WAR
GAMES

M

WAR GAMES

The Secret World of the
Creators, Players, and
Policy Makers Rehearsing
World War III Today

Thomas B. Allen

McGRAW-HILL BOOK COMPANY

New York   St. Louis   San Francisco
Toronto   Hamburg   Mexico

1 2 3 4 5 6 7 8 9 D O C D O C 8 7

ISBN 0-07-001195-8

LIBRARY OF CONGRESS CATALOGING-IN-PUBLICATION DATA

Allen, Thomas B.
   War games.
     1. War games.   I. Title.
   U310.A49   1987   355.4'8   86-27498

ISBN 0-07-001195-8

To Aaron Allen Witte and his world

CONTENTS

CONTENTS

ACKNOWLEDGMENTS

My first move in the game was to talk to Norman Polmar, a friend and my original guide to the military world. Norman read the manuscript, and, thanks to his wise suggestions, improved it immensely. He read as a friend but edited as a pro. For him, special thanks.

Jim Hessman, editor in chief of *Sea Power*, started me off by assigning me one day to do an article on war games. The result of his assignment is this book.

I also wish to thank Al Lefcowitz, friend, writer, scholar, and professor of English at the U.S. Naval Academy, who gave me the first research clues toward automated war gaming.

Every person I interviewed, on and off the record, helped me make more moves. I wish to thank all publicly listed on page 387—and those who cannot be publicly named because of the rules of the game. I am especially grateful to Lincoln Bloomfield, Colonel Trevor Dupuy, Ted Atkeson, and Jim Dunnigan.

I received a great deal of help from David C. Humphrey of the Lyndon B. Johnson Presidential Library, who led me to war games that had never been made public and then helped me through the maze phase of the game, the Freedom of Information Act.

Others who helped me around the board were Frank Uhlig, Michael Goodman, Dov Zakheim, Ike Kidd, Bill Lind, Rick Kessler, Bill O'Neil, and Michael R. Doyen.

Connie Allen Witte, James Witte, and Roger MacBride Allen helped me with research, and Roger and his brother Chris made my computer extremely friendly.

The last move is to home, where my wife Scottie gave me encouragement and support that I have taken for granted—until this moment when, thinking of this book, I think of her and how much she means to me.

WAR
GAMES

The Spectrum of War: An Introduction and Scenario

It is early in 1985, about a year and a half after the assassination of Benigno Aquino, political rival of Philippines President Ferdinand Marcos. The Reagan Administration still publicly supports Marcos. But on this day, in a long, low-ceilinged room in the basement of the Pentagon, U.S. military officers and senior government officials have gathered to decide how to hasten Marcos's downfall. "While we do not have much . . . hope of controlling a deteriorating situation," an official says, "I think we can step in at various points and screw it up."

A high-ranking State Department official refers to Marcos's regime as "a political process in the state of disintegration." The diplomat urges the others at the paper-littered table not to make the mistake the United States has made in the past with other such "clients" as Iran, Iraq, Libya, and Ethiopia, when "we have failed to make clear our detachment from the sinking ship, and therefore have gone down with it." The Philippines will sink, not the United States.

Michael H. Armacost, U.S. Ambassador to the Philippines, has begun secret negotiations with Imelda Marcos to see if she is interested in succeeding her husband. Another speaker calls this "a fatal mistake," just like the idea to support a takeover attempt by Philippines Defense Minister Juan Ponce Enrile. The official admits, though, that "we would be willing to let him have as much—play as fast and loose with whatever remains of democratic institutions as we've been forced to allow Marcos to do. So these are the kind of things I think we have to, you know—for getting value out of this certain process—this has to be it. . . ." (No one mentions Corazon Aquino, Benigno's widow, who will run for President against Marcos, lose in a rigged election, and be swept into power by a popular revolution in February 1986.)

Someone wonders about the effect of congressional involvement in the worsening situation in the Philippines. An intelligence officer says, to laughter and applause, "Any one of us who has been in a situation like this has experienced the visiting Congressman who gets up and makes vociferous anti-Marcos statements on the floor of the House and then comes out to the Philippines, and you say to him, 'Well, Mr. Congressman, I hope you'll let Marcos know how you really feel about democracy in this country.' And he says, 'Well, that's your job.' And he goes in to Marcos and tells him what a good leader he is and how important the Philippines is—and your leverage is shot. . . ."

These quotations—frank, hostile words about a man who, in public, is still a valued ally—are from a Pentagon war game. The secret negotiations between Armacost and Imelda Marcos, just like the talk about helping an Enrile plot, sprang from a scenario writer's imagination, not reality. But the words are real; they are on a tape recording I obtained while doing research for this book.

The Pentagon games are as real as imagination can make them, and, as in the Philippines game, they often take yesterday's events, merge them with today's planning, and file the results away for tomorrow's action. These are not lieutenants and captains playing at war. Armacost, for example, soon was to become the Under Secretary of State for Policy. Three- and four-star generals sometimes go into that room to plan the action for tomorrow or the day after.

The National Security Council, which in the Reagan Administration often acted as a vest-pocket CIA and State Department in the West Wing of the White House, is a frequent sponsor of political-military games. So many active-duty and retired military officers, trained by gaming, have been assigned to the NSC in recent years that some observers knowledgeable about war games have detected a gaming mentality behind some of the NSC's more lurid operations. At best, war games give players a chance to make crucial decisions without having to endure the sweaty palms and chaotic communications of a real crisis. At worst, war games can give a Marine lieutenant colonel or a Navy admiral a sense that the smooth and rational moves of a NSC game can be used in the real, messy world where nations play for keeps.

Modern war games stress reality, not theory. The crumbling of the Marcos regime and the fast growth of a Communist insurgency in the Philippines were real, as was the threat to two strategic U.S. outposts, Clark Air Force Base and the Subic Bay Naval Base. Reality was just outside the doors of the game rooms, and the new generation of Pentagon game players wants to keep reality that close.

Computerized war games can produce military contingency plans virtually on call. Players include the commanders in chief (CINCs) of military commands throughout the world, hooked into a complex computer system called SINBAC (System for Integrated Nuclear Battle Analysis Calculus). And nuclear wargaming with automatons instead of human beings is now a reality.

"The Joint Chiefs," a former Pentagon civilian official told me, "want to give the CINCs high-grade tools. The war planning picture is not pretty. It's not exactly monks poring over manuscripts by candlelight. But that's close. So the scenario equipment could be real war-fighting equipment at the same time."

This is one of the eerie new realities of wargaming. In today's computerized games, players look at video displays whose artificial images often are exactly the same images that would appear on a real video display during a real war. In the combat control centers of modern war, commanders see electronic symbols of distant targets, not the targets themselves. Electronic wargaming is preparing generals and admirals for warfare that, to its managers, will look like a video game.

Wargaming covers a spectrum of reality and make-believe. At one end is the pale abstraction of analytical gaming. Moving through the spectrum from that faint shade of reality, gaming next appears as electronic simulations, then as people-around-a-table games like the one in the Pentagon basement. The spectrum next begins darkening with the reality of exercises—games played with real

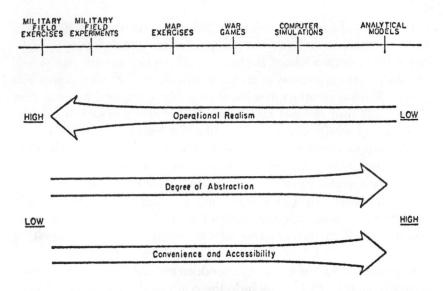

The spectrum of war gaming begins with the flesh-and-blood of military exercises, continues through games involving players and computers, and ends with analytic games. The realism of military maneuvers is countered by the economy of abstraction. (Credit: J. G. Taylor, *Proceedings of the Workshop on Modeling and Simulation of Land Combat.*)

warriors and real weapons, real warships and real warplanes. At that end of the spectrum, right next to the exercises, is the dark of war itself.

Wargaming is a simulation of war, a horror show without props. Actors, chosen for the real war, walk the stage in rehearsal, ad-libbing to a script that keeps them bound by reality. The play goes on in war colleges, and in some forty universities and think tanks that stage games for the Department of Defense—and other customers, including foreign countries. BDM International, a Washington-based military consulting firm, in 1986 was working on a game that would be sold to Iraq for use in its war against Iran. The heart of the game would be a model of Iranian transportation and oil-pipeline systems, a kind of dart board for Iraqi pilots to practice on before flying off to bomb the real thing.

BDM is one of several major think tanks that develop highly sophisticated games that play out crises involving nuclear-war control and communications. Recently, the Strategic Defense Initiative, better known as "Star Wars," has inspired a host of new games. From many of these games come ideas that find their way into American foreign policy and military contingency plans. But little is known about gaming beyond the game rooms because of the secrecy that is imposed upon the identity and the moves of the players.

Finding details about the games is itself a game of hide-and-seek. The history is both secret and ephemeral. The historian's office of the Department of State, for example, knew of no records of an ultrasecret "games center" working for the Department of State as early as 1948. The center was described in a book by Miles Copeland, a State Department official who said he played in games from the summer of 1955 to the spring of 1957, when he worked with the Middle East Policy Planning Committee. Copeland wrote that a "diplomat to whom I showed the original draft of this book chided me for 'revealing a lot of information which had best be forgotten' and for 'needlessly' puncturing a view of our Government 'which it is best for the public to have.' " Something of the same attitude still prevails in gaming circles.

There are few game records in the archives where declassified government records usually are kept. In the days before widespread use of xeroxing and filming, the making and distribution of transcripts was considered a luxury, and so many games were not preserved. Games of the 1970s and 1980s, which are better preserved than their predecessors, are top secret or secret. So, while the accounts of past games are unknown because of carelessness, the accounts of present-day games are unknown because of highly protective care.

In wargaming stadiums there are no bleachers. The public is never invited to the games that help to determine how the Cold War will be waged, how events could kindle World War III, and how that nuclear war would be fought.

When our leaders speak on radio and television, we hear the speeches that enunciate the nation's declared strategies. But we do not hear or see the words that are made in the heat of simulated crisis—the words that offer sometimes chilling glances into future realities. Sometimes the simulations inspire raw emotion. At a nuclear war game run from the Reagan White House, for example, administration officials applauded when, at the climax of an escalating crisis, U.S. missiles destroyed Moscow.

The visible stadiums for wargaming are the colleges of war. Each has a speciality. For the venerable Naval War College, in Newport, Rhode Island, it is the Global Game, "a test bed of current strategic wisdom." The Army War College in Carlisle, Pennsylvania, has a computerized game called Janus, which tests tactics for battlefield-grade nuclear weapons. At the Air University near Montgomery, Alabama, games are dominated by air power and, lately, space power. The National Defense University and the Industrial College of the Armed Forces share the riverside greensward of Fort McNair, a tranquil redoubt in the District of Columbia. The Industrial College, watching over the plain matters of logistics and mobilization, practices the moves a nation must take to go to war. The National Defense University plays counterterrorist and special-operations games, along with full-scale strategic games.

Beyond the stadiums, wargaming exists, almost unseen, on a plateau of the nation's defense establishment. Never at the peak where the Secretary of Defense is, and never at the bottom, where the soldiers and the sailors are, wargaming is an abstract concept in a Pentagon dominated by the hardware of high-cost weapons.

The abstractions include the concept of fighting a nuclear war. Military analysts may hesitate to predict the outcome of conventional warfare. But, writes Fred C. Iklé, Under Secretary of Defense for Policy, "Such modesty is missing for nuclear war, where pretentious analyses and simplistic abstractions dominate and blot out the discrepancies existing between abstractions and possible reality—a reality that for so many reasons is hard even to imagine."

Games give policymakers a cheap and quiet way to go to war for the mundane purposes of planning budgets, for tinkering with the size of Army divisions and Navy fleets, and for putting nonexistent weapons and outlandish tactics onto mythical battlefields. Numerous games have been developed in obscurity by the Defense Advanced Research Projects Agency, the Army Concepts and Analysis Agency, and the Army Research Institute for the Behavioral and Social Sciences.

The least known and most important stadium is the one in the Pentagon basement, directly below the National Military Command Center—and that close to the real world. The games once were run by a Joint Chiefs of Staff organization called SAGA, whose felicitous acronym stood for Studies, Analysis, and Gaming Agency. SAGA is now JAD, the Joint Analysis Directorate.

There seems to be no precise date on which SAGA became JAD. But several officers and civilians familiar with gaming told me that the word *game* was out of favor by the early 1980s, when *simulation* and *modeling* became the preferred terms. Many academicians have also shied away from the term not only because "gaming" does not sound like a serious occupation but also because they believe that *simulation* more accurately describes today's computerized studies of conflict. A couple of military gamesters shrugged and smiled when I asked whether the eradication of *wargaming* from the defense lexicon had anything to do with *WarGames*, one of the most popular films of 1983. In *WarGames* a teenager computer wizard almost starts World War III when he connects with another computer and plays what he thinks is a war game, unaware that he has engaged an Air Force computer that controls U.S. missiles.

The fiction of the movie was linked to reality in media accounts of a rash of computer break-ins by young computer hackers. Because many of the breached computer networks contained defense information, military officials became sensitive to public perception of wargaming, which is looked upon as a matter best kept inside what players often refer to as the gaming community.

The Naval War College, unlike the Pentagon, still clings to the old word, probably because the Navy has been the oldest and most persistent user of gaming. Many of the basic rules of American wargaming were developed at the college during the years between World War I and World War II. And in modern times its Center for War Gaming became a major stadium for the planning of World War III.

Between the wars the college selected America's probable future enemies and plotted warfare against them. In a large game room players moved fleets of little lead ships on the gridded tile floors, where eight inches represented two thousand yards of ocean, where the Blue Fleet was always American, and where other colors were given to opponents. In the 1920s the enemy was the Red Fleet— the Royal Navy. In the 1930s the enemy was the Orange Fleet—

the Japanese Navy. Now Blue is not only the United States but also its allies, including Japan, and the enemy is not a colored fleet but simply Red—the Soviet Union and its allies.

The Naval War College, which commands a bluff overlooking Narragansett Bay, was founded in 1884 as a place where Navy officers could "study their profession proper—war." The war college perhaps should be called a war graduate school, for its students are seasoned naval officers sent there for a year of study to round out their professional careers and put them in line for promotion. Many of them will become admirals, a few will rise to the Navy's highest ranks. All will have spent many hours waging make-believe wars, for wargaming is an old pursuit at the college.

Traditionally, the games pitted ship against ship or fleet against fleet. But in recent years the games have pitted nation against nation, and the officers studying their profession proper have been joined by civilians studying, if not a profession proper, then a task they may someday have thrust upon them: the prevention or the waging of war. As an official description of its mission says, the college "designs, plans and executes games for the highest decision-making levels, civilian and military, in the government."

Every year since 1979 high-ranking military officers, senior officials of government departments and agencies, together with defense contractors and think-tank analysts, have assembled at the Naval War College for the Global Game. There is a football-stadium excitement about the game. And, during intermissions and after the game, the players get together in a kind of locker-room camaraderie. (The towel-snapping image is not overly drawn. Women players are as rare in the male world of the war game as they are in football locker rooms. To further the jock image, players call their postgame analysis a "hot washup.") In their figurative locker room players can make friends, line up contracts, and enjoy the clubby privacy provided by their security clearances, the common dues of membership.

Although the global war game invariably involves NATO forces, NATO allies are not invited—because, I was told by a frequent player, allies would inhibit the game, which looks at the world and at war strictly through American eyes. "Imagine what the Germans would say," the player told me, "if they were around the day one of us said, 'Let's write off Germany.' "

Very little has been written on wargaming outside the little community that sponsors and plays it. The Pentagon has a great deal of data on the games that have been played since the 1960s. But several people familiar with the games told me that the highly classified information has never been digested and historically assessed. Many of the reports that do exist are fragmentary and classified top secret.

One of the few unclassified assessments of political-military games was made in 1977 by Robert Mandel, a political scientist whose sources included the records of several games run in 1960 through 1969 with teams of players drawn primarily from the Central Intelligence Agency, the Department of Defense, and the State Department. One of Mandel's important discoveries was that a team's perception of its own intentions is much clearer to that team than it is to the opposition. In one game the U.S. team consistently underestimated the intransigence of the Soviet team and naïvely supported a Soviet proposal "which might have temporarily stopped the fighting in the area, but which would have given the U.S.S.R. control of oil in the region." Military players, Mandel found, consistently perceived the United States "as being significantly more cooperative and trustworthy" than the Soviet Union.

The games seemed to show that a nation's perception of its own strength and weakness depended upon whether the nation was trying to preserve the status quo or change it. A nation seeking to change a situation would see itself as stronger than the nations standing in the way. But, at the same time, a nation that was not seeking change also usually perceived itself as a relatively strong nation. These perceptions go to the heart of the theory of nuclear deterrence, itself a product of gaming.

Deterrence, as Mandel pointed out, "depends entirely on perceptions of capabilities rather than on the capabilities themselves." The games showed that what strategists call signaling is not always clear and does not necessarily get the results the signaler wants. A knowledgeable adversary can figure out what led to a signal and realize what it may conceal.

Mandel's study also indicated "a severe inability by the United States to understand the expressed intentions of another state." This in turn complicates the opposing nation's reaction because as far as its policymakers are concerned, *they* knew what they said.

In a crisis, as in a game, Mandel reported, U.S. policymakers may shift toward military solutions because of a magnified image of Soviet belligerence.

Well-run games resemble real-life crisis management more than many people realize. "In a government crisis," Mandel pointed out, "the same narrowing of attention on the issue takes place; the information available may be scanty and may therefore prevent a realistic depiction of the situation by participants; and sessions may be similar in length to game moves."

Using the leverage of the Freedom of Information Act, I managed to obtain wargaming records that underscored many of Mandel's findings—not in theoretical terms but in the very real conduct of the Vietnam War. The secret games I found and got declassified were being played during that war by the same White House and Pentagon officials who were running the real war.

Had these records of Vietnam-era war games been retained by the Pentagon, they probably would have been lost or, at best, kept secret through restrictions designed to protect the records against disclosure even through the Freedom of Information Act. But, because of an archival fluke, the game records somehow had found their way to the Lyndon B. Johnson Presidential Library in Austin, Texas. As presidential records, the games were subject to mandatory declassification review by the Department of Defense, and I was able to get them.

The dialogue in the games showed again and again an arrogant, unquestioning belief in signaling; a reliance on gaming-table analysis over GIs-in-jungles reality; and an incredible faith in the usefulness of strategy worked out against a "Red Team" that consisted of American military officers and civilians playacting Viet Cong and North Vietnamese leaders.

Vietnam-era gaming, which attracted the rational war planners who first appeared in the McNamara Pentagon of the 1960s, has continued to influence policymaking in every presidential administration since. The U.S. strategic community is strongly influenced by gaming-inspired analysts who believe that international conflict can be viewed essentially as a game with rules. Nuclear war-fighting doctrines have a chesslike quality because they were created by war gamers. Many of the doctrines developed in the 1970s endured and were built upon by the Reagan Administration in the revival of the Cold War, itself a kind of game in which, with luck, no one is killed.

In appeals that went as high as Secretary of the Navy John F. Lehman, Jr., I requested an opportunity to observe all or part of a Global Game. I even offered to allow that portion of my manuscript pertaining to the game to be subject to review for possible disclosures of classified information.* I had hoped that my presence would perhaps add a touch of realism to the game, which typically is played in a military vacuum: no media, no congressional reactions, no civilian panic over war-threatening or war-waging moves. (I was not accepted into a Global Game, but you will read about another in which I participated.)

Although I was rejected as a spectator at the Global Game, I was accepted as a player in an unclassified game at the War College. The war was waged in Room 32 on what Navy people call the second deck of a fine old stone building. The small room's spotless window framed a cloudless sky and deep-blue water flecked with sailboats darting before blustery winds.

My teammates and I straggled into the room from a stand-up breakfast of juice, coffee, and doughnuts. Also heading for similar rooms were five other ten-member teams. The war game was part of the college's annual conference on international security. One of the speakers had set the theme for the conference by declaring that we live in an era of "violent peace."

Most of the players were military officers, including a few allied officers who, like me, were barred from the Global Game. The highest-ranking officer was a two-star U.S. Marine Corps general. The other players, several of them retired officers, included political scientists, military and civilian members of the War College faculty, representatives of think tanks devoted to defense matters, defense industry executives, middle-level Pentagon civilians, an analyst on Soviet affairs from the Central Intelligence Agency, several military intelligence officers, a couple of congressional staffers with defense interests, a military historian, and a State Department official who was listed in the roster of players as an official concerned with international security matters. Except for one woman,

---

*No part of the manuscript was ever submitted for review.

a former middle-level Department of Defense official, all the players on the six teams were men.

Our team included a Norwegian officer who did not say much and nine Americans. For our game we would not need the computers used in the Naval War Gaming Center, stadium for the Global Game. We sat around two tables pushed together and covered with green baize. In front of each of us was a white, lined pad, a couple of perfectly sharpened yellow pencils, and a Crisis Simulation Game Packet, which contained a statement on the game's objectives, a schedule, and a few pages entitled "The Persian Gulf Scenario." At precisely the same moment, in our small room, and in five others in the building, everyone started reading.

The scenario began with a factual background paper that reviewed the three-and-a-half-year-old Iran-Iraq war up to early 1984, when the Ayatollah Khomeini and other Iranian officials "warned repeatedly that Tehran would close the Gulf if Iranian oil exports were impaired." Then the scenario added a fictional development:

In April of 1984, approximately 350,000 Iranian troops launched the long-awaited spring offensive and severed the main highway between Baghdad and Basrah. Iraqi defenders in the area were isolated, with little prospect of being reinforced or resupplied. However, in an attempt to relieve the pressure around Basrah, Iraq launched around-the-clock strikes against Khark Island. These strikes included the use of Exocet missiles by Super Entendards against ships en route to and from the island.

On 25 April an Iraqi Super Entendard launched an Exocet missile against a super tanker outbound from Khark Island with a full load of oil. The 357,000-ton tanker, leased by a Swedish firm but registered in Saudi Arabia, was eventually lost. Because of the loss, Western insurance companies made substantial increases in insurance rates and threatened further increases should the attack against ships in the Gulf continue. Khomeini subsequently stated that the "ultimate confrontation with Western imperialism in the Gulf must now be met."

By late April, intelligence sources reported that Iran had moved aircraft to bases in southwest Iran and increased naval patrols in the vicinity of the Straits of Hormuz. There were also reports of additional small aircraft on Iranian islands in the vicinity of the Straits. Khomeini vowed to close the Straits and also warned the Gulf states that Iranian retribution will not be limited to the Straits.

Then we were given another piece of paper. With this sudden bit of information, the game was on. The paper said:

SANA (Saudi Arabian News Agency) has just reported that a Japanese tanker was severely damaged by an explosion in the Strait of Hormuz and is in danger of sinking. The ship was en route to Saudi Arabia and may have struck a mine. Iran has issued a statement claiming responsibility for the incident and declaring that it has the capability to close Hormuz by putting mines in the Strait. The statement further warned that only ships en route to Iranian ports will be provided safe passage through the Strait and any attempt to threaten Iranian sovereignty will be met with attacks on oil facilities in other Gulf states. The Pentagon has reported that U.S. AWACS* have detected an increase in Iranian fighter aircraft patrols and the redeployment of some Iranian forces.

This incident follows what has already been a sharp increase in insurance rates. Exports of Persian Gulf oil are projected to fall to 25 percent of current levels should the situation continue.

Our leader was a tall, black-haired CIA specialist on the Soviet Union. I will not use his name. (Under the usual rules of a military-sponsored game, players are not identified—so that they can play without worrying about how their moves might affect their real-life careers.) As soon as we had read the scenario, the CIA man went to an easel holding a blackboard, picked up a piece of chalk, and asked us what American objectives would be in this crisis.

The ranking U.S. Navy officer present, a commodore, spoke up: "It's 'Don't let Iran be a winner.' " A retired Navy pilot, now an executive of a leading military think tank, said, "Keep the Soviets out and keep the Straits open." The commodore, a slim, soft-spoken man, added, "And don't get us involved in combat."

On the blackboard the CIA man wrote *preference* under *OBJECTIVES*. He seemed to feel that we needed to see the crisis as an event that could give the United States something it wanted. But the subtle difference between *preference* and *objectives* was lost as we began discussing and sometimes politely arguing over what ought to be written on the blackboard.

A man from the Congressional Budget Office asked, "What is the national interest here?" An Army historian said, "Maintenance

*Airborne Warning and Control System aircraft.

of U.S. stability there." Then the State Department man spoke for the first time. He said he wanted to put economic pressure on Iran and he wanted the United States to "get shirty" with the Soviet Union, although it was not clear what reason the United States had for any irritation toward the Soviet Union. The commodore, who would become the dove to the State Department hawk, repeated his prime objective: "Don't get us involved in combat."

One of my few contributions was a question about whether U.S. Navy ships should attempt to rescue the crewmen of the Japanese tanker. The CIA man and the State Department man gave me my first lesson in war games realpolitik: the object of the game was to solve a crisis, not risk Navy assets in a rescue mission. I later remembered that there had also been a debate over the Iranian hostage rescue. The military had argued for a large-scale military operation, not a rescue mission. The civilians had got what they wanted—and the military advisers had got what they had warned against.

The majority of the people around our table and around the other tables were civilians, but, because of their past or current occupations, they were militarized civilians. No one had needed a special explanation of the scenario's technical terms, such as Exocet missiles and Super Entendards. They all knew that in the Falklands War a single Exocet missile fired from an Argentine Air Force Super Entendard sank HMS *Sheffield*. "What's the Exocet count?" asked one of the militarized civilians soon after the game opened. (*Control*, the godlike umpires on the third deck, did not give us an answer. *Control*, in a game like this one, is everything—enemy and neutral countries, fate and the weather—that is not on our side.)

We spent nearly an hour looking for words to put on a blackboard. By now we had begun acting—playing—as if we could control events. We were trying to think real thoughts about a fictional situation that somehow no longer seemed fictional. Finally, the CIA analyst had this on the blackboard:

OBJECTIVES

Minimize East-West confrontation.

Keep Strait open and the oil flowing.

Get Allies involved.

Seek to end the war in stalemate.

Avoid combat.

Enhance GCC [Gulf States Confederation].

Now, as we began to decide on what our CIA leader labeled *MOVES*, a civilian faculty member of the War College's Strategy Department arrived to help the CIA analyst run the game. Gradually, the game began changing. The CIA man remained our leader, but the newcomer was more like a player-coach. Under his professorial prodding, a team member asked Control whether the imaginary intelligence officials on Control's staff could answer these questions: Did a Japanese ship *really* sink? Were there *really* mines? (The commodore had professional doubts about the Iranian Navy's ability to sow mines in the straits.) Where were the nearest minesweepers? (The only friendly naval force in the area, according to a map on a wall, was a U.S. Navy battle group, which included an aircraft carrier but no minesweepers. British and French minesweepers were three weeks away; it would take about seven days to deliver U.S. Navy helicopters equipped for minesweeping, and it was unclear where they could operate from.)

The think-tank executive said he had heard that Saudi Arabia had bought four minesweepers some time ago from the United States. "Why can't we get them involved?" he asked.

"We could buy them back," I said, "and put U.S. crews on them—"

"And paint out the numbers," the CIA man added.

While the team waited for Control to deliver intelligence reports, the professor said impatiently, "It doesn't matter if it was a mine. We have to start making minimal moves."

Several men who seemed to have Navy backgrounds started referring to the United States in a war-game term: *Blue NCA*— Blue from the old Blue Fleet tradition and NCA from the much newer concept of National Command Authority, the entity that contains and directs American strategy: the President and the Secretary of Defense, their alternatives, deputies, or successors.

"Blue NCA has a contingency plan with Saudi Arabia," someone said, as if he absolutely knew. "Blue NCA should start consulting with the Saudis."

Another suggestion was being made—"We should try to encourage the British and French to move up their minesweepers"— when a Navy captain from Control entered Room 32 with the news that the people in intelligence did not have the answers to the team's questions. Several players made joking remarks about this. They seemed used to not getting answers from people in intelligence.

"Intelligence never solves your problems for you," the commodore said, smiling. A couple of other team members also smiled and nodded. I assumed that they had been participants in crises that were not simulations. It took a certain inner-circle knowledge to talk about Blue NCA contingency plans and a certain panache to throw around the names of real countries as if they were Boardwalk or Marvin Gardens in a Monopoly game.

Like the professor, the State Department man was getting impatient. Although technically not a chain smoker—he lit his cigarettes with a lighter, not the previous cigarette—the effect was the same. He was wreathed in smoke. Leaning forward, out of his haze of acrid smoke, he said, "This blockade is unacceptable. There must be an element of *threat*. We have to find out who's in—and for how much."

The CIA analyst agreed that more information was needed. He suggested a probe through Sweden or Algeria. "The Algerians were reliable during the hostage crisis in Iran," he said. "We want an honest broker. Or how about the Paks?"

After a few minutes of diplomatic talk among team players (no one raised his voice; everyone was polite), the team settled on the idea of talking to U.S. allies and simultaneously opening a channel with Algeria. The game-wise man from State put forth what he called "The Line" that should be pushed: "This is unacceptable. The civilized world will not tolerate it. We speak for several nations. We will attempt to persuade Iraq to back off—no, desist—on Khark Island. We will also say—and we'll have to shade this—that military force may have to be brought to bear. Tell Iraq our allies can't accept attacks on Khark Island and tell them we will keep up military supplies to them."

The man from State, the War College professor, and our CIA

leader began making moves with the practiced skill of players who had been through this kind of exercise before. They convinced the rest of us that the team should try a double diplomatic approach: Tell Algeria to tell Iran not to carry out its threat to close the Straits; tell Saudi Arabia to tell Iraq to stop bombing Khark Island. At the same time, England, France, and the United States would start moving minesweepers toward the Persian Gulf because there still was no solid information about the mines. By now our *MOVES* on the blackboard were:

Diplomatic pressure.

Limit arms sales—

longer-run impact France→Iraq. Israel→Iran.
British→Iran.

Increase public awareness

Add economic incentives.

The think-tank executive suddenly asked a businessman's question: "What happens now if the chairman of Exxon calls the White House and says he will triple oil prices? We need a coordinated strategy by major users as to how to handle producers." Before anyone could answer, Control signaled that it was time for lunch.

When we arrived back in Room 32 after an hour-long sandwich lunch, we each found at our places a new situation paper that said sixty days had passed while we were eating. During those sixty days, Shiite terrorists calling themselves the Knights of Ali had been attacking tankers at night by attaching limpet mines to the hulls or by firing at the tankers from dhows. The flow of oil from the Gulf had virtually ceased. The per-barrel price of crude oil had gone from $28 to $45. Our think-tank executive had been prophetic. The situation paper went on to tell us what else had happened in the past sixty days:

Oil reserves held by the U. S. and other non-communist countries
. . . have been drawn down to the point where there is some concern
for their relative margin of safety in days of supply. . . .
The economic impact of the oil disruption was severe, and it led

to recession, unemployment and a downturn in world trade. National and worldwide recession was being felt in the international trade arena. The recession was exacerbated by a sharp rise in worldwide unemployment, rates of inflation and interest. It was evident that the Western world was on the brink of a downturn in world trade which would cause a severe loss of GNP in the U.S., western Europe, Japan and the lesser developed countries. . . .

Khomeini has stated that the Strait of Hormuz will remain closed until President Sadam Hussein of Iraq is deposed. The Iranian coastal area and islands near the Strait have been reinforced, and Iranian air and naval forces have been redeployed to bases near Bandar Abbas on the Strait.

We all reacted with a sense of shock, *real* shock, not just a reaction to a bad break in a game. We were really feeling upset about what was happening in our imaginary world. "What is happening to our institutions?" someone indignantly asked, as if real institutions were really going through what the situation paper had described. I had an unreasonable feeling of helplessness and failure. Some of us spoke softly to each other about having failed.

Then Control sent word that the "sense of world opinion" was that the attacks on the ships were being carried out by "government-sponsored terrorists." Control also reported, "The Soviets are expressing great concern about the deployment of Western forces." But we had kept the American battle group, along with a few Royal Navy and French ships, outside of the Gulf. At this low point for Blue, our CIA leader scolded us. "Nothing yet," he said, "has met our primary objective to keep the oil flowing." The man from the State Department urged action. "Paramilitary operations seem to have been doing all this business," he said angrily. "Against this kind of thing, counterforce is required—patrol boats with .50-caliber machine guns, if you're dealing with dhows. There's a certain amount of counterterrorism expected."

Our team was veering sharply from the morning's course of diplomatic probes. Now force was on the table. After a discussion about the possible use of escort ships, the State Department man spoke again. "Piracy," he said, "is the best recognized international crisis. A dhow that attacks and is chased should not be allowed to make it back to port." He looked around the table for support and information. "Will the Omanis allow us to put planes there? Using air cover out of Oman, or small combatants as escorts—"

The War College professor interrupted with an academic question: "What do we want the world to look like in six weeks?"

The man from State ignored the question. "Counter paramilitary activity," he said. "We have to get the oil out."

"How about another diplomatic initiative?" asked the historian from his cone of silence.

He, too, was ignored. "Since we're dealing with a man who is somewhat of a nut," the man from State said, "we are taking a risk. He may escalate this to a military problem."

Now it was the commodore's turn to become impatient. "We're a helluva long way from our supplies," he said. "You can *gradually* attrite his resources. We can't end this by having U.S. forces go into Iran and start a war."

"You say we can keep taking cough syrup," the suddenly belligerent professor said, glaring at the commodore. "You're just putting off the day of reckoning. We should be looking for ways to twist his arm behind his back."

The CIA man looked around the table. "Is there any incentive that will get the Iranians to open the flow of oil or end the war?"

He got several suggestions: Blockade Iranian ships going in and out of Iranian ports; increase arms to Iraq, perhaps through France; attack the ports from which the dhows sail.

"Taking out sites means a dhow for now, a plane for now," said the commodore. He had been a carrier pilot. "What bothers me is everything we're doing is military. . . ."

The professor asked how long it would take to get fighter planes and missiles to Iraq. Before anyone could answer him, the game ended. We had used up all our time and we had not found an ending for the game. As I walked out of Room 32 I looked out the window. The wind had picked up, and the sailboats were knifing through the bay, heading home.

Exactly five days after I had read and played out the Persian Gulf scenario, I read this real report in *The Washington Post*:

> A Saudi Arabian supertanker was set afire in an air strike off the Saudi coast in the Persian Gulf today, and U.S. officials said the attack almost certainly was carried out by Iranian warplanes. . . .

Five oil tankers have been attacked in the gulf this week, two by
Iraq and three apparently by Iran. The development has significantly
broadened the hostilities and raised concerns among the world oil
and shipping industries about continued operations in the Persian
Gulf. . . . Major maritime insurers announced that higher "war-risk"
premiums have been imposed for the northern third of the Persian
Gulf.

Then, one month after we finished a war game without finding
an ending, there was, as in the game, a new situation report from
The Post:

A Pentagon spokesman said yesterday that the United States "would
do all possible to protect U.S. assets and interests" in the Persian
Gulf in the face of an Iranian threat to stop and search commercial
vessels in the Strait of Hormuz.

"We would attempt to protect our assets and our vessels by appropriate means," Michael I. Burch, assistant secretary of defense
for public affairs, said. . . . Burch also said the USS America has
joined its sister carrier, the USS Kitty Hawk, in the Arabian Sea for
"routine" exercises.

The Persian Gulf Scenario, which had produced a war game,
had become an exercise, and was perhaps on its way to the next
place on the spectrum—war.

CHAPTER 1

War Stadiums:
Then and Now

Most of my fellow players in the Persian Gulf war game were amateurs. But not the State Department man, I would learn later, when I talked to him in my own hot washup—a series of interviews with former teammates. He was a middle-level career executive with a corner office containing a cluttered desk, a small safe, filing cabinets with combination locks, and a table bearing a couple of seemingly moribund plants. On the wall behind his desk was a large map of the Soviet Union. His State Department address was INR/SEE: Bureau of Intelligence and Research, Soviet and Eastern Europe. The bureau produces studies written by analysts who draw from intelligence sources, academic research, and State Department cables; the bureau's reports, which are written for decisionmakers, are starkly factual and secret. They are also anonymous, which is what he wished to be.

In his view modern gaming traced its origin to a time, "somewhere in the post-World War II era, when we began to think about nuclear authorization and how agonizing it was. And we began to develop policy for it. You think about decisionmaking processes. Even if you want to do something for a purely military purpose, you have to understand that when you come to these mountain points, there is going to be a Presidential decision."

He paused, and I realized I had to fill in the spaces between his thoughts. He was used to letting others come to conclusions about his information. He was trying to explain why grown men in the State Department, the Pentagon, and the White House played games, war games, as a way to move toward those mountaintop decisions. Like many players, he seemed to dislike using the word *game*.

"You do these exercises for a variety of reasons," he resumed. "One, maybe the simplest one, is procedural. You want everybody to run through certain drills: What kind of messages do you send? When do you get the President out of bed? You're going to do exercises for procedures in communications, for formats—that sort of business. Another kind of exercise would be to train people in thinking about these problems. In other words, you want analysts, commanders, staffs to sort of get used to the idea that you have to do some strategic thinking.

"And then there is the notion of using the war game as an analytical tool. There I myself would be a little skeptical. Once, when we were playing Red, we did something that we thought made sense, something totally unanticipated by the sponsor of the game. The results were kind of catastrophic for Blue, and the game went on along an entirely unanticipated tangent." He smiled, and I could see he liked remembering how he had upset the game board. "Well, life is like that," he said.

"You know," he said, "it is not an awfully economical way to think through a problem. And, depending upon who he is and how much he has been engaged, a player might get game-wise, kind of the way some students get test-wise," he said, smiling again. He knew he had been game-wise at Newport.

The pickup teams at the Newport games are like those at a baseball clinic where devotees of the sport get a chance to go a few innings with the professionals. When the clinic is over the amateurs feel they understand the game better, and the professionals go back to their real games and real stadiums.

One of those stadiums is the National Defense University, which runs the newest game in town. The university's War Gaming and Simulation Center opened in 1982 under the direction of Sterling Hart, a former designer of toy-shop war games. He is one of several people who have worked, sometimes simultaneously, on war games played for play and war games played for real. The center, like the

Naval War College, discourages public spectators. But there is a pride of newness about the place, and people there will talk, off the record, about the innovations that the center brought to the old games of war.

The NDU gaming center once conducted seven major games in eleven months. The freewheeling games, sometimes involving more than two hundred people for twelve days, were pared back by Hart's successor, Air Force Colonel Kenneth J. Alnwick. But the National Defense University had established itself as a new and highly visible wargaming stadium.

"Some of us went to the global game at Newport in the summer of 1982," an official told me. "The Blue side was interested in Special Operations. No one would play it. Ray Bell [Colonel Raymond E. Bell, Jr., a psychological warfare officer and then deputy director of the NDU wargaming center] stepped in and tried to plan a Special Operations mission." Apparently the idea was to counter Soviet rear-echelon *Spetsnaz* units.*

Out of this play came Bell's SUWAM, Strategic Unconventional Warfare Assessment Model. Next came SUWAM III, which Hart carried in a Kaypro II portable computer to Hawaii and South Korea to demonstrate U.S. tactics to U.S. and South Korean officials. The model shows how to plan and carry out raids, ambushes, and intelligence gathering "in a high-risk environment."

The center's interest in psychological warfare led to a game called PSYOP. "It can target different groups within, say, Poland during a European crisis, where it would be to the United States' advantage to know whether certain groups would be helpful to the United States," I was told. PSYOP designers identified seventeen power groups within Poland, estimated the likelihood of their revolting or going on strike, and developed a "loyalty index," based on the probability of support of the United States in a war.

The university's global war games, which compete with the Navy War College for high-level players, have explored ideas that more traditional games have avoided. One, played in 1983, drew nine admirals and generals, two ambassadors, and fifteen members of the staff of the Joint Chiefs of Staff. The players looked at the most

---

*These are "Special Purpose" or "Special Designation"—*Spetsalnaya Naznacheniya*—troops directed by the Soviet security and intelligence services.

fundamental question of war: Should it be declared? "The difference between a legal state of war and mere hostilities," a player
later wrote, "triggered a debate as to whether a declaration of war
would impede termination of hostilities."

Another regularly played NDU game examines terrorism, counterterrorism, and assassination—topics that strategy-minded war
gamers usually avoid. Real-life examples of public and private diplomacy are folded into the game, as are the pressure of the media,
a haphazard flow of information, and the conflicts generated in a
nation when a group of its people is held hostage. (There is an
expert on terrorism right on campus; Bruce Laingen, one of the
Americans held hostage in Teheran, is vice president of NDU.)

NDU students also play the Force Deployment and Sustainment
Game, in which officers struggle with the varying demands of as
many as five competing wartime theaters. Players can summon up
information on available manpower and on the readiness of various
industries as far as ten years ahead. These and several other games
use actual information wherever possible. Thus, as Lieutenant
General Richard D. Lawrence, president of NDU, has pointed
out, games can be played alongside reality. Like many other military game enthusiasts, he wants "to develop the capability to 'war
game' actual crises as they unfold so that real-time alternatives can
be explored and problems anticipated" through "quick-response
models and games that are easily adaptable to fit specific crisis
scenarios."

The wargaming of reality is also being explored in the basement
of the Pentagon, where SAGA became JAD. For nearly six years,
beginning in 1973, Frank Kapper worked in that stadium as the
scientific and technical adviser to SAGA.

Kapper has spent much of his federal career amid acronyms. He
was chief of the mathematics and computer division of SHAPE
(Supreme Headquarters, Allied Powers in Europe) and at DOD
(Department of Defense) he was deputy director of the Air Force's
DDA (Directorate of Data Automation). In 1983 Kapper became
Assistant Deputy Under Secretary of Defense for technology transfer. The appointment took him out of direct involvement with
gaming, but he remained part of what is known as the gaming
community, an assortment of perhaps 2,000 retired and active-duty
military officers, Department of Defense officials, think-tank ana-

lysts, academicians, and, surprisingly, government game contractors who are also creators and purveyors of the kind of war games sold in upscale toy stores and hobby shops.

Kapper has always been a warrior at heart; he joined the Marines at fifteen, earned a judo black belt, and became a champion marksman with rifle and pistol. He also has a doctorate from St. Louis University and, as a specialist in quantitative methodology and organization research, is the author of more than thirty-five technical publications. He is a civilian scholar, but a thoroughly militarized one.

Like most of the people in DOD, he liberally sprinkles his conversation with acronyms. He usually explains them to the uninitiated, sometimes as he goes along, sometimes later. When he talked about his days at SAGA, he often uses two acronyms: SIOP, pronounced *Sy-op*, and RSIOP, pronounced *Rice-op*. SIOP is the United States' Single Integrated Operation Plan for total nuclear war; RSIOP is the hypothetical Soviet Union SIOP, with R as in Red Team and O for "offensive" instead of "operation."

"SAGA," he told me, "had, and still has, the responsibility for creating the RSIOP. They take all the best intelligence they can. Then they construct likely scenarios. What SAGA did was create a [strategic] weapon-by-weapon lay-down of the plan under different scenarios, or you could call them different options. Then they take the actual SIOP—the real-world plan that the United States has for a strategic nuclear war—and then they game those two. They do a simulation using different computerized capabilities. And then they determine what the consequences of execution are under different scenarios and different responses to different scenarios, a series of outputs depending upon the outcome.

"The consequences of execution, as it is really called, is one of the most important things that is done by SAGA in gaming the world. It has the broadest impact on the survivability of the free world, especially the United States. The consequences of the simulations feed directly into your current SALT talks, and so forth, and the fact is that the 'window of vulnerability'* was determined as a consequence of these simulations."

*The window was a reputed gap in U.S. nuclear deterrence, primarily the vulnerability of Minutemen silos to Soviet intercontinental missiles. Advocates of the MX missile extended the metaphor when they said the MX would "slam shut" the window.

When the Soviet Union makes an arms-control proposal, the offer is cranked into a war game using what players call real numbers: the actual U.S. arsenal and the best estimate of the Soviet arsenal that U.S. intelligence can provide. Then a number of "nuclear wars" are played to see just what effects the proposal would have on the balance of U.S.-U.S.S.R. forces. The results of these wars are passed on to arms-control policymakers.

Gaming the U.S. SIOP against RSIOP, Kapper said, "is just a matter of parametric variation, and all that means is you play games with the key parameters to the scenario or to the model of the phenomenon. On the one hand, for example, you can increase the number of weapons, you can change their performance characteristics." When I asked for examples, Kapper looked perplexed, for he rarely dealt with people who did not have security clearances. "Specifics are classified," he said.

But he did talk about this ultimate Blue-Red game in general terms to give me "an understanding of how a crisis develops and how you can control it"—because, he said—"once you punch the buttons and everything takes off, you are no longer in control."

"We would play it both conceptually and with people for a variety of objectives," he recalled. "When you normally do the RSIOP and SIOP gaming together, you assume you have already gone to war. Typically, they attack you first. Or you attack them first. Or it's simultaneous. Then you can make policy, such as, 'Let's say they attack us first and we decide to ride it out.' So you can go simplistically. You can 'fractionate' the crisis or conflict in a number of ways. And what you are really talking about is the operational reality of a political decision: 'What the hell happens if I decide to ride it out? I ain't going to launch.' There have been studies for years on this. For instance, take the phrase 'draw down.'

"Let's say I decide I'm going to take all the hits they can throw at me on my missile field X. Let's say I have a hundred missile fields, and on this particular one, I want to find out what the hell it's going to cost them to kill my capability, and that means C-cubed* centers as well as the missiles themselves. Well, effectively the Soviets start out with so many missiles. Well, as they start to

*C-cubed is Pentagonese for C³: Command, Control, and Communications. C-cubed-eye (C³I) adds Intelligence.

launch and attack, there is a draw down of their resources availability, and that's called a draw-down curve."

He then did what analysts always do. He drew a graph. It looked like this:

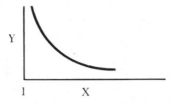

"If you can imagine you have a Y and an X, and this [Y] is the number of missiles and this [X] is time, then effectively what you are seeing"—he made the firing of Soviet nuclear missiles a soft, sloping line—"is a curve that draws down. A draw-down curve. It's resource allocation activity. You're looking at his resources and your resources, and you're looking at your draw-down curve versus his draw-down curve, and two Ys and Xs: his missiles and your missiles and what does his curve look like relative to your curve? What happens if you take the political decision to hold fast? If he draws down his [missiles] to zero, yours may go down from 100 percent to 30 percent, maybe 50 percent. This is something anyone can understand. A major strategic concept."

What Kapper was showing me was part of the wargaming spectrum, with its dark reality of war and its shades of reality, beginning with the dim, nearly pure abstraction of analytical games and their graphs and soft, sloping curves. Beyond, in the spectrum's twilight of reality, is the Pentagon's wargaming stadium.

Entry to the Pentagon's gaming stadium is tightly controlled. A visitor gets a red identification badge and an escort officer. The escort takes the visitor past the command center's guard post, down a flight of concrete stairs, and into a low-ceilinged, ill-lit basement corridor. We walk a short way and then stop below what looks like one of those big, internally lighted white-glass boxes that hang in the windows of small-town restaurants. Instead of EATS, the sign says SAGA.

"For years people have used it as a landmark," my briefing officer, an affable Air Force colonel, told me. "The basement is not laid out the way the rest of the Pentagon is, so people who give directions down here in the basement say, 'Go to the SAGA sign and turn left at the next corridor' and things like that. The other landmark down here is a purple water fountain."

He handed me a written description of SAGA soon after I passed under the sign. The paper, like the sign, still called JAD by its old name.

Veteran gamesters still say "SAGA" when they talk about the gaming in the Pentagon's basement, but the officers who briefed me there never slipped. They invariably referred to the activities by their official name, political-military simulations. That is also what they were called when Robert S. McNamara, as President Kennedy's Secretary of Defense, ordered the Joint Chiefs of Staff to inaugurate them in 1961.

McNamara was reacting to the President's anger over the debacle of the Bay of Pigs. Kennedy believed that the abortive invasion of Cuba had been badly planned, that he had been badly advised, and that the military had little understanding of how the political world worked. To give Kennedy's military advisers a better grasp of realpolitik, McNamara turned to universities, particularly Harvard and the Massachusetts Institute of Technology, where political scientists had been using gaming in crisis-management studies. The academicians welded their new ideas about political gaming to traditional military wargaming.

The result was the Joint War Games Agency, each of whose three divisions—Limited War, General War, and Cold War—ran top-secret games. The name of the agency was later changed to the Studies, Analysis and Gaming Agency, but the mission, and most of the methods, never changed. Pentagon gaming as described in a 1967 official paper is quite similar to what is described in a paper I was given in 1985. Both, for example, point out that the games are not intended to be predictive. And scenarios in both eras describe situations occurring in the future. "We are using the same format that we were using in 1962," the briefing officer told me. "We modify it, but basically it's a disciplined structure."

A notice to participants in one of the early games says, "Any resemblance between these hypothetical events and the real world is *not* coincidental. In preparing the scenarios from which BETA I & II will be initiated, a number of real-life considerations have been introduced. They have, however, been intermixed with some highly imaginative and speculative material intended to provoke thought and to stimulate politico-military discussions. BETA is neither a war-game nor a technological analysis. The aim is to explore a wide range of plausible, (if not necessarily probable), contingency situations in order to derive broad insights into current and potential problems. Participants are reminded that play is limited to the TOP SECRET NO FOREIGN level and discussion of RESTRICTED DATA is not authorized."

The statement can probably be generally applied to most of the games played before and since. The games traditionally *have* been based on reality. One for which obtainable records exist was called Epsilon I-65. Played in 1965, it looked ahead five years at NATO— presumably without France, which did withdraw its military forces from NATO in 1966. Ingredients for its scenario included transcripts, full of anti-American rhetoric, of a General Charles de Gaulle speech and press conference. Participants in Epsilon I-65 came from the State Department (including representatives from the Paris and Bonn embassies), the CIA, U.S. Information Agency, the White House, the Joint Chiefs of Staff, the Army Chief of Staff, the Chief of Naval Operations, the Chief of Staff of the Air Force, the Commandant of the Marine Corps, and the Defense Intelligence Agency. In the thin files on Epsilon there is a paper festooned with baroque, star-centered doodles, along with a note: "Details are not too important."

When I told my briefing officer about the game records I had found in government archives, he seemed surprised that I had been able to see records of games, even though they were nearly twenty years old. In the Pentagon what happened in a game in 1965 can still be top secret in reality two decades later.

My formal introduction to the secret little world of Pentagon gaming began with a film that opened with clips of the Soviet invasion of Afghanistan and Cambodian troops on the march in Vietnam. "Because of United States assumption of leadership in the West," a narrator said, "crises throughout the world have a profound effect on this country. The reaction of the United States to international crises is vital to the continuation of world peace and stability. Many forums are available to make full use of informed judgment in international affairs, particularly in crisis situations. . . . Political-military simulations are one such technique."

Occasionally a player's familiar face—Henry Kissinger, Alexander Haig, James R. Schlesinger, Admiral Bobby R. Inman, former director of the National Security Agency—flashed the screen while the narrator explained how the games ("simulations") were played. For anyone who tries to know the world through the specifics of history and culture, an attempt to understand Pentagon wargaming can be a frustrating enterprise. The language of the game is dense with bureaucratese. The narration of the film, like the briefing paper I had been handed, was excruciatingly generalized.

Neither SIOP nor RSIOP was mentioned by my briefing officer. Much about gaming is highly secret, and many officials involved in gaming treat the subject as if nothing has ever been published about it. Even the existence of the Pentagon gaming center, in its first incarnation as the Joint War Games Agency, was considered a secret. The public's first knowledge of its existence came in 1962 in one of the oddest newspaper leaks in Pentagon history—disclosure in a comic strip.

Milton Caniff, creator of the comic strip "Steve Canyon," was a great favorite of Air Force officers, who saw to it that Caniff was given inside knowledge of defense matters. He was especially briefed about Air Force bases that could serve as backdrops for the adventures of Steve Canyon, a firm-jawed Air Force officer. Caniff's friends added to his knowledge by letting him sit in on a SAGA game. The idea was to give him background. But one Sunday in 1962, to the chagrin of Caniff's sponsors, there it all was: beautiful industrialist Copper Calhoon, a nasty though patriotic adversary of Steve, sitting at a table in the Pentagon basement and being told, "The politico-military game is played in this manner. . . ."

Caniff's comic-strip description of the game in 1962 was clearer than the one I was shown in the official movie and briefing paper in 1985. Copper Calhoon and the others at the table were told, "You are persons of importance in your own fields. From you we hope to receive new slants on problems we live with here in the Pentagon every day. You have been divided arbitrarily into Blue and Red teams. The Blues will be the United States and its allies. The Red will pretend to be the command staff of the Red bloc nations. You will be given a series of hypothetical situations. Each team, in separate rooms, will decide what next to do. You will have all the actual forces of your own group to throw into the mock crisis. . . . We on the Control Team are umpires. We do not wear uniforms, but we are from all branches of the Department of Defense. We will give you true answers as to known strengths on both sides—which is one reason these games must be classified as secret—and no documents you use may leave these rooms."

Ever since Copper Calhoon's day, industrial executives have been invited to the games. So are, in my briefing officer's words, "the same kind of people who will be summoned to the White House Situation Room to deal with crises" and "the kind who can go back to their offices and have some impact on policy." Players have included not only cabinet members and ambassadors, but also academic specialists on subjects from Afghanistan to Zambia, and ranking officials from the State Department, the National Security Council, and the Central Intelligence Agency, and from every other executive branch agency (with the possible exception of Health and Human Services, for which there are rumors but no record). Congressional representation is relatively rare; one veteran Capitol Hill analyst of national security matters told me neither he nor any of his colleagues had played in a Pentagon game, though all of them had played at various war colleges. All Pentagon players must be American citizens and hold security clearances. Even representatives from NATO countries are excluded.

The thirty to thirty-five players—officially, "participants"—are usually divided into two player teams and a control team. Each team is put in a room that contains little more than a long table, chairs, and equipment for viewing video tapes. On the walls are maps (usually those published by the National Geographic Society for its members) and whatever printed material that needs to be displayed. There is also a television camera attached near the ceiling and aimed at the table.

The game scenario—"believable, real and projected world tensions, activities, and policies"—sets up the crisis situation, which is generally a year or more in the future. The JAD game staff, which numbers about forty-five, may work for as long as six months developing the scenario. Researchers often go overseas to interview U.S. ambassadors, senior military commanders, and experts on the region that is the setting for the scenario. JAD staffers may even gather information on the performance of specific weapons—"to the trench level."

As an example of one of those information-gathering missions, retired Army Colonel Alvin D. Ungerleider told of a trip to Israel, Jordan, and Lebanon in 1968 in preparation for a Middle East war game. Ungerleider, then chief of the Limited Warfare Division of the Joint War Gaming Agency, recalled, "We wanted to gain as much information as we could on types of forces, deployment, and all that goes with land warfare. This was after the 1967 war, and we wanted to know the answer to, 'What is one Israeli soldier worth compared to how many Arabs?' We set up a ratio of one to eight—one Israeli to eight Arabs. When we tried this ratio, nothing happened. The Arabs couldn't move. So we started going down. Seven, six, down to three. The scenario was that the Egyptians would strike across the canal, not on Yom Kippur but, we said, on *Shabbat*, the Hebrew Sabbath. We had them to almost the exact spot they did reach in the 1973 Yom Kippur War."

The typical game scenario is built up from such real-world information and intuition. The final game scenario is in two parts, a "world scene," in recent years usually presented documentary style on a video screen, and a crisis, in written form. The world scene thrusts the players into the future and lays out the situation in which the crisis takes place. The staff strives to make the video presentation realistic. "We get support from the Defense Communications Agency, which as a matter of course records all the news programs, and they save that, and we can go and raid those clips," I was told. "The world scene creates synthetic history to bring participants forward. We project into the future to change history a little bit and to get participants away from current-day policy restraints and let them freewheel a little."

The chairman of the Joint Chiefs of Staff, the nation's senior military officer, selects the subjects for the four or five major games played each year. His candidates for subjects include those solicited from the State Department, the Central Intelligence Agency, and every major military command. The solicitation letter, which is tightly circulated, asks for issues that may be developing within the next two years. "We like to think that there is no *no*—that if you can't address the subject in this medium, then I don't know where you can," my briefing officer said. Subjects are "chiefly geographical—regions of the world that have specific problems," although there have been games about terrorism, strategic balance, and arms negotiations. Sometimes a game will examine contingencies, such as a naval quarantine of Nicaragua or a dry run on dealing with the Philippines during the collapse of the Marcos Administration.

"It's not a 'war game,'" my briefing officer insisted. "We have gone into conflict, but we are not equipped to give any confrontation analysis. We do no analysis here. In a political-military simulation, the focus is on the role military forces play in international policy. Indeed, for a country like the United States, they are a major component. We emphasize policy issues. We want them to talk about the broad policy issues and get some insights into implications of an event and how the world might unfold and what that means for the United States. There's no role-playing—that's important to me. A lot of people wonder why we don't role-play, because that's commonly done in war games. In war games where you know you're going to get into a confrontation, role-play moves the game ahead more rapidly."

To look at the transcripts of games, as I have, and say that the players are not "role-playing" is to claim that an actor in a play—or a quarterback in a game—is merely part of what is going on. Of course people role-play, at least unconsciously. When a person has the right in a game to use *I* and *we* and *our* and mean *an entire country*, then an odd form of role-playing is obviously happening.

Consider this exchange from a 1964 game, played to see how to "prevent the loss of South Vietnam to Communism." In the dialogue between two "Blues" can be heard not only impassioned role-playing but also an early example of the hawk-and-dove debates that would rage over Vietnam for years to come. Several Blues speak in the 1964 critique. They are not differentiated in the transcript, in keeping with the concept that *teams*, rather than individuals, are playing the game. Nor, in keeping with the traditional anonymity of the game, are they identified.* But they patently do not speak with one voice.

---

*Although I was granted reports on the games under the Freedom of Information Act, Pentagon censors "sanitized" the reports by blacking out names.

Control, during the critique of the 1964 game, mentions that, as a result of some aggressive Blue moves, "world opinion" was going against the United States in the United Nations. Blue speaks up, reacting not to the game moves but to his own view of reality. "Who says world opinion is going to go against us?" he asks. "I admit we have been trying in the last several years to alienate every friend we've got, but we haven't alienated them *all* yet."

A Blue says, "Everybody knows we are at war with Communism," and then begins a litany of countries "we've lost"—Latvia, Lithuania, Poland, Czechoslovakia, China, Cuba.

Another Blue questions whether people think that "we *lost* Lithuania. They don't think we ever had it."

"Lots of people do," the first Blue says. "We had Czechoslovakia. . ."

"*We* had Czechoslovakia?" someone asks.

"Yes. It was a free and independent country," a Blue replies.

"You have to be a little careful, I think, of equating our opinions with world opinions," the dovish Blue begins. But the hawkish Blue resumes speaking.

"If we back off," he says, ". . . we've lost Vietnam. We've got to use some other type of tactics. . . . We could put North Vietnam into the Stone Age."

The modern JAD games try to focus on teams, not individuals. But players often come away talking more about insights than major discoveries. In a Middle East game, for example, Blue's problems included a shortage of aviation fuel. When Red started setting kerosene fires in a large city, a player from the CIA remembered that many Middle Eastern cities had large kerosene supplies—and kerosene could substitute for jet fuel in an emergency.

A game does not move rapidly; it can last four days, from about eight in the morning to about four in the afternoon, with occasional overtime. At lunchtime on the first day orders are taken and sandwiches are brought into the rooms. On the other days there are lunch breaks. The staff does it that way, the briefing officer said, "so we can get a full cycle in the first day. They just sit there at their table and keep going. The other days, we need to get them out of our hair. . . so we can get the materials typed up properly and get the rooms straightened up."

A typical move takes about three or four hours to make and report to Control. "When they begin to address the new situation, time is stopped. The four hours fit on one tick of the clock," the briefing officer said. "There are no late-breaking news flashes, no demarche from the Soviets. The reason is because we want them to achieve consensus and discuss the policy aspects of it. We don't want them to prove how they can respond with the right actions."

Moves are conveyed to Control in writing, according to a strict format that resembles the outline of action I had seen on the blackboard in the Persian Gulf game: assessment of the situation, national objectives, strategies, specific moves, contingencies. When a team makes a move, Control "projects" the scenario in response to that move. "Since I've been here," the officer said, "I've seen projections as short as two hours and as long as one year. Some of our simulations have started out in a hot-war situation. So then you are going to have short-term projections. On others we've gone out a number of years. It seems to me the further out in the future you get, the longer the interval between moves can be without destroying your game. I've also seen a game—not done by us— where they had three-year intervals between moves, and that was too much. It just took away too many options, too many prerogatives of the teams."

Control may also inject twists—wild cards, some of which may have come from a stockpile of secondary crises. (The Federal Emergency Management Agency, for example, may add a natural catastrophe, which strikes simultaneously with the international crisis.) There may also be a Green Team representing neutrals and a Yellow Team of friends who are not official allies. In a Red-Blue game, where Blue is the U.S. National Command Authority and Red is the Soviet's, "Red moves are seen by Blue through a filter," I was told. "Anything that Blue's intelligence would not normally receive, Control is going to filter out. And if Control adds anything to it, he has to add that back and tell Red what he added in terms of what the other countries in the world might do. Control is the weather, the calendar, the other countries, all other factors. They can have earthquakes in their pockets."

A move might have several facets, he said, giving a cautious example: "Move the Seventh Fleet from point X to point Y. Notify the Joint Chiefs of Staff to alert European Command to prepare for Z. Dispatch a demarche to whatever country it might be, warning them not to do this. Schedule the President to go on national television."

The briefing officer's aide, an Army major, talked about the art of creating a scenario. "The type of assumptions that you put into the scenario has a direct bearing on the types of outcome that you will have and, as a result, you try to focus more on the plausible than on the worst case. You want plausibility which, in turn, gives you credibility. The longer you project the scenario into the future, the more synthetic history you have to build in. You have to be sure you are projecting current trends and events that follow a plausible line.

"We try to provide the player teams with actions and information that tend to replicate the National Command Authority's. We may not have fast-breaking defense news—film clips and so forth—but we do try to provide them with the tools that would be ordinarily available to them, primarily in the way of information." Usually the team members themselves have enough knowledge and experience to make feasible decisions. If, say, they decide to set up an airlift, someone on the team will probably know how, when, and where to do it.

"The public affairs aspect of a crisis are often very, very important," the colonel said. "And they are difficult to capture in our medium because we don't have the clamor, the telephone, the demand for a statement. . . . We tend to think of the executive branch. But it's very useful to get a Congressman or a legislative staffer to say, 'I don't think Congress will buy that.' "

In an adversary game the player teams oppose each other. In a parallel game each pair of teams is given a problem and responds to it without knowing how the other pair responds. In both versions the Control team monitors the player teams and makes sure that the game is moving along. Control keeps the player teams under constant surveillance through a closed-circuit television system to get a "better understanding of team intentions and rationale." Control represents so much reality and sees so much of the world that Frank Kapper calls Control "the Hand of God."

A player team is, collectively, the National Command Authority—the leadership of whatever nation the team represents. No one member role-plays an actual job, such as president or chairman of the Joint Chiefs of Staff. The team does, however, select a captain who serves as moderator in discussions leading up to the collective decision on a move. When the players are assembled my briefing officer said he tells them to be themselves, not somebody else: "We don't want to know what you think George Shultz would do in this case. We understand where you're from, that you have certain biases and certain ways of looking at things. You're in uniform, or you're State Department or Treasury or whatever. But it's your ability to articulate, to persuade your colleagues, and try to achieve a consensus—that is what is important. We want you to bring into the room what you have in your own kit bag. Don't try to adapt to what you think some role would be."

The game goes on for three moves, sometimes four. The game is played so that the players will make moves, not to see if the moves are inherently good or bad, wise or stupid. "You can have a seminar and discuss things for days and days, but until you have to make a decision based on it," the colonel said, "you haven't completed that last ten percent of the thought process."

Whatever is said by a government employee, civilian or military, is considered off the record. "The reason I think we're still around after all these years is non-attribution. We have espoused and practiced a successful policy of non-attribution. People can say, 'I know my Secretary may have taken this position. But I don't agree.' And he can feel confident that it's not going to be in *The New York Times* or *Washington Post*."

A high-ranking Navy officer who had played in three JAD games when I interviewed him called the system "no-fault" because "you can't be judged for what you do in there." But unofficially—and with unknown future career reverberations—a military or civilian superior may not like what he sees on the closed-circuit television or on the videotapes that are sometimes circulated among high-level officials. "One of the guys in Control asked me why I didn't recommend a military solution," the officer recalled. "He called me a wimp because our team didn't go and fight. They called us all wimps. I told them they could learn a lot out of our *not* fighting."

A game usually ends with a critique or, borrowing from the Naval War College, a hot washup. "That's the first time the teams get a chance to talk with each other, and the first time they see what the other team did, how they approached it," the briefing officer told me. "Since 1981 we have modified this, and either as part of Move Three or as an additional move, we ask for the policy recommendations. It's a difficult twist because, after you've gone for two or three days and the people are really into a situation, and then you say, 'Stop. Leave the game behind. Come back out of the future, come back to today and tell us, based on whatever insights or greater understandings that you have, of all the implications and dimensions of this situation, what recommendations do you have for U.S. government policy today?'

"The kind of recommendations we get out of a simulation—we may have twenty to thirty." He turned to an aide. "How many did Tau generate?"* "About thirty-five," the aide answered. The colonel nodded. "They ranged all the way from very general to very specific. They could be as broad as—the United States government should initiate a policy review of this region of the world or it could be very specific—the U.S. government should put an additional diplomat in a certain capital." Another officer told me that every game has produced "enough ideas for them to get into the policy picture."

JAD-generated ideas are getting into that picture because JAD in recent years has been closing the time gap between JAD games and reality. Dr. William G. Lese, Jr., Frank Kapper's successor as scientific and technical adviser to JAD, has presided over this change. An organization long renowned for what many called "seminar gaming" has been changing into a computerized source of "rapid-response analysis," in Lese's words. "The chairman [of the Joint Chiefs] needs advice in a day or two," he said. "We felt we had to fix this.

"For war games we want computer systems *and* the big dog-eared maps and the grease pencils. At the CINC level, suppose there's a threat in his theater. How is he to fight it? Has he the right forces *now*? Are the right forces headed his way at the right time? The war-planning community—the CINCs and us—must have the same kind of hardware and software."

Contingency planning is an old military occupation, which com-

---

*JAD gives Greek-letter names like *Tau* to its simulations because of a lingering academic tradition. That professorial labeling, which clashes gently with the military's tradition of hefty acronyms, goes back to the earliest days, when academic game creators and the Pentagon discovered each other.

puters have speeded up. Traditionally planners have used war games as the basis for contingency plans and then filed the plans away for future use, possibly years later. Now if the Joint Chiefs or the National Security Council warns a CINC that something may be happening in, say, the Horn of Africa, the CINC responsible—who in 1986 happens to be a Navy admiral—may ask the Naval War College for a scenario that shows Soviet forces, down to such details as how many and what kind of aircraft are operational and on average how many are down for repairs. This kind of information is routinely stored in the college's wargaming computers. Lese's tools will make the parochial request to the college unnecessary.

Lese's SIOP versus RSIOP scenario begins, he says, by "taking about four people and christening them Soviets. We make them think like Soviets. We give them all we have on the Soviets. We ask them, 'If you were a planner in the Soviet Union, how would you go about doing it?' They look at targets: Nuclear systems? Cities? Leadership? They use tactics, allocating their weapons against us. Then we send that plan to Omaha [headquarters of the Strategic Air Command] and they war-game it in the larger strategic, homeland-to-homeland nuclear exchange. Should we sit and ride out the first attack? What would be the effects of launching with warning? With launching under attack?"

Playing Red is the most challenging aspect of Blue-Red gaming and, among war simulators, the most controversial. Even enthusiastic advocates of gaming cannot easily explain the worth of having Americans, many of whom are not experts on the Soviet Union, make believe that they are Soviet strategists. Members of the Strategic Studies Group at the Naval War College, for example, just shift sides from Blue to Red. A commodore who spent a year in one of the groups told me, "The Med group would play the Soviets for the Pacific group and vice versa. We worked on ideas like what is the thinking of the Soviets, how do their weapons systems work, what are the ranges and capabilities." Much Blue-Red playing is at that level of sophistication.

In the heart of the warrior is a patriotism that cannot be easily muted, in play—or in study. At the U.S. Naval Academy, for example, an associate professor of Russian history said, "I've had many students ask me if I were a communist." The midshipmen think that "somehow if one either implicitly in some way criticizes the United States or says something favorable about the Soviet Union, that makes you unpatriotic. . . ."

If an American warrior cannot find it in his heart to be a Red, then his presence on a Red team skews a game and so distorts its results that it is difficult to see their value to realistic policymakers. This is one reason that diplomats like my skeptical teammate in the State Department tend to look upon gaming as a military pastime with limited worth to civilian policymakers.

Many civilians who possess the necessary clearances to play war games harbor such animosity toward "the Reds" that they think bad-guy instead of Red-guy. I know a civilian naval analyst who, chosen to play a Soviet submarine squadron commander in a Naval War College game, "started thinking bad and mean," projecting a patriotic attitude rather than his professional knowledge of a Soviet submariner's training and doctrine. "I wanted to put my men on long shifts, treat them harshly, be rotten," he said. "I thought more of that than I did of tactics." Another civilian who tried to be a wily Red in a JAD game made an Army general so suspicious, he sought out the game director and demanded to know whether the Red player's security clearances had been thoroughly checked— because a real American would not have been as convincing.

Academicians seem to be safe choices to play Red and heed the gaming imperative—"think Red." Lincoln Bloomfield, an MIT professor who was a pioneer SAGA game director, told me about a game in which a player was a chillingly convincing Red. The game, played in the late 1960s or early 1970s, focused on Guatemala, then "very much like Nicaragua or El Salvador today." Bloomfield, "in a sort of puckish mood," asked Richard M. Bissell, Jr., to play a guerrilla leader.

Bissell, as CIA deputy director for operations, had been an architect of the Bay of Pigs and had resigned after the invasion. "He knew as much as anyone in the world about how the United States responded to insurgencies," Bloomfield said. "He played the most brilliant hand any guerrilla leader has ever played in screwing the United States government nine ways from Sunday because he just knew instinctively what the American weaknesses were and how to exploit them. When the game was over, some general came up and said, 'You've just got to lock away all these files because it's a blueprint for how to diddle the U.S. government.'"

The insistence on keeping games locked up has done more than keep them from "the Reds," however. Cool, objective analysis of the games, particularly those during the Vietnam era, would give Blue an opportunity to see how realistically Red was Red and whether swaggering, bomb-'em Blues truly understood the rules and goals of real American policy. Bloomfield, for one, has been urging such an analysis.

Gamers sometimes try to solve the Red player dilemma by establishing "a Red cell," staffed by intelligence experts from the CIA, the Defense Intelligence Agency, and universities. The problem—"How to Think Red," as a National Defense University workshop put it—is complicated by the fact that the U.S. intelligence community tightly holds what is authoritatively known about Soviet decisionmaking and begrudges sharing this information with the gaming community. "Defectors are also used," a Pentagon official told me, "but you have to remember that whatever they know was at the time the individuals were working for the Soviets. That might be years ago. We know we are constantly changing, and we assume they are, too."

But all games, including war games, are inherently conservative and resist change. Players preserve strategies from one season to another. New rules are resisted. Team colors rarely change.

The tone and orchestration of games played in the 1960s will have familiar strains to listeners in the 1980s. The nations in the games that follow are the United States, the Soviet Union, China, India, and Pakistan. The games are about nuclear weapons.

CHAPTER 2

"It's Hard to Start a War": The Nuclear Factor

The letters from Air Force Colonel William Thane Minor, chief of the Cold War Division of the Joint War Games Agency, to McGeorge Bundy and other high-level Johnson Administration officials sounded like an invitation to a seminar. Professor Thomas C. Schelling, of Harvard's Center for International Affairs, would be the moderator at a Senior Review, with Refresher Material, of a political-military game involving India, the Kashmir Issue, the Sino-Indian Border Dispute, Indian-Pakistan hostilities, and activities in Sikkim, Nepal, India, Burma, and Pakistan.

By the time of the 1966 letter the pattern of the Pentagon games had changed little since Minor's predecessor, Air Force Colonel William Jones, had overseen the founding of the Joint Chiefs' war-gaming operation. Jones gave the games credit for helping to bring together the "whiz kids" of McNamara's Department of Defense and the suspicious, hostile officers of the Pentagon. The games had given the civilian newcomers and the military veterans a chance to meet in a friendly but competitive atmosphere. Jones called the games "the damnedest salesman's conventions you ever saw." In 1967, when Earle G. Wheeler, chairman of the Joint Chiefs of Staff, invites Walt W. Rostow, special assistant to President Johnson, to play in a game, the camaraderie of politician and soldier is manifest in the letter, addressed to "Dear Walt" and signed "Bus."

The subjects of both the 1966 and 1967 games foreshadowed imminent events. The Air Force Television Facility recorded the Senior Review of the India-China-Pakistan game, and the Joint War Games Agency staff made a film of it to show officials as part of the preparations for talks between President Johnson and India's new prime minister, Indira Gandhi, in the White House in March 1966. The game, played in February 1966 but set in October 1970, had a grimly prophetic scenario. Famine, foreseen in the script, was hovering over India when Indira Gandhi arrived in Washington. The United States had rushed food to India, as in the scenario, and Gandhi had dispatched troops to put down uprisings in Kerala, just as the games agency script writers had written. But the scenario's threat of superpower confrontation had never come.

The 1967 game focused on how U.S. and Soviet strategic weapons and antiballistic missiles figured in a crisis set in 1972. The game ended little more than a month before Johnson and Soviet Premier Aleksei Kosygin met at Glassboro State College in New Jersey on June 23 and 25, 1967. The principal subject of their summit meeting was antiballistic missiles.

In the annals of these games can be seen the glint of a Red-Blue mirror image that distorts the "Red" thinking of true Blues and flares into distrust in the real world of India and Pakistan, of Johnson and Kosygin. Then, as now, players soar beyond themselves—*you* and *I* become *us* and *them*: *Blue* and *Red*—*the United States* and *the Soviet Union*. These anonymous men bestride the world. Reading their words is like watching a shadow play and wondering if the players will ever become flesh and blood, wondering if their gestures—an army hurled here, an aircraft carrier hurled there— will ever become the hard, real stuff of war.

"One of Control's problems is to introduce plausibly the behavior of the countries," the Game Director says.* He is talking at the Senior Review about Nu, the 1966 India-Pakistan-China game. But he could be talking about all games before or since. "Plausibility usually comes up for a little criticism at these sessions. Let me say two things about it. First, most of life seems to be a sequence of

*The Game Director's name is still deleted in the scant files about Pentagon political-military games. Obtainable records indicate that Minor and Schelling participated. But, due to the vagaries of game files, neither their names nor the names of anyone else appear in discussions of the game that were released under the Freedom of Information Act.

implausible events. . . . The problem is to choose among implausible alternatives and even if one can interpret these games as true history, rather than synthetic history, one would still, as the historian does, have to say, that's just one way things could have gone. . . .

"The other point about plausibility is the Control Team often finds itself groping for something that is fairly plausible, chooses something, works it over for a while, and it becomes very, very plausible through a process of getting familiar with it. I think, frequently, what these games can accomplish is to demonstrate that what often appears on the surface to be implausibility or improbability is merely unfamiliarity. It's hard to work with any sequence of events in a game for several hours without its beginning to seem either real or as one that could be real."

Assessing games in general, he says, "Some games are particularly good at focusing on the process of decisionmaking, of planning or estimating an adversary. Some games stir up substantive problems and policy issues. Some games are especially rich in by-products. Most games are a splendid cram-course in local geography and politics."

The primary problem that the game attempted to stir up was the possession of nuclear weapons by China (which is referred to as "Communist China" or "Chicom" to distinguish mainland China from U.S.-supported Taiwan). The game looked at what could happen in 1970—the time frame of Nu—if China had "a limited nuclear capability." This consisted of thirty-two medium-range ballistic missiles with a range of about 1,000 miles and some nuclear-carrying Badgers, a Soviet-designed medium-range bomber. A purpose of the game, the Game Director said, was to discover whether China's nuclear capability "would either lie like a shadow over the situation or bring about some kind of nuclear action."

The possible use of nuclear weapons dominates Nu, which was the twenty-fourth game run by the Joint War Games Agency since 1961, and the first concerned with a Chinese-Indian conflict. Most of the previous games had explored policies dealing with Berlin, the Middle East, and Southeast Asia; there also had been games on disarmament and one on revolt in East Germany. Chi I-63, played in October 1963, explored a favorite Kennedy Administration subject—counterinsurgency, with a complex game involving the long-term trouble spots of Angola, Indonesia, Iran, and Venezuela. Three games—two on Southeast Asia, one on the "internal problems" of NATO— had been made into "documentary-style" films that were shown to Secretary of State Dean Rusk and other high-level officials.

Lower-ranking officials from several agencies had played the first round of Nu in the days before the Senior Review. The "action teams" had been given stacks of information—"Rainfall varies from 150–200 inches along the Arakan and Tanasserim coasts" of Burma; China's "freight-car inventory is fairly modern and is not being used to capacity"—and a chronology beginning with the second millennium B.C. and ending with "16 Oct 1964 Explosion of China's first atomic device" and "14 May 1965 China's second nuclear test."

Nu was essentially two games in one. One action team made moves in reaction to a starting scenario that had China issuing veiled nuclear warnings to India and then pushing Indian forces back to Chinese-claimed borders, with Pakistan meanwhile invading Kashmir and threatening India. The other scenario began essentially the same way but had the Soviet Union intervene with an air strike at a Chinese nuclear missile site.

The game revealed a nuclear secret that would haunt future war games: Players hate to cross the nuclear threshold. For example, a player, cryptically identified as *CPR II (OSD)* [a member of Chinese Team II, from the Office of the Secretary of Defense] says, "I was the nuclear hawk on our team, and I did in fact try and get our team interested in a small nuclear explosion, ambiguous to its nature, so that we might even get the Russians and the U.S. confronting one another over who may have been responsible. But I got no enthusiasm out of the Chinese for that gamble."

At that point the CIA member of the team remarks, "I might say, Mr. Chairman, I was the dove on the Chinese Team, and all the way through my thinking was, 'Let us not shake those nuclear weapons because we have far, far too much to lose.' "

The game's scenarios and Control itself drive the players again and again toward nuclear confrontation. Later the Game Director admits, "It was hoped in this game that we might push things to the point where at least some kind of nuclear intervention would be seriously considered." By way of explanation, he says, "If any generalization comes out of these games—and it may not apply to the world, I think it does—is that it is very, very hard to get a war started."

The countries—that is, the Americans *playing* countries—continually resist nuclear escalation. One player, part of a China team, complains about the emphasis on nuclear issues. All the teams, he says, ignored "a collapse of India and what impact this would have on a worldwide basis. We ignored the impact of these large Chinese ground forces that were in position and able to move in practically any direction. We ignored the impact of the loss of Sikkim, Bhutan, and the Nagaland—again always coming back to the nuclear question. Actually the thing I have some trouble with here is, what is this tremendous emphasis on something that even today's world will not face?"

U.S. team members "had quite a lengthy discussion on at least two occasions as to whether we shouldn't consider taking out the Chinese nuclear facilities," but the team concluded that the facilities "were probably more of a liability to the Chinese than they were a real concern to us." Another U.S. player says his teammates wanted the Soviets to knock out China's nuclear capability—but "not necessarily by the use of nuclear weapons." At one point, he says, "we discussed the question of how far, for example, the President of the United States could put himself in the position of egging on two major powers, which was really what we were doing, if by no other action than by removing any restraining influence that we might have otherwise observed in normal circumstances. The extent to which he could satisfactorily do this—up to and including a point short of a major war—was discussed."

The White House man on a U.S. team, speaking "from the point of view of United States interests," tells the director, "I think that if one were drawing lessons, one might say that there's a pretty sizable argument in terms of persuading the Indians not to go in for a nuclear capability. We concluded that [nuclear weapons] didn't mean much to the Chinese. . . ."

The Game Director, who seems to be snappish about the players' nuclear reticence, describes U.S. avoidance of confrontation as "Let's hold their coats and see how far they'll go." But he appears to endorse someone's suggestion that coat-holding "might be a rather cheap way for both the U.S. and the U.S.S.R. to achieve a common objective, and that's the elimination of this needling Chinese nuclear threat. . . .

"If the Soviets took out the Chinese nuclear capability," he later says, "both would be weakened. And the United States would gain relative to both. The Chinese would be weakened by having lost their nuclear weapons and lost a share of influence and power. . . . At the same time, the Russians would certainly have lost a great deal in terms of any long-run expectations of being able to reestablish some degree of influence in China by this move."

A Chinese team member admits that they had "found our own capability to be an embarrassment. It seemed to emerge that nuclear weapons weren't a very good weapon." A China team player from the Defense Intelligence Agency says that from his team's point of view, "thinking in as Chinese a manner as possible," the team felt China's objectives "were to secure the frontier countries without arousing a major war."

The Game Director asks members of one of the Indian teams whether they had been worried—"either about a Chinese attack or about the U.S. provoking an attack by being too wayward with nuclear weapons."

The India II man, from the Office of the Secretary of Defense, says, "I don't think that we were terribly worried about either of these. I think we felt that the threat by the U.S. to attack China was a very fine deterrent indeed. We thought it was rather splendid and we felt comfortable under the nuclear umbrella. . . . [It's] easy to play a line of getting rather angry at the U.S. We Indians were very self-righteous. I would say that the problem of the U.S. attacking the Chinese with nuclear weapons wasn't very serious."

A member of one of the U.S. teams says, "The team really wanted to get into a situation that was positive to us—Sino-Soviet confrontation. We recognized there were dangers, and one of the things that concerned us most (it particularly concerned one member of the team) was the fear that bringing the Soviets and the Chinese into a confrontation might open the door to greater Soviet influence in the subcontinent. This bothered us, but our conclusion on this point was that the risk was worth taking. . . .

"From my point of view, we were not at all reluctant to see the Chinese and the Soviets move in this direction. We recognized that this might even raise the question of a nuclear engagement.

We weren't much concerned about the Chinese nuclear capability, except that we were interested in seeing it disappear, and preferably disappear by Soviet action rather than on our own."

In a summation of the nuclear issue, the Game Director says of the failure of the game's "Chinese" to use nuclear weapons, "While that doesn't prove that the Chinese wouldn't, it proves that the Americans [on the Chinese teams] didn't find this a plausible situation. . . . It seems to me . . . that to the members of the Chinese team, it does not look as though a threat to retaliate against the Chinese first-use of nuclears in 1970 with these weapons makes any real difference, because it isn't plausible for the Chinese team that they should initiate nuclears. . . ."

The Agency for International Development man on a U.S. team says, "I don't know whether I speak for the whole team or not, but I know the team well enough that they'll say so if I don't. Insofar as I'm concerned, I went on the assumption that the U.S. wouldn't ever use the nuclear bomb first. It's a paper bomb from that standpoint.

"I would guess that the Chinese would come to the same conclusion. They'd be foolish to use the bomb first themselves, when they've only got a few of them and we have an incredible capability. It seems to me that nuclear bombs for this kind of exercise might just as well be shut up in a cupboard somewhere. They don't really enter into the picture at all. . . .

"But still, wouldn't some valuable insights be derived if people were willing to try things where the risks are clearly not as great as they are in the real world? If you can assure them that trying these things does not brand them with the name of 'hawk' and 'dove' outside of the game and you also urge them in a preliminary briefing and discussion to be willing to gamble—to play the game with the purpose of deriving the insight rather than to see whether or not to keep the United States from losing or whatever—wouldn't this be important?"

"This is a war in which if you lose it's *not* for keeps," the Game Director answered. "We usually try to solve this by designing what we think is an absolutely foolproof starting scenario which requires hard decisions on both sides."

The discussion suddenly veered to the idea of the game itself, with the "Indian" from the Office of the Secretary of Defense

saying, "I'd like to add to your comments about the way these
games seem to go, not only that the aggressor can take a small bite
and get away with it because the 'defender' is willing to settle for
a small loss, a very small loss, a loss which he considers negligi-
ble. . . . If I were playing a thousand games in succession and each
time I would only lose a very, very small bit, maybe after about
three hundred games, I'd start worrying. . . .

"I guess this carries over into the real world in the sense that
one *does* have to look at the question, Where does one stop giving
in? That's a point which Mr. [Secretary of State Dean] Rusk makes
very often nowadays about South Vietnam, I personally think quite
accurately. If you don't stop aggression there, then you have the
other choice of either in Laos or Thailand or Malaysia, or somewhere else.

"I wonder if one could design games that run far enough in time
of the moves where lots of small losses by one side become no-
ticeable, aggravating, and a cause for a rather strong countermove
even if it's somewhat riskier than one feels like doing."

He gets no direct answer. In what passes for an epilogue to the
game, the Game Director sidesteps the nuclear issue: "Now that
the game is over, and now that we have several years until 1970
to think about the problem, maybe we should look very seriously
not only at the question of how serious it would be if India collapsed,
but what difference it would make if it collapsed: (a) with China
taking the credit; (b) in spite of American efforts to help; (c) because
the Americans were unwilling to fight Pakistanis, and therefore let
it go; (d) for internal reasons that, had they been anticipated, might
have been somewhat remedied; (e) for things that were so wholly
beyond our control that all we could do was get ready for it."

If the dialogue of the India-Pakistan-China game revealed a detached, academic, game-obsessed view of the world, the words also
disclosed that men around a table, consciously or unconsciously,
can easily transport themselves to Olympian heights, wielding imaginary power with an ease that Zeus would envy. The temptation,
from the perspective of the 1980s, is to smile at those men of the
1960s, playing at being nations. But many of the men who played
these games of war went back to their offices and really did manipulate, or attempt to manipulate, the ways of nations. And men

like them still gather in the basement of the Pentagon, playing games that presumably will remain secret for decades.

The route from the game table to the world of real power politics is not well marked. Secrets keep the way dark, and those who have commuted between play and power are reluctant to concede that they have shaped policy by playing games. As an anonymous officer from the Cold War Division of the Joint War Games Agency told the players of Nu, "None of this will go outside for any type of distribution attributed to anyone, as is normal in these games." Anonymity is guaranteed forever, as is demonstrated by the blacked-out names in the game records released under the Freedom of Information Act.

Political-military games have value. They would not be in their third decade of existence if they did not. Nor would busy men waste their time at play if the busy men were not gaining insights from the games. Nor would *secret, secret noforn* (no foreigners), and *top secret* labels be stamped on these pages of synthetic history if they were, indeed, only that. In those games hawk dueled dove, nations fought wars and responded to crises, policymakers squabbled over policy ideas, especially about Vietnam. We know that the games produced results, but, as Frank Kapper put it, "Specifics are classified."

Perhaps the most valuable legacy of the games are the few scenarios doled out by the Pentagon censors. Just as literature acts as a guide to culture, so do the scenarios track strategies and throw light on secret, enduring fears. I thought of the offhand remark by a Newport gamer—"Let's write off Germany"—when I read the scenario for the 1967 Pentagon game to which Walt Rostow had been invited.

In the scenario it is April 1972. A global crisis is rapidly building because the world has just learned that West Germany is making nuclear weapons under the leadership of Franz Josef Strauss (who in real life had been West Germany's Defense Minister). Strauss, according to the scenario, "categorically refused to give ground on West Germany's right to defend itself. . . . It was apparent that he had strong popular support as the spirit of nationalism reached a fever pitch throughout West Germany." Intelligence sources, according to the scenario, "indicated that the program was advanced, that the FRG [Federal Republic of Germany] had em-

ployed 'break through' centrifuge techniques and fissionable materials
provided by Argentina."

I wondered, as I read that scenario, why its writers had made a
nuclear irresponsible West Germany the starting gun for a game.
Given the anonymous Game Director's penchant for "plausibility,"
where had the scriptwriters found the plausibility for this bizarre
bit of synthetic history? Was it possible that responsible officials
in 1967 actually harbored fears of a nuclear-armed Germany? Did
U.S. policymakers fear the rise of Strauss?

The scenario continues with a reference to a story in the German
magazine *Der Stern*. The U.S. Ambassador to Bonn, according to
Der Stern, has told Strauss that the United States had no choice
but to "engage in basic reexamination of U.S. policy toward Germany if the Germans persist with their nuclear program."

The scenario continues with France threatening complete withdrawal from NATO and demanding inspection of German nuclear
facilities by the European Atomic Energy Community. A senior
French official states that the French government "will no longer
support the FRG [Federal Republic of Germany] against external
aggression and denounces strongly 'German perfidy' in breaking
its word to the Western European Union regarding manufacture
of nuclear weapons."

The Soviet Union detonates a massive nuclear weapon just beyond the atmosphere, at the edge of space. The President of the
United States, in a television address, calls the explosion "a vio-
lation of the Nuclear Test Ban and Outer Space Treaties." The
President also denounces the dispatch of Soviet troops into Rumania. The President ends the speech by emphasizing U.S. desires
for peaceful exploitation of space. He hints that the U.S. is about
to launch a spaceship for a manned landing on the moon.

Soviet and Warsaw Pact forces are brought to full alert and begin
moving to forward positions. Reserves are called up and mobilization begins. The Soviet Union sends diplomatic notes to NATO
countries, France, the United States, West Germany, and the
Senate of West Berlin.

The European notes denounce the German nuclear program as
"a grave threat to the peace and stability of Europe and the world."
The note to the United States asks for U.S. pressure on Germany
and points out that hopes for détente, resumption of talks on stop-

ping antiballistic missile deployment, and reaffirmation of the test ban treaties hinge on stopping the Germans from building nuclear weapons. The note to Bonn specifically says that if the German weapons-making continues, the Soviet Union will be forced to consider "appropriate actions to protect its national interests and security."

The United States increases its Strategic Air Command airborne alert.

A Polaris submarine carrying nuclear missiles has been lost. Preliminary reports discount Soviet involvement and indicate the disappearance was the result of an accident.

In a private note to the Soviet leadership, the President intimates that the United States can scarcely apply hard pressure on the Germans when the Soviets are acting so provocatively. The note holds out the prospects of talks on world security problems, including Germany's nuclear program, if the Soviets show more cooperation.

The U.S. Secretary of Defense announces significant redeployment of U.S. forces "from Southeast Asia," indicating improvements in political stability in that area and of peaceful U.S. intentions.*
The United States begins preparing for the resumption of nuclear tests at Kwajalein, steps up production of submarines, and decides to build sea-based antiballistic missiles.

At this point in the game Red II† makes an estimate of the situation, which is sent to Control. The estimate appears to be more the work of a wishful-thinking, hard-line American playing Red than an actual Red.

Military: "Strategic. No politically useful superiority exists for either side. United States has superiority in delivery vehicles but not enough for first strike because:

"a. It cannot eliminate our assured destruction retaliatory capability.

"b. Its decision-making process makes first strike unlikely. USSR cannot destroy U.S. retaliatory capability . . . .

"Existing ABM deployment is irrelevant to issue. With four U.S.

---

*When the players read this scenario in April 1967, American troop strength in Vietnam was about 470,000 and U.S. planes were bombing North Vietnam.
†Presumably there was also a Red I, for the game designation, Beta I II-67, indicates a double Blue-Red game.

and three Soviet cities protected, ABM does not affect retaliatory capability. . . . U.S.S.R. is superior to central front in Europe. . . . Trend of arms race is against U.S.S.R."

Political: "Liabilities. U.S.S.R. has suffered a series of defeats:

"a. A communist state has been defeated in Vietnam.
"b. A communist government has been overthrown in Cuba.
"c. Soviet protegé has been defeated in Middle East through initiating use of nuclear weapons.
"d. Need to use troops in Rumania."

Now Red moves. On May Day the Soviet Union imposes "a selective blockade" of West Berlin. "Supplies necessary to prevent starvation, disease, etc., will be supplied by East Germany upon request from the Berlin Senate." At the same time the Soviets begin a diplomatic campaign aimed at isolating West Germany. The Soviet Union warns that the Germans are "playing with fire and by fire they will be burned." Support develops for the Soviet attempt to stop the nuclear arming of West Germany.

"There were increasing indications," the scenario says, "that the West was no more enchanted with the prospects of a nuclear armed Germany than were the Russians. *The Economist* in London glibly called for [Prime Minister Harold] Wilson 'to start laying the groundwork for a renewed shuttle bombing of Germany.'. . .The publication was severely castigated in Parliament."

Soviet leader Aleksandr N. Shelepin—a real but rather obscure member of the Politburo—in his two-hour May Day speech in Moscow "pointed out the ineffectiveness of diplomatic approaches" and makes "a thinly veiled threat of direct action." The President, via the Washington-Moscow hot line, warns Shelepin that the United States will "invoke its guarantees under the NATO Treaty" if the Soviet Union uses force on West Germany.

On the morning of May 5 the Soviet Union bombs West German nuclear facilities. Some thirty covered trucks are seen leaving the installations just before the bombings. The Allies refuse West Germany's request for an immediate retaliatory air strike against Soviet air bases in East Germany. ("The U.K.," the scenario says, "did not consider the Soviet action to fall within the definition of an aggression calling for automatic NATO response.")

An atmosphere "of intense frustration and even despair," the scenario says, "is building up in West Germany." The Allies' refusal to take retaliatory action against the Soviets "has created a cabinet crisis in Bonn and near-riot conditions in Munich and the Ruhr. The editorial tone of most of the German press appears to support the Strauss government's position and even liberal publications carry a note of isolation and great apprehension."

Meanwhile, on the other side of the world, massive concentrations of Chinese troops have moved into North Korea in what seems to be a buildup for a Chinese-North Korean invasion of South Korea. This is every real strategist's nightmare—and the scenario writer's favorite move: a two-front crisis. On May 18 North Korea and China launch a coordinated ground and air attack into South Korea. "Defenses which had been considered impregnable were penetrated in four places and an estimated three Red Chinese armies plus North Korean elements" threaten to overwhelm the two U.S. divisions "guarding the traditional invasion route to Seoul."

Now the scenario breaks off from its narrative of imaginary events and makes what appears to be one of the points of the game: The decision to use nuclear weapons is affected not only by the U.S.-U.S.S.R. strategic balance in terms of nuclear weapons but also in terms of antiballistic missiles and the mix of nuclear and nonnuclear weapons.

"Since 1968," the scenario says, "South Korean forces had been drastically reduced because of US military aid cuts and pressures to strengthen the civilian economy. The twelve ROKA [Republic of Korea Army] and two U.S. divisions in South Korea had, since 1970, keyed their defense plans almost entirely to the early use of nuclear weapons. This doctrine had been widely discussed in military journals and apparently [had] not been overlooked in Peking."

The U.S. commander in South Korea transmits an urgent message to the Commander in Chief of U.S. forces in the Pacific (CINC-PAC): A decision must be made within twelve hours on which of three choices he should make:

1. Arrange for a naval evacuation of all U.S. forces from Inchon "with probable high casualties and loss of virtually all matériel."
2. "Stand and fight and be overwhelmed by what appears to be a five to one numerical superiority and growing enemy air strength."
3. Authorize the use of "tactical weapons to relieve the pressure."

The day after the Soviets bombed West Germany, China fired one missile over India into the southern reaches of the Indian Ocean and a second over Japan and into the Pacific Ocean. Blue intelligence estimates that the Chinese have twenty-five missiles able to carry a 4,000-pound payload 6,000 nautical miles.

Blue asks intelligence for an assessment of the strategic balance between the United States and the Soviet Union. Intelligence cites the shortcomings of the Soviet missile system and "the newly refined test data" for U.S. hardened warheads, electronic penetration aids, and individually directed multiple independently targeted reentry vehicles (MIRVs). The conclusion: "The United States enjoys a major strategic nuclear advantage."

On "the most conservative basis" a first strike by the Soviet Union would result in 30 to 50 million Soviet casualties and 20 to 30 million U.S. casualties. A first strike by the United States would result in 100 to 120 million Soviet casualties and 5 to 10 million U.S. casualties.

Then Control's scenarists add another twist: Through what appears to be a leak, six days after the Chinese missile launchings, U.S. newspapers reveal that a high-level inspection of the nation's antiballistic missile system has "cast doubts on its overall effectiveness." The stories say that a Soviet *second* strike—apparently in reaction to an American first strike—would produce 20 to 30 million U.S. casualties, not the 5 to 10 million previously expected.

And, if China launches a first strike, the result would be "no less than" 10 million U.S. casualties. A preemptive strike against Chinese missile sites might cut down the prospective casualties from that quarter, but "there was no assurance that all of their capabilities could be located and destroyed." So even a preemptive strike against vulnerable China with her relatively small nuclear arsenal would probably result in "several millions of casualties."

The available game records end at this point. The cliff-hanger game seems to have been designed to get players into a crisis of decision and leave them there. The players were given a lesson in a political-military algebra whose answers could be barely imagined and never tested. Nuclear weapons make it hard to start a war.

The ABMs that flit in and out of the Control-driven scenario probably got there as a result of a simmering debate over the worth of an ABM system. It is easy to imagine the players, at the hot washup, talking about ABMs, developing or hardening ideas that they would take back to their offices and their superiors. Or, biding their time, taking with them to the next plateau of their careers ideas that were born during the game that bombed Germany.

The ABM debate carried over into the Nixon Administration, and earth-based ABMs evolved into space-based weapons, the Star Wars of the Reagan Administration. The debate often involved the citing and the denouncing of numbers—levels of accuracy, casualties, ranges, dollars, how high the stack of bargaining chips. Numbers long have been used to keep score at the game tables of war.

CHAPTER 3

Measuring War: The Men, Methods, and History

When William Lese sat in his Pentagon office and described the missions of SAGA's game-playing successor, the Joint Analysis Directorate, he did not lean back in his chair and tell anecdotes the way academic game directors do. He is an analyst and has flip charts and a table of organization arranged inside the usual boxes. He methodically singled out each box and tersely described what was in it: *Nuclear Plans Analysis Division*, "Nuclear planning." *Capabilities Assessment Division*, "Conventional side." *Contingency Simulation Division*, "Political-military simulation and real-world crisis political-military simulation." *Joint Force Allocation and Analysis Division*, "Wrapping it all together. What is the best total force? What is the best mix?" *CINC Support Division*, "The war fighters." *Technical Support Division*, "The look-ahead guys, modelers. They look for the best computers. And we meet with Trevor Dupuy."

Trevor Dupuy, mentioned in the same breath as the Pentagon's best computers, is a soldier, an old soldier with a parade-ground posture, a precisely trimmed white mustache, and an air of command. When he asks a question of one of his associates—inevitably, a retired officer—*Colonel* Dupuy gets a brisk answer in a military accent: A date is given day first, month second; eight o'clock in the morning is *oh-eight-hundred*; a car is a vehicle.

Dupuy is the founder and executive director of HERO, the Historical Evaluation and Research Organization, "dedicated to promoting the cause of historical analysis and improving the national security of the United States." He has fought in Burma and taught at Harvard. He is the author or co-author of more than eighty books. Speaking or writing, he can put more into a declarative sentence than most Department of Defense analysts can manage to put into a dozen paragraphs. Historian, soldier, student of war, he spans past and present gaming, real-life battlefield and computerized games. And, by creating a calculus of battle, he has joined mathematics and traditional gaming in a way that makes warfare—from tin soldiers and board games to NATO and nuclear weapons—a military continuum.

"The senior decision makers of the U.S. military establishment," Dupuy wrote in 1985 in the respected professional journal *Army*, "are increasingly basing their decisions on the outputs of computer models and simulations of combat which are widely recognized to be unreliable and unrealistic." Some Army officers and Pentagon policymakers could perhaps shrug off criticism from Dupuy as the sermons of a gadfly and an irascible character with ideas more Napoleonic than nuclear. Dupuy is not universally beloved.

But Dupuy was not alone in condemning the heedless use of simulated facts. He was endorsing the thoughts in another *Army* article, one that had denounced "the undisciplined nature of the overall analytical community." The name of the author was not revealed, apparently because, as an officer on active duty, he risked punishment for stating his views. The use of "Anonymous" as a byline was an extremely rare departure from *Army's* editorial policy.

The author was Major General Edward B. Atkeson, former director of the Army Concepts Analysis Agency, the Army's built-in think tank that makes contributions to the SIOP and produces most of the Army's war games and models. Atkeson wrote that many wargaming simulations "were as perforated with logic holes as a sieve." In war games, he said, NATO tankers kill three or four tanks for every one lost, but doubters can search in vain for the hard facts that back up these paper victories that "emerge from the analytical process via the bowels of a computer."

Like Dupuy, Atkeson wanted analysis to look beyond numbers.

"You can play the Iraq-Iran war, where one side is using just a sort of rabble in arms and the other side is harvesting them," Atkeson, recently retired, said in an interview. "But if one side outnumbers the other, the numbers will count. Well, of course, Israel couldn't survive that way. So there are some quantitative differences that someone has to take into account that perhaps the purest mathematicians—particularly the Rand types—wouldn't be comfortable with.

"I don't know anyone who can do it in such a way that his colleagues would agree, with the exception of Trevor Dupuy's approach. Trevor says, 'Look and see what actually happened.' War wasn't invented yesterday. There have been a lot of them, and most of these kinds of issues have come up in one way or another.

"The Soviets will take a campaign or a battle and they will dissect it to its individual parts: what happened to an engineering company on the third day of the engagement if they had to cross a stream and it only had fourteen guys to build a bridge that normally takes sixteen to build? And the stream was going at this certain rate of flow? All those little factors. They will take history and break it down into its particulars, file them away in computers, so that when you come to a situation, and you know that part of your campaign involves an engineering company crossing a bridge with only fifteen guys to do it, the computer will search through these compilations and aggregations of previous experiences and seize them and say, 'Here is a norm for you.' You don't have to follow that, but if you deviate very strongly and you're a colossal failure, you better go home and shoot yourself.

"We consider that there is a scientific aspect to warfare, but fundamentally it's an art. And we train artists to manage their resources. We tell the artist to do as he pleases. He can follow doctrine. It's easier if he follows doctrine because then he won't screw up his neighbors. If he's successful, we won't mind what he does. If he's unsuccessful, we'll fire him and get somebody else. The Soviets have a different view. They tend to emphasize the scientific dimension, and they look for historic principle as sort of the first determinant. Of course, they will take into consideration technological advances and the great complications of nuclear weapons and aircraft speeds. But they have essential norms. They know how fast they expect their forces to operate under certain circumstances and they know essentially how much it takes to accomplish something."

In a quest for a "scientific theory of combat," Dupuy has enlisted Atkeson, other retired officers, active-duty officers, gamers, analysts, and strategists, including several from Canada, Great Britain, France, West Germany, Israel, and Switzerland. In 1984 Dupuy banded them together in the Military Conflict Institute. Dupuy's battle cry is *Validate!*—and not merely with numbers.

"If you fall back on math," he said in an interview, "you're kidding yourself. You have to fall back on history. Intellectually, the process of validation involves the ability to say that you have come up with a theoretical explanation, in a model, of how the real world works. Any model is some sort of a simplification of the real world for some particular reason. And, having simplified it, or having come up with this version of this particular aspect of the real world, then you want to know whether this model will predict what will happen or explain what has happened. And if it can't be related to what actually happens in the real world, then, in my opinion, it is worthless. There are people who will say, 'We can't validate our models because we are dealing with weapons that weren't used in the past.' That's very true. None of us, unless the divine hand of God has been laid upon someone and made him a prophet, none of us can predict the future.

"In my opinion, the purpose of a model is to try to be able to predict what is *likely* to happen, for whatever reason the model has been designed. And the only way you can do that is to go back and see whether, if you applied this conceptual entity in the model to a set of circumstances in the past, it would give results similar to what actually happened. And to say you can't do it because your model deals with future weapons—if your model is dependent upon the design of the weapons—then there is something seriously wrong with it. Because it isn't weapons that run wars. It's people who run wars."

A model, as Dupuy's friend and supporter Frank Kapper defines it, is "an objective representation" of some portion or aspect of the real world. "It may," Kapper's definition continues, "be a representation of an object or structure, or an explanation or description of a system." What most of us think of as a model airplane is an iconic model. The models Kapper has in mind are descriptive symbolic models, which use words or diagrams, and descriptive mathematical models, which use numerals and mathematics symbols to represent such abstract realities as quantitative relationships. At this point in the vocabulary of the gamer, model definitions become hazy. As Kapper explains it, "Math models are further subdivided into *analytic* math models for which an exact numerical solution can be determined, and *simulation* math models that may be used to converge on solutions to very complex problems involving uncertainty (probabilistic occurrence) and risk."

To a modeler, the English language has "good descriptive capability" but is very ambiguous, and cannot be manipulated, while mathematics, though not much help in descriptions, is not ambiguous, and can be manipulated. In the inevitable charts that modelers use for communications, ambiguous English is at one point ("Primary Function: descriptive explanations and directions") and unambiguous mathematics ("Primary Function: problem solution and optimization") at the other. Theorists tend to group at the abstract end of the gaming spectrum, where, as Kapper put it, gaming becomes more "convenient, accessible, and reproducible in outcome . . . because the actual set of variables which can affect the war game proper have become much more limited, and have in turn been simplified."

Mathematical models can deftly illustrate practical problems that occur in battle—such as the breakdown of equipment or the varying skills of soldiers and sailors. Such problems do not come up in political-military war games, which are aloof to the sound and smell of battle. A model is more "scientific" than a war game because a model, operating purely without real human involvement, can be precisely replicated.

Two competing fighter planes, reduced to specifications in a model dog fight, can be put into a computer instead of into the sky. They can be run through thousands of duels, with a slight change in some control factor for each duel. Then computers can analyze the duels and see what factors helped one plane win and one plane lose. It is all mathematical, all logical.

A war game, full of human players, can be repeated but not replicated. People do not act like equations. In a submarine-fighting-submarine model, for example, there may be many quantifiable variables, from water temperature to the intensity of background noises. Each of these variables can be changed (or, in modeler language, "manipulated") through innumerable replications of the duel. But that kind of model does not have human players; there are no competing submarine skippers using their luck or cunning.

A "free-play" war game will include those skippers and their human foibles. To some analysts, however, that kind of game is dangerously unscientific. "People tend to want something to hang their hats on, and numerical results, like two CVs [aircraft carriers] were sunk, are convenient hooks," a Center for Naval Analyses study says, touching a sensitive Navy nerve. Even in games or models the Navy does not like ships sunk. "It always helps," the study advises, "to express game results in terms of the human interactions as opposed to modeling prognostications; for example, a report could say 'when the losses were heavier than expected a decision was made to withdraw,' instead of 'after losing 12.65 ships, the decision was made to withdraw.' "

As precise a man as Dupuy could not differentiate *game* from *model* from *simulation*. The words hover over imaginary battlefields like a mysterious, ever-shifting concept of the Trinity. "The terms models, wargaming, simulation are used synonymously," he said, a trace of irritation in his colonel's voice. "And yet," he pensively added, "we are not too sure what we mean when we use them. For instance, the political-military *game* is one of the things that JAD does. At contingency planning, then you're starting to get down to purely military *planning*—the interaction of two opposing military forces.

"I believe that models can be used for military gaming. But political-military gaming? You can't predict. You can't even predict what's likely to happen. You may get some insights, but unless you have enough data that can be quantified and enough data that will give you what I call actuarial comparisons, you can't be sure of what is typical and what is atypical. The model is useful only if it can represent a *typical* situation."

Dupuy obviously had his own definitions in his own mind when he made that comment on gaming and models. Such mono-grammed definitions characterize the speech and writings of most gamers, modelers, and simulation makers. As far back as 1963 a report asked "the defense analysis community" to develop a com-mon vocabulary "or remain in the Alice in Wonderland world of project management—'when I (Humpty Dumpty) use a word [or model], it means what I choose it to mean—nothing more or less.' "

Members of the defense analysis community tend to use graphs, models, and diagrams rather than words when they attempt to define what they do. The concept of the wargaming spectrum comes from this tendency to be visual rather than verbal. One gamer's spectrum, for example, puts *military exercises* at the deep-purple end and *analytic games* at the deep-red end, with *manual war games, computer-assisted war games, interactive computer games*, and *computer games* forming the rest of the spectrum.

Military exercises involve flesh and blood troops; analytic games are so abstract that they do not even have to use English. A manual war game is simply a group of people making believe they are engaged in a crisis; this is the kind of game I played at the Naval War College. If we had had a computer at our elbows, we would have been playing a computer-assisted game, and if the computers had responded to us, perhaps questioning our decisions or offering options, the game would have been interactive—a kind of chess match with an automated, no-nonsense, data-drenched chess player. If we were not in the room but the computers were more or less playing each other, then that would have been a computer game. Mathematically minded gamers dwell on the right of the spectrum amid the abstractions; mathematically illiterate gamers (such as most policymakers) cluster to the left, with the *real* soldiers.

As wargaming became a civilian occupation as well as a military one, attempts were made to formalize definitions so that everyone concerned with gaming would be talking the same language. These standardized definitions, published in a 1979 study sponsored by the Defense Advanced Research Projects Agency, were written by two academicians associated with Rand, the ultimate defense think tank. Civilian gamers, especially those who are working under Rand contracts, tend to make the distinctions created by the definitions:

Gaming: "A gaming exercise employs human beings acting as themselves or playing roles in an environment that is either actual or simulated."

Wargaming: "A war game is defined by the Department of Defense as a simulated military operation involving two or more opposing forces and using rules, data, and procedures designed to depict an actual or hypothetical real-life situation." Such a game may be for training, operational, or research purposes. A war game may be "manual," meaning without computers. It may also be computer-assisted or completely played by computers.

   Simulation: "The representation of a system or organism by another system or model designed to have a relevant behavioral similarity to the original."

   Model: "A representation of an entity or situation by something else that has the relevant features or properties of the original."

Such definitions have little more than academic interest to Dupuy, who, in his efforts to make war scientific, has tried to use a soldier's common sense, which he believed is enough to span the gulf between the mathematical modelers and the warriors who see warfare as a very real, deterministic, historical process. In 1985 he wrote an article that questioned the validity of mathematically based models designed to produce dry runs of battles for Army officers. The professional journal *Army* entitled the article "Criticisms of Combat Models Cite Unreliability of Results." Dupuy wanted the title to be "Human Factors Without Human Experiences."

"The operational research analyst will try to take experience and analyze it and see what it might tell him about the future," he said not long after the article was published. "But because most operational research analysts are mathematicians, not historians, they don't understand history and they don't think it's relevant. So forget the past, just take what weapons can do theoretically and reach out into the future and predict things.

"Any military model, force-on-force model, uses some method of quantification, and some method of quantifying weapons. Now one of the issues that you will come across is firepower scores. What are our firepower scores? Should we use firepower scores? Whenever people avoid firepower scores, they are still trying to find some way to quantify weapons, which ultimately goes back to some kind of firepower scores because this is what weapons are for—to hit, kill, maim, damage, wound. And if you are trying to take into consideration their effects, you have to find some way of quantifying what you think their killing effect is, or their destruction-dealing effect, or their damage-dealing effect."

I asked if the reluctance to get such figures into models stems from squeamishness on the part of policymakers. "It's not squeamishness," he said. "It's the Lanchester problem—the raw data is based on the interaction of a number of things which are very hard to quantify: the effects of terrain, the effects of weather, the effects of leadership, the effects of mobility, the effects of defensive posture. And recognizing that you can't get answers that are like reality if you just take the firepower of the weapon."

In my conversation with Dupuy he frequently mentioned two names—Lanchester and Clausewitz. They are keys to an understanding of gaming and an understanding of Dupuy and what he and his band of combat theorists are trying to do.

Frederick William Lanchester, born in 1868, was a British engineer who, early in the twentieth century, became interested in aerodynamics and the use of aircraft in war. In 1914 he published several articles on the subject and in one of them examined what had happened to "the principle of concentration" in warfare. "In olden times," he wrote, "when weapon directly answered weapon, the act of defence was positive and direct, the blow of sword or battleaxe was parried by sword and shield. . . ."

In modern times, Lanchester said, the equation drastically changed. He expressed the difference in what has become known as Lanchester's Law: "The *fighting strength* of a force may be broadly defined as proportional to the *square of its numerical strength multiplied by the fighting value of its individual units*." He reduced this to an algebraic notation, with r and b representing the numerical strengths of Red and Blue and N and M representing the fighting value: $Nr^2 = Mb^2$.

Karl von Clausewitz, the nineteenth-century Prussian general and writer on military strategy, also tried to find the answer to success in war, which, in his famous definition, he called "not merely a political act, but also a political instrument, a continuation of political relations, a carrying out of the same by other means." For Dupuy, Clausewitz's most important quotation is on "friction," and Dupuy bases much of his work on what Clausewitz wrote:

> Everything in war is very simple, but the simplest thing is difficult. The difficulties accumulate and end by producing a kind of friction that is inconceivable unless one has experienced war. Countless

unforeseeable minor incidents . . . combine to lower the general level of performance, so that one always falls far short of the intended goal. Friction is the only concept that more or less corresponds to the factors that distinguish real war from war on paper. . . . None of [the military machine's] components is of one piece: each part is composed of individuals, every one of whom retains his potential of friction [and] the least important of whom may chance to delay things or somehow make them go wrong. . . .

This tremendous friction . . . brings about effects that cannot be measured, just because they are largely due to chance. . . .

Action in war is like movement in a resistant element. Just as the simplest and most natural of movements, walking, cannot easily be performed in water, so in war it is difficult for normal efforts to achieve even moderate results.

Dupuy believes that Clausewitz, who "argues with himself the twin questions of whether war is a science or an art" nevertheless perceived, as Lanchester was later to perceive, a quantifiable theory of combat. Dupuy writes the "Clausewitz formula" as $P = NVQ$, "in which P stands for combat power, N stands for numbers of troops, V stands for the operational and environmental variable factors which indicate the circumstances of a battle, and Q stands for the quality of the troops."

Dupuy's ideas pervade modern gaming, from board games in hobby shops to the contingency-planning games in the Pentagon. His Quantified Judgment Model is known throughout the world of gaming as the QJM. Both professional and amateur war gamers use the QJM, or adaptations of it. The QJM encompasses all of what gamers call the gaming community: the military establishment, which uses games in the Pentagon and the war colleges and for other types of training; academe, which uses political-military games in international studies; commercial companies that sell board and computer games to the public; and defense firms that use games to develop and sell weapons systems. The QJM is also well known outside the country. Soviet military authorities have alerted their war planners to the QJM, describing it, quite accurately, as a combat analysis model used by the Defense Intelligence Agency and Pentagon force planners.

The QJM dates to the mid-1960s, when Dupuy was analyzing

World War II air-ground operations for the Air Force and air-support tactics for the British Defence Operational Analysis Establishment. For both projects he "decided to try to measure the combat potentialities of opposing forces by quantifying their total weapons firepower" through an Operational Lethality Index. For this measurement of death in battle, Dupuy examined American, British, and German weapons and soldiers in sixty engagements in Italy in World War II.

Out of that research first came a concept Dupuy called the Power Potential. "If the ratio of the two Power Potentials—P_f for the friendly force divided by P_e for the enemy force—was greater than 1.0, we postulated that this meant that the friendly side should have been successful; if the ratio gave a value less than 1.0, then the enemy side *should* have been successful." To find a formula for quantifying the outcome of an engagement, he examined battles as far back as Austerlitz in 1805 and looked at the casualties.*

Dupuy called the result of his studies the Quantified Judgment Method of Analysis of Historical Combat Data. "When discussing the two major formulae with an operations analyst friend," Dupuy later wrote, "I was somewhat surprised when he referred to them as 'models.' He assured me that these were indeed the kind of formulations which OR [operations research] analysts called models. I felt a pride akin to that of Molière's *Bourgeois Gentleman* when he realized that he had been speaking prose for forty years."

In an earlier modeling of war Dupuy had begun with weapons, going back to the javelin, sword, and sarissa, the long pike of Macedonian warriors. Assembling statistics on such matters as relative incapacitating effect ("the likelihood that an individual blow if it hits a target will incapacitate the target"), number of potential targets per strike (ranging from "one-for-one lethality" to "multiple casualty possibilities"), and effective range ("will increase proportionately as the reach of a weapon extends beyond that of a man's arm"), he produced a theoretical lethality index of weapons. His formula:

---

*He also looked at such arcane statistics as rate of advance and discovered that a British cavalry unit in World War I had traveled faster than Israeli armored forces did in the Six Day War in 1967. (Critics, though, questioned the value of this statistic, claiming that differences in terrain and opponents were too important to ignore.)

Lethality index = rate of fire per hour × targets per strike × relative incapacitating effect × range factor × accuracy × reliability.

A sword's theoretical lethality index (TLI) is 23, based on the formula (60 × 1 × .4 × 1.03 × .95 × 1.0).

For a one-megaton nuclear weapon exploded in the air the TLI is 695,385,000 (1 × 8,500,000 × 1 × 101 × .9 × .9).

He compounded other factors, including mobility, ammunition supply effect, and a ceiling effect for aircraft, to get a total TLI for a modern "mobile fighting machine" and added a dispersion factor, which shows the spread of soldiers from the tightly packed phalanxes of the Roman legions to the deployment of a modern 100,000-man army across 4,000 square kilometers. On the battlefield he found seventy-three Combat Variables, some calculable, such as weapons' effects, and some intangible, such as morale. He assigned values to specific variables—terrain, weather, air superiority, posture (attack, defense, and other operational factors), mobility, vulnerability—and then did the same with intangibles; besides morale, these included leadership, training, and relative technological development. He also correlated such factors as weather related to artillery and weather related to armor and created more and more complex formulas for aspects of battle.

Dupuy also found ways to quantify the chaos of battle. By three "measures of effectiveness"—accomplishment of mission, gaining or holding ground, and casualty rates—he came up with a Result Model. Testing all these formulas against the battles in Italy in World War II, Dupuy was satisfied that he had found a measurable theory of combat. By 1973 Dupuy and his HERO associates could show that he could reproduce combat in World War II with his formulas. (The formulas also worked for the battles of Waterloo, Antietam, and Gettysburg, and the German Somme offensive of World War I.)

In that same year the QJM entered the Pentagon, where war gamers used HERO models to calculate the effectiveness of NATO weapons against the weapons of Warsaw Pact nations. Later HERO modeled thirty-eight hypothetical battles, including seven in which U.S. forces faced Soviet forces. The attacker—the Soviets in five of them—won all of the battles. On the basis of these and other studies, Dupuy has questioned NATO doctrine and, in an age of growing dependency upon technological weapons, has steadfastly maintained that superior strength always wins.

"I happen to believe," Dupuy said of his QJM, "that numbers are very useful because you can demonstrate practical relationships with them. But not everything can be quantified. I have come to the conclusion that war is a lot more deterministic than people think. And that most things in military affairs can, in one way or the other, be reduced to numbers even if you have to give a general dimension to things you can't quantify."

As an example he cited combat effectiveness. "I am able to give a general value for combat effectiveness, historically, in comparing military forces. It has pretty consistently demonstrated some validity historically. I can't tell you how much of that is due to leadership, how much is due to training, to chance. But a number of writers on military affairs have pointed out that luck seems to favor the guy who is aggressive and follows up on it."

Dupuy had likened some of the numbers in his QJM to "the technique insurance specialists use to determine insurance risks." Speaking for himself and his associates, he said, "We do not believe that it is possible, or ever will be possible, to predict individual events which are dependent upon human behavior, whether this be the behavior of one person or of a group of people."

To see the QJM in action, I watched over the shoulder of Bob Hardy as he summoned from his Apple computer a battle on the central front in Europe on a foggy day in May. Hardy works as an analyst for the Vought Missiles and Advanced Programs Division of the LTV Corporation, which in 1985 was the sixteenth largest U.S. defense contractor. Soon after the company acquired a computerized version of the QJM, Hardy said, "We changed a few things. There was little tactical air or air support. So we modified it." LTV's defense products include aircraft, airframes, and a multiple-launch rocket system. With the modifications, QJM became a selling tool.

A computer menu appears on the monitor screen. The menu asks Hardy whether he wants to use the previously stored data on U.S. and Soviet units or make new units? Does he want the attacker to be the United States or the Soviet Union?

The menu then lists the U.S. units available—from armored or infantry divisions to aircraft that can be armed with specific weapon packages. For each unit there is a number showing total personnel and an operational lethality index. There are also numbers for the factoring of damage and repair of trucks, tanks, aircraft.

```
Do you want more U.S. units?
```

Hardy asks for an A-10 flight and new ordnance.

```
Do you want more Soviet units?
```

He asks for three Soviet tank divisions and six flights of helicopters.

For terrain he picks *mixed woods* from a list of fourteen possibilities. (The lethality index will automatically adjust for casualty rates in each terrain.) Hardy ponders the weather: from *dry, summery, extreme heat* through eleven variations to *wet, extremely cold.* For *season* he chooses spring and for *month* May.

Hardy scrolls the other parts of the battle equation:

```
Who has air superiority?
```

A *Posture Table* asks him to decide whether the defense is hasty or prepared, and another table inquires about levels of "weapons' sophistication." (The QJM gives top rating to U.S. weapons, followed in sophistication by Britain/France, Germany, the Soviet Union, Israel, Arabs, China.)

```
Length of engagement.
Do you want surprise?
What kind of surprise? Complete?
Substantial? Minor?
Road quality: Good, medium, poor.
Road density: European standard, moderate
density, sparse.
Rivers/streams: Width? Fordable or not
fordable?
Do you want to add reinforcements after the
first day?
```

```
Mission factors: The attack is
            all-out
            secondary
            holding
            general
 Defenders are delaying
            withdrawing
            holding
            giving all-out defense. . . .
```

In four minutes Hardy had made all the choices for his battle.

Turning away from the screen, Hardy began talking about how he got to be an analyst and a computer warmonger. He had been described to me as one of the few analysts who had actually heard the whine of bullets on a real battlefield. After combat service as a Marine in World War II, he was sent to Tianjin, China. "I learned to play chess in China," he said. "I used to play with the British consul general. He looked at each piece in terms of war. The pawns were infantrymen, the knights were armor, rooks were artillery, the bishops, a heavy force strong enough to move along diagonal lines. The queen was a general, and the king was The Nation or The Cause.

"The Chinese sat on the curb and used chalk and made pieces out of bottle caps and drew characters on the caps. So one day I finally got across that I wanted to play. There must have been fifty people around. A Chinese friend came along and said that I had raised a big question in the minds of the other Chinese: How can he play our game and not know our language?"

He turned and looked again at the screen. I had expected to see an animated battlefield, something like the ones that pulsate in a video arcade. This one was full of numbers. Hardy pointed to the numbers and recalled manual games that had taken him two weeks to play. "Now," he said, "you can play the same game in ten to twelve hours because the bookkeeping is being done by computers. In an uncomputerized game, there is lots of data and very little time to play. Not many 'what-if's. Now I can fight a battle through a day of time—or through seven days—in fifteen minutes."

The battle was raging in the form of marching numbers. Hardy explained some of them. "Things like fatigue slow down your rate of advance," he said. "Troops are more vulnerable because they are less alert. And they take more casualties." All of these factors were being processed and coming onto the screen as numbers.

Hardy scanned the numbers for *Day 1*. "Perhaps," he said, "the defender is pushed out of his position. Fatigue is already setting in, perhaps more on the attackers than on the defenders." *Day 2*: "A new standard rate of advance because the power potential ratio has changed."

There was no *Day 3*. Hardy seemed politely bored. "I can fight it one hour at a time," he said, again turning away from the screen. "I can make fifteen minutes be a day for up to thirty days. At thirty days, I pick off data from the thirtieth day and start it again. It's an automatic routine."

I asked him about the operational lethality index. "To find out the OLI," he said, "you start off with a rifle, and I know if I put it in a vise and line it up with a target one thousand yards away, I'll hit the bull's-eye every time. Then I give the rifle to an eighteen-year-old kid off the farm. I give him six weeks training and he hits the bull's-eye four out of six times. That is operational training lethality. Now I feed him cold rations for four days and make him seasick, put him on a beach, and have people shooting at him. And maybe he gets one out of ten hits. That's combat effectiveness value.

"To figure out what happens, you adjust factors, such as fatigue. You evaluate tactics and strategy. What is the real impact of the road network? What is the best weapon? How many casualties are you willing to take to give up a kilometer? We're playing a game, but in conflict if an enemy attacker has a five-to-one advantage, well, it's not guaranteed he'll win, but statistically the odds say he will achieve his objective—push you back, inflict more casualties on you than he takes." He paused. "You can go nuke with this game. I can overwhelm him with heavy lethality with a nuke." Hardy said it as if he would not think that would make a good game.

"When we first got this," he said, pointing to the QJM on the computer screen, "both sides were using World War II weapons and had World War II lethality rates. We changed things. The Soviets changed from primarily towed artillery to self-propelled, and so we had to change a data base. Things like that.

"I can evaluate an LTV [Vought Corporation] new weapon that we are trying to sell. I can show the generals. All aerospace companies accept this at some level. If you have good, qualified people, it works. People can use a wood chisel for a screwdriver, but after a while they don't have a good chisel. You can fool around with the numbers, but eventually the game will catch up with you."

One of the prime users of Bob Hardy's numbers wars is Fred Haynes, a retired Marine major general, a combat veteran of three wars, and a Vought executive. He called the QJM "a good little system" that can be readily used for "an industrial approach" to war. "Industry has recognized over the past few years a need for the capacity to do several things in a much better fashion: to look at what the battlefield may be like five, ten, or twenty years from now—and systems that may affect that battlefield can be found anywhere from the bottom of the sea to geosynchronous orbit.

"A number of companies, ours included, have begun to invest a fair amount of money in what we call mission-area analysis, which is often shortened to mission analysis." Such analysis, he said, enables engineers and managers to understand Pentagon needs. It also helps U.S. companies to understand what allied countries like Thailand or South Korea may need to defend themselves.

"In approaching the marketing side of the weapons industry," Haynes said, "a U.S. firm that is interested in sales to a friendly foreign country will not infrequently set up a scenario that will look at that country's capability in terms of defending against some potential enemy. It will then look at what kinds of combinations of weapons might improve that country's ability to provide for its own security. A good case in point is Thailand, which faces Vietnam, which is now somewhere near the fourth-largest military force in the world."

Haynes analyzed Thailand's defensive needs with the brisk professionalism of a Marine general. "Thailand does not have the firepower by itself to defend against Vietnam. They just flat don't have it. Fortunately, the Chinese can muster maybe twenty divisions in a hurry on the northern border of Vietnam, which can pin down a very large chunk of the Vietnamese forces even without hostilities.

"The trick is then to figure out how the Thais can best handle the Vietnamese forces that remain on the border. You do this initially by working less detailed computer war games, like QJM, that will give you some idea of the relative capability of each side. You'll game a battle that lasts anywhere from six days to two weeks, depending on what your common sense says is the best timeframe. And you ask yourself, what could you do if you had an extra squadron of X aircraft with X type of weapons available for X number of sorties a day? And so on and so forth. You'll game in as many of those aircraft or other systems as it takes to neutralize the threat."

QJM is only one of many warfare models used by a major defense contractor like LTV. A small model like QJM, relatively cheap and quick to run, can give its user some ideas of what is likely to work in a specific warfighting situation. To test out these preliminary results, however, LTV must resort to larger and more detailed combat models that run on much more powerful computers. On a map of Korea, Haynes showed how one of these larger models had gamed a North Korean attack on South Korea. The situation had not changed much since a 1965 game (Chapter 9), in which the North Koreans poured across the border. "By the seventh day, you can see that the enemy forces can make a significant penetration," Haynes said. "That says you've got to do something, and what do you do? Well, using the same set of circumstances, we run the game again, adding three batteries of the Army's multiple-launch rocket system . . . ." Vought's Missiles and Advanced Programs Division manufactures a multiple-launch rocket system, referred to as MLRS. The basic rocket warhead contains six hundred and forty-four bomblets. A single launcher load of these will saturate a target area the size of ten football fields.

Readouts from the model showed, essentially, that the more MLRSs deployed, the slower the FEBA [forward edge of battle area] moved south toward Seoul. "FEBA means the frontline, where the really tough fighting is, where the soldiers are face-to-face," Haynes said, pointing to the map. "If you took a battery of MLRSs in the U.S. Second Infantry Division and added another three batteries of MLRSs, you then would create a condition that would put the FEBA more where you want it at the end of seven days." The FEBA stabilizes near the Demilitarized Zone. "And then we said, 'What if you added some more? What if you added another three batteries?' Well, running it up that high doesn't really make all that much difference, although it helps. It reduces the penetration somewhat more." The South Koreans can punch whatever numbers they want to use into the model and see for themselves what effects various combinations of MLRSs or other weapons systems would have.

Vought has also run games between NATO and Warsaw Pact forces. "We have determined that probably the most comfortable thing, from a professional military point of view, would be to buy ten more armored or armored and mechanized divisions for NATO's central front. That, of course, is politically and financially impossible. But gaming has shown a cheaper, more practical way: If we attack Warsaw Pact follow-on forces with, among other weapons, the Army's new ATACMS [Army Tactical Missile System], NATO can delay and disrupt Soviet reinforcements." "This," Haynes concludes, "is a defensive strategy which raises the nuclear threshold, which lowers the risk of nuclear war because it lets conventional, non-nuclear forces do the job."

The QJM has proved it can produce numbers that satisfy customers like Hardy and Haynes. For Dupuy, the QJM has settled Clausewitz's dilemma about the nature of war. But by deciding that war is more of a science than an art, Dupuy intensified the attacks of war-game critics who say that numbers cannot make a science of what is at most a minor art form, a military pastime that is expensive and possibly dangerous.

There was an obvious air of make-believe about the sand-table games that Napoleonic-era generals played. In the computer age, however, the equipment a commander uses to play at war often resembles—or actually is—the equipment he will be using to direct real war. The images on the screen veil the reality of battle. More important perhaps, the detachment from battle intensifies the notion that war-fighting has become a science. As a behavioral scientist put it in 1982, "Until recently, strategic decisions regarding what information to seek and what actions to take were more art than science. Today the resonance, at least, is more scientific. Whether such exercises of rational thought are truly a contribution to national security is the question."

In a 1980 report on the Pentagon's use of analytical tools, including war games, the General Accounting Office said, "Many of the problems encountered in the classical sciences, engineering, or accounting, are *rigorously quantifiable*. Their structure is well-understood, and their mathematical formulas provide a clear-cut representation of the 'real world' problem. A *squishy problem*, on the other hand, may be given a mathematical form that looks like an unambiguous representation of the real world problem; but the appearance is only superficial, and evaporates rapidly when probed to any great extent."

The gaming of NATO, the report said, is a "very squishy problem."

CHAPTER 4

World War III: Playing NATO Games

For nearly forty years Western strategists have created scenario after scenario describing the clash they envision as the beginning of World War III—the Red hordes of the Warsaw Pact invading a Western Europe defended by NATO's thin Blue line. Numerous military and civilian analysts have spent most of their careers studying the NATO-Pact battle. From this army of analysts have come tens of thousands of studies and the plots for annual war games on wheels, gigantic NATO maneuvers that churn up West German soil and produce more reams of printouts for still more studies.

The theoretical encounter between Warsaw Pact invaders and NATO defenders has been analyzed more than any other anticipated battle in military history. The Technical Centre of SHAPE (Supreme Headquarters Allied Powers Europe) in The Hague runs what is probably the largest gaming operations outside the United States. Since the founding of NATO in 1949 planners have run simulations looking at the battle from every possible angle. These are the combat models that the General Accounting Office selected to demonstrate a "very squishy problem" involved in trying to make a science of war. The scenarios and the models of the NATO-Warsaw Pact battle must go beyond numbers into people and policy, and it is here that the squishy problem begins. What, the report asks, defines victory? Is it "casualty levels, ground gained, or the control of strategic objectives? Over what time period. . . ? What is the purpose of NATO's conventional military force—to defend in a major war, or to defend in lesser conflicts that are below some predetermined nuclear threshold? Will France participate? What are the budget realities?"

There are also the problems of people. I asked Joel Resnick, a veteran analyst of warfare, how he defined the very unanalytic word *squishy*. "People," he said. "A squishy problem is hard to get hold of and has undefined borders. It has lots of human beings in it, and what they do is important, especially individuals." "NATO," Resnick said, "is particularly squishy on many levels." He ticked off several that are hard to model: "Weather. Leadership. Morale. Will the French come in? What is the role of conventional weapons?" Each of these questions can be individually modeled, but when all of the models are put together, the result is squishy.

Even if policy questions are ignored in a model, many seemingly measurable aspects of war may get shaky numerical values. One NATO model, for example, assumed that ammunition would run out every twelve hours and troops would need twice-a-day replenishment. But there was no way for ammunition to get to the guns because the model lacked a logistics network. The report also noted that the model's break point—the number of casualties that forces a battle to end—"is simply a mathematical convenience."

The mathematics of war can give strategists dubious and sometimes ludicrous answers. Looking at the Lanchester equations, the report said, "Lanchester's original work described a greatly simplified war—a force carried only one type of weapon, and the effects of terrain, tactics, supply, reinforcement, etc. were all ignored. The theory has been enriched since, and is very complex today." The report looked at a duel to the death between sixty Warsaw Pact tanks and twenty NATO tanks under Lanchester square-law rules. When the NATO and Pact tanks are equally effective, the NATO force is annihilated and 95 percent of the Pact tanks survive. When NATO tanks are nine times more effective, both sides are wiped out at the same time. Under the square law, the outnumbered NATO tanks must be *more than nine times effective* to win the duel.

Another problem stems from parochial scenarios in which U.S. services not only give no meaningful roles to NATO allies but also ignore the part that the Navy and Air Force will play in a real war. The Army looks only to the battles on the land. The Navy, centering the NATO war on a "maritime strategy," sees aircraft carriers and amphibious landings. The Air Force focuses on strategic bombing rather than vital, though less glamorous, close air support. In a recent game, run by the Navy, NATO naval forces left themselves open to Soviet Backfire bombers, which sank twenty-two unprotected allied ships.

Hovering over the simulated NATO battlefield is more than Clausewitz's fog of war. This is a battlefield that may be struck by tactical nuclear weapons, that may be sprayed with lethal gas, or given an instant epidemic of plague. Radios, telephones, computers, even the engines of vehicles may be knocked out by a high-altitude nuclear burst. The three following scenarios, each covering one of the potential ingredients of a NATO-Pact battle, were written by people who studied the effects of a specific ingredient and then applied that knowledge to a scenario.

U.S. Army Colonel Jean D. Reed, the author of the first scenario, in fact, had some direct experience in setting up the field-level U.S. policy for the use of nuclear weapons. In the late 1970s, according to Reed, the United States military doctrine in Europe had been this: If the conventional defenses "were in danger of failing and the integrity of the corps position were in jeopardy, tactical nuclear weapons fired in 'pulses' of perhaps as many as 200 warheads in a corps zone would be used to halt the enemy's attack decisively and create a situation conducive to negotiations and termination of the conflict on terms favorable to the United States and its allies." Reed was directly involved in "informal staff talks" with West German officers who were concerned about the doctrine. As a direct result of the talks, U.S. Army field manuals no longer published a specific number of "pulses."

Scenario: Unleashing Tactical Nuclear Weapons

The United States and the Soviet Union are "in a face-to-face confrontation in Southwest Asia" in this scenario written by Reed, a senior research fellow at the National Defense University. "The Soviet Union and the Warsaw Pact decide to divert Western attention from the Middle East and to launch limited objective attacks in Western Europe. In addition to their intent to limit the ability of NATO to shift forces to the Arabian Gulf, the attacks are aimed at separating West Germany from the NATO Alliance and neutralizing forever any German potential for again making war on the Soviet Union or the nations of Eastern Europe."

The scenario then launches, "on 1 April 198X," the classic textbook by Warsaw Pact forces: an attack "along the full extent of the border between NATO and the Warsaw Pact countries." The Third U.S. Armored Division* and the remainder of the U.S. V Corps deploy to their battle positions along the border. Reinforcements are on their way from the continental United States, but it will be at least ten days before they can reach the front with their stockpiled equipment.

"Although NATO is able to inflict heavy losses upon both the first-echelon and second-echelon Warsaw Pact regiments," the scenario continues, NATO forces are pushed back after a three-day battle. Then, as the two Soviet divisions in the follow-on echelon threaten the Third Division, the NATO corps commander estimates that within the next twelve to twenty-four hours the Warsaw Pact forces will open a 20-kilometer hole in NATO's lines. He expects that tank-led armies, now some 100 kilometers to the rear, will aim at this opening and lead the enemy on a blitzkrieg race to the Rhine.

"The corps commander calculates, however, that with the forces available and limited use of tactical nuclear weapons, he can strike the flank of the enemy penetration before the tank army is committed." The tactical nuclear weapons will stop the Warsaw Pact invasion long enough for the reinforcements from the United States to reach the front. The corps commander requests authorization "to use a limited number of nuclear weapons."

The scenario says that U.S. and NATO military and political authorities have "anticipated the development of such a situation in war games prior to the conflict, and plans for the required nuclear weapons support were included in one of the contingency plans developed before the war. . . . Release authority for the weapons to support the counterattack is granted in short order, but with a

---

*For information on the nominal composition of U.S. forces, see page 356.

clear understanding by Allied political and military authorities of the possible consequences."

The corps commander launches his counterattack with non-nuclear weapons, including precision-guided missiles. He uses his nuclear rounds against the enemy's command and control units to disrupt communications. "Simultaneously," the scenario continues, "the Central Army Group and 4th Tactical Air Force launch conventional and nuclear strikes against the following tank army and severely disrupt its advance, throwing its command and control into chaos. Again, the improved effectiveness of NATO's new classes of nuclear warheads achieves a high level of military effectiveness.

"The reduced collateral damage [death and injury of civilians] of these warheads couples with the West German civil defense program and keeps civilian casualties relatively low," the scenario says, sidestepping one of the obvious results of using nuclear weapons on the soil of a NATO country. The scenario ends triumphantly:

> The attack of the U.S. 11th Armored Cavalry Regiment reinforced by the VI German Panzer Brigade is successful. Forces in the penetration [area] were defeated, destroyed, or captured, or they withdrew to reconsolidate and reorganize. The NATO line forward of Highway 8 is restored and V Corps reconsolidates its position along the interzonal border.
>
> The Army group commander receives word that high-level discussions are taking place between Pact and NATO authorities relative to a cessation of hostilities. Throughout the period he had been advised by SHAPE of continuing communications between the United States and the Soviet Union over the Washington-Moscow hotline and knew that the United States threatened to escalate to strategic nuclear weapons if the Pact responded to NATO's use of tactical nuclear weapons with theater nuclear weapons of their own.

Scenario: Loss of Command and Control

"The most remarkable thing about the Soviets' crossing the nuclear threshold was that it went virtually unnoticed," begins another scenario written by a senior research fellow at the National Defense University. "The first recorded sign of something unusual occurred when a division logistics center reported a total failure of the commercial computer that processed all the requisitions for spare parts of the division logistics center. . . . Had anyone been in a position to gather the data, a survey would have shown that almost every piece of data-processing equipment in the corps area—and considerably beyond—had failed. But commanders had fixed their attention elsewhere, and disruption of spare parts resupply ranked low in priority.

"Of more immediate concern was the widespread, sudden, and infuriating loss of communications coupled with the indications, gained mostly from satellite resources, that the Warsaw Pact's predicted renewed attack was about to take place. The detonation of several nuclear weapons across the corps front some hours later removed all doubt and made the analysis of the effect of the first two high-altitude nuclear bursts on equipment a matter of interest mainly to historians and technicians."

The electromagnetic pulse (EMP), generated by the high-altitude bursts of Soviet nuclear weapons, had knocked out telephones, switchboards, and computers, eradicating their stored data. The pulse badly damaged all radio sets that had sizable antennas and destroyed numerous FM radios, radars, and receivers used primarily for eavesdropping on enemy communications to gather intelligence.

In this version of the NATO-Warsaw Pact clash, the war had been going on for less than a week, with indecisive results. Then, as in the previous scenario, NATO field officers had requested release of tactical nuclear weapons to stop, or at least slow down the enemy advance. But the request was "strangled in its own procedures."

SACEUR (Supreme Allied Commander, Europe) had already requested release of what the scenario calls Selective Employment Plans (SEPs) even before the Pact attack began. By the time the actual requests were made from the field, however, the SEPs were outmoded—and political leaders, agonizing over the decision, had not even authorized them. Then came the EMP-triggered loss of NATO communications. "The fog of events left SACEUR in the position of requesting release for air-delivered weapons to be fired on targets of opportunity, an option that ultimately had minimal effect on the battle's outcome."

In the aftermath of the EMP, Pact forces unleashed nuclear weapons that all but destroyed NATO command elements and communication links. The scenario ends grimly:

> At the unit level, commanders faced numerous other problems. Attempts to determine areas of significant radioactive contamination moved slowly. The lack of prompt information about contamination made commanders reluctant to move across unsurveyed terrain. . . . Unanticipated psychological consequences further reduced unit effectiveness. Almost no one knew who had fallen victim to radiation sickness. Few soldiers had dose measuring instruments, and every soldier on the frontlines thought he might be a radiation casualty. Lack of training on symptoms sent droves to various medical treatment facilities under the impression they might be suffering from radiation sickness. Rumors and misinformation passed among individuals. . . .
>
> As a result, the Warsaw Pact attack at the seam of the two Army groups succeeded, allowing the enemy to wedge them apart and push deeper into Germany. The Pact then succeeded in exploiting the assault phase with second-echelon units. With forces divided and flanks menaced, Allied leaders found themselves facing an intolerable dilemma: escalate the nuclear battle and possibly trigger a strategic war, or open negotiations with the Warsaw Pact while in a weak bargaining position.

Scenario: Gas Attack

Petty Officer Sam Barnes, an air controller at a naval base in southern Europe, becomes one of the "individuals" in a squishy problem in this melodramatic scenario, produced for the Pentagon by the Institute for Defense Analyses and a group of retired military officers.

"My God, it must be gas! A chemical attack!" Barnes exclaims as the melodramatic scenario begins. His "eyes began to burn and he felt a tightening in his upper chest and throat. In only a second or two more, he began to have difficulty breathing and felt extremely nauseous. In his final moments of consciousness he took one last look around the field from the tower. . . .

"Fire and smoke were everywhere, and the gas emanating from the chemical warheads had begun to take a terrible toll. Air operations ceased. The wounded and dying were everywhere and the survivors from the explosions were rapidly succumbing to the chemical agents. . . . Very few of the base personnel had managed to get into their protective gear in time. . . .

"Rescue crew in protective chemical gear began to move about the field. The toll included hundreds of men and women, caught without protective clothing, who died within thirty minutes following acute stages of nausea, loss of muscular control and then unconsciousness. Others suffered nearly the same symptoms but escaped death following a long period of incapacitation. Petty Officer Barnes was one of the latter. He was found an hour after the attack unconscious but alive. He had been exposed only briefly to the chemical toxins as they settled in the cloud just above the elevation where he had been working. . . . Medical supplies of all kinds were quickly used up. In many instances medical packages themselves became contaminated with the chemical residue which lay everywhere."

The three scenarios illustrate not only the squishy problem of modeling as complex a phenomenon as a NATO-Warsaw Pact war but also the problem of using a model—in this case a scenario— as a form of advocacy. The first scenario, which the author, an Army officer, admits is "simplistic," is "designed to show a possible means of limited employment of tactical nuclear weapons." The second scenario, also by an Army officer, is used to show that current military doctrine "inadequately addresses the fighting of a tactical nuclear war." The third, produced by a small think tank under a subcontract from a major one, was the work of twenty-one retired military officers, all of them flag officers, who said in their report that they were "extremely concerned" about the potential use of chemical weapons by Warsaw Pact forces.

In the military lexicon a theater is where a war is fought. NATO has become another kind of theater, a place for imaginary wars based on scenarios that seem unlikely to be the scripts for a real war. "I think I can say without any contradiction," former Secretary of Defense Robert S. McNamara has said, "that there is no piece of paper in the world that shows how either the Soviets or the

U.S., Warsaw Pact or NATO, can initiate the use of these warheads with any advantage to itself. They are militarily useless."

"The debate about NATO's defense options has changed little in more than twenty years," Paul K. Davis, an analyst for Rand, the pioneer strategic think tank, wrote in 1985. His paper stemmed from a speech he gave to the German Strategy Forum in Bonn-Bad Godesberg, West Germany, the land that is part of the squishy problem, for, according to NATO scenarios, it is upon this soil that much or all of a real war would be played.

West German concern about tactical nuclear weapons dates to a secret NATO war game played at the United Kingdom Defence Operational Analysis Establishment at Byfleet, Surrey, around 1960. The nuclear battle involved weapons with a total explosive power of between twenty and twenty-five megatons. If the weapons had been fired so that they burst in the air, the homes of about 3.5 million people would have been destroyed and about 1.75 million people would have been killed or gravely injured. A ground burst would have killed about 1.5 million people immediately or through radiation and left another 5 million suffering from the effects of exposure to high levels of radiation.

Another authoritative estimate came from Admiral Noel Gayler, who saw nuclear scenarios from several perspectives—as Commander in Chief of U.S. Pacific forces, as Deputy Director of the Joint Strategic Target Planning Staff (which develops the SIOP), and as director of the National Security Agency. Gayler said that he had seen estimates that a million civilians, most of them West Germans, might be killed as a consequence of using NATO nuclear weapons to stop "four nominal tank breakthroughs."

The political reality of unimaginable death and destruction fuels skepticism about the value of NATO scenarios. "The point is arguable," Davis said, "but I will assert that I have *seldom* heard a senior or mid-level official of the U.S. government base policy-level conclusions on the results of a theater combat simulation. To the contrary, I have *often* heard such simulations derided and arguments based on them dismissed out of hand. The reasons for this lack of influence are many and varied. The most fundamental reason is that the combat simulations lack *credibility* among experienced policymakers. . . ." (Davis was already at work on a revolutionary approach to gaming [Chapter 16] designed so that

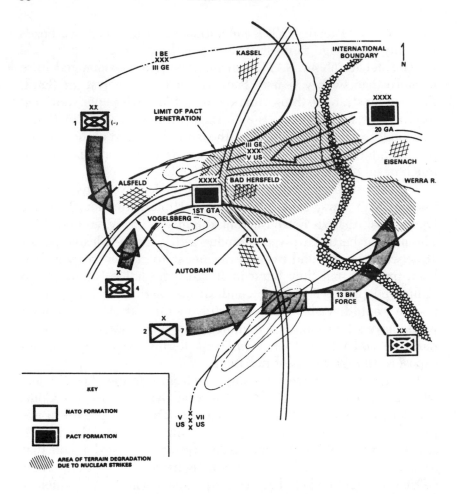

Nuclear explosions devastate some 830 square miles of land (shaded area—mostly in West Germany) during a study of NATO gaming sponsored by the Defense Nuclear Agency. The games, featuring a battle near what NATO strategists call the Fulda Gap, were played in 1980 with a scenario projected into 1986. Rectangular NATO and Warsaw Pact symbols, also used in hobbyist war games, indicate the size and type of units in the battle. (Credit: *Proceedings of the Workshop on Modeling and Simulation of Land Combat.*)

analysts and policymakers "would no longer be able to characterize strategies with 'one-liners.' ")

Retired military officers look at NATO from a viewpoint far different from that of an officer still on active duty, advancing up the career ladder and attending the National Defense University. As retired Army Colonel Leslie G. Callahan, Jr., briskly summed it up, "We've got our whole spectrum of NATO warheads down to one-tenth of a KT [kiloton]. The Russians have no nuclear warheads in Europe less than twenty-five KT. Now how the hell do you make a game out of that?"

Retired Brigadier General Edward B. Atkeson—the "anonymous" author of the article criticizing simulation—was the U.S. Army Deputy Chief of Staff for Intelligence in Europe and later the CIA's National Intelligence Officer for General Purposes Forces. He has two major criticisms of NATO-Warsaw Pact scenarios: unrealistic estimates about Pact armies and unrealistic concepts about the wartime worth of tactical intelligence.

He sees "a basic 'fault line'—not unlike a seismological fault—running between the Soviets and their allies, which, if subjected to great stress, could rupture, with serious ramifications for Soviet fortunes." By his view the nations that the Soviet Army supposedly relies on would be extremely unreliable comrades in a real war. NATO scenario writers and analysts, he says, have "largely ignored the fault-line phenomenon." Atkeson developed his fault-line thesis with the aid of think-tank games (rather than NATO ones). The games played out various kinds of defections by Warsaw Pact combat units—and nations—and translated the effects into what could reasonably be expected on the battlefield.

Distrust of our own allies is reflected, he says, in the U.S. attitude toward the sharing of intelligence information. "If there was one gaping sore in the intelligence system we had," he recalled, "it was the fact that we have designed our intelligence coordination apparatus so that it can only go to American forces. And we forget that most of the guys on the Blue side are non-Americans.

"We have all sorts of sophisticated equipment—satellites and so forth—that, say, the Dutch don't have, and so we can tell our forces about all kinds of things that they can't tell their forces about. And we haven't built the links for providing their forces with the information we have. There's no way to pass the information. In peacetime it's not apparent, because it doesn't matter if the information doesn't get to them for a while."

As for plans that are carried out in combat as smoothly as they were in that *Unleashing Tactical Nuclear Weapons* scenario, he said, "Once a year we have a big order-of-battle conference. We decide what the threat is, and it turns out to be whatever the Greeks and the Turks can agree to." At least one major NATO war game was abruptly ended because of a Greek-Turkish disagreement.

Lieutenant General A. S. Collins, who served in World War II as an infantry commander, commanded the Fourth Infantry Division in Vietnam, and from 1971 to 1974 was Deputy Commander in Chief of the U.S. Army in Europe. He called the use of tactical nuclear weapons to offset heavier enemy conventional forces "a dangerous myth." Hiroshima—which he saw with a soldier's eye two months after it was bombed—"was destroyed by what is considered to be one small tactical nuclear weapon in today's arsenal." He quoted an Army field manual as saying that a single nuclear burst in one battle could "produce enough losses to make whole units ineffective." During World War II, he said, "over a period of approximately twenty-eight months the Third Infantry Division had 4,992 men killed in action, and the Fourth Infantry Division lost 4,097. What happens when that many men can be killed in only a few minutes. . . ?

"People talk about keeping a nuclear war limited as if all this violence, movement, and dying will take place in some antiseptic laboratory. . . . Once the nuclear barrier is broken, it is hard to believe that commanders, steeped in the US military tradition of concern for the lives of their troops, will respond with moderation, especially if their units have suffered severe losses in the first attack. . . .

"Nuclear strategists often describe nuclear attacks, in which millions would be killed and wounded, as though they were great chess games with cities, aircraft carriers, great industrial areas, and other sources of national power as the pieces. As each one is destroyed, that piece is removed from the board and the match goes on as usual. . . ."

It is this perception of war as chess that has produced an odd and little-known marriage of fantasy and reality. Recreational war games on the shelves of toy stores and hobby shops are produced by the same thinking—and sometimes by the same contractors— producing games for the Department of Defense. The board games and the arcade-style video games are often dazzling conjurers of reality. And the strategic thinking behind these games is often brilliant. But they are games. And it is in recreational games that serious war games are rooted.

CHAPTER 5

War as a Game

One day in the fall of 1980 James F. Dunnigan, a wisecracking, cigar-puffing New Yorker who made a living creating hobbyist war games, approached the Pentagon office of Andrew W. Marshall, who assembed secrets for the Secretary of Defense. Someone inside the small suite of offices pressed a button and the crypto-electronic lock clicked open, admitting Dunnigan to a sanctum of secrecy.

Since 1973 Marshall had been the director of Net Assessment, the Department of Defense's tightly guarded citadel of knowledge about the military strengths and strategic doctrines of the United States and the Soviet Union. Marshall was the assembler, custodian, and judge of some of the nation's most sensitive information— U.S. war plans, the targets of U.S. nuclear weapons, Soviet defense secrets that agents had stolen at the risk of their lives. Net assessment was considered so vital that Marshall could call for aid from anyone in the U.S. government. Now he was asking Dunnigan, an unlikely version of Pentagon consultant, to help him. "We're sitting there," Dunnigan recalled, "and Andy, whom I respect, is asking me to do a game on how to keep a nuclear war going after a limited exchange."

The subject did not bother Dunnigan. "I've been dealing with that—I've been blowing up the world—for years. What bothered me was that we—SPI—were called in at the last minute." SPI is Simulations Publications, Inc., which by 1980, with Dunnigan as president, had produced nearly three hundred hobbyist board games, including Jutland, Normandy, and The Origins of World War II, all of which Dunnigan had done, and such fantasy board games as Dragonslayer, Swords and Sorcery, and War of the Ring, based on Tolkien's *Lord of the Rings.*

The meeting between Andy Marshall and Jim Dunnigan symbolized the little-known connection between what might be called *real* war games that help a nation prepare for war and *recreational* war games that help hobbyists pass the time. Real war games are the blueprints of real war. Recreational war games are the playthings of war. Yet, the good playthings are more than that. They are accurate, fascinating reenactments of war. They teach the thinking of war, and, like autopsies, they have great value in showing the living why certain actions can be fatal.

Dunnigan had a passion for war as history, as something more than the sand-table exercises of traditional military wargaming. He had, for example, designed The Origins of World War II as a game of politics, and the way the game had been played had given him an insight into war and human nature. The object of the game was to avert World War II. "Each player," he recalled, "had different values attached to Poland, and so forth. There were also different points for securing control. The idea was to get the largest number of points. Everybody knew what to do to avoid World War II. Most of the games started World War II."

Dunnigan was a student of war and strategy, not an analyst with think-tank credentials. By the time he had his talk with Marshall, Net Assessment already had contracted with the Rand Corporation and another defense research firm, Science Applications, Incorporated, for a revolutionary, multimillion-dollar research project aimed at improving methods of strategic analysis (Chapter 17). So Rand and SAI, among the nation's most influential think tanks devoted to national security matters, had, as Dunnigan put it, got there before him. "But I walked out with a $40,000 contract. The game is still going on. It is called SAS." (SAS, for Strategic Analysis Simulation, is a global war game played at, among other places, the National Defense University. Lieutenant General Richard D. Lawrence, president of NDU, says the game "allows the players to analyze decisions at the National Command Authority, commander in chief, and fleet commander levels.")

Dunnigan said Marshall had called in a commercial* war gamer because Marshall was not satisfied with what he was getting from game designers working for the Department of Defense. "Whatever gaming was done was command-directed," Dunnigan said. One day, while working on SAS, he asked a military game designer, "How do you validate these things?" The designer said, "We don't. We are under constraints. We are tasked by our decision-makers to do certain things, and quite often that includes verifying a decision that has already been made."

"Well, obviously, I didn't have that constraint, and so the fact that our games were done in a more open atmosphere allowed them to develop more fully," Dunnigan said as he reminisced over a cigar and a tall glass of lemonade. We sat in the small, walled backyard of his Greenwich Village flat. "I'm sure the people in the government could have done just as well if they hadn't been working under the bureaucracies, and, as a consequence, they began taking [commercial] games—whole—and turning them inside out and giving them another name. A lot of things you see now can be easily traced. Sometimes they are very blatant about it."

One of Dunnigan's games, Firefight, appeared first as an official U.S. Army training device and then as a hobbyist game. He has since said that the game for hobbyists was more realistic than the one for real soldiers. Dunnigan designed Firefight for the Army in 1974 to train platoon leaders and company commanders. The game, based on U.S. and Soviet tactics and weapons specifications, pitted four-man U.S. and Soviet fire teams against each other on a game map of either Fort Benning, Georgia (site of the Army's infantry school), or a training area in West Germany. The Army's Training and Doctrine Command used a later version of Firefight to evaluate weapons under development.

As Dunnigan tells the story, Army-ordered changes in Firefight made the game unrealistic. "Our historical research indicated that the small-unit commander's ability to tightly control his troops was a sometime thing," Dunnigan told a conference on gaming in 1982. "Historical actions at this level could not be accurately simulated without taking into account the problem of 'command control.' In

*Game-makers prefer this term rather than "amateur," "hobbyist," or "recreational."

game terms, it simply meant that a certain percentage of the commander's orders would be ignored or misinterpreted. In 1976 someone in the chain of command did not want this subject addressed in the game. . . . I could never understand how one could properly train troops with a simulation that did not cover the effects of poor communication.

"The other factor suppressed was underbrush. Our terrain maps were quite accurate, taken from detailed survey maps and personnel reconnaissance by our staff, at least of the Fort Benning site, where our guy almost disappeared in a swamp." The result, Dunnigan said, was that soldiers would have trouble finding and aiming at tanks. But, at the time, "much was being made of the long-range killing power of anti-tank weapons." And so the swamps on the terrain maps officially vanished.

A hexagonal grid system, first used on the map boards of commercial war games, appears on the computerized maps used in many high-level Department of Defense games. Hobbyists liked the hexagonal grid and saw it as an improvement over the traditional squares of the chessboard. On a hexagonal grid the movement and massing of military units can be better approximated than on squares. Many of the game rules are built around the hexagons. The damage caused by a bomb, for instance, usually is calculated according to a hexagonal pattern, with the destruction attenuating away from the impact point by a formula based on the hexes.

Military gamers picked up the hexagon idea, commissioned hexagon-based games from designers like Dunnigan, and incorporated the hexagon into computerized games developed by the Department of Defense. On an Army game board or computer monitor screen (the modern versions of the sand table) games are played with rules derived from the hexagonal design. A move across a hexagon on thickly wooded terrain will cost the player more "movement points" than, say, a crossing of a hexagon on a plain. Units sitting on hexagons in wooded areas are governed by rules about obstructed "line of sight." A unit is represented by a flat piece of cardboard that is moved, like a checker, about the map. A full-size map, twenty-two by thirty-three inches, has about two thousand hexagons gridded upon a stylized terrain.

Trevor Dupuy, a friend of Dunnigan and a fan of commercial games, has criticized the use of hexagonal grids in military games.

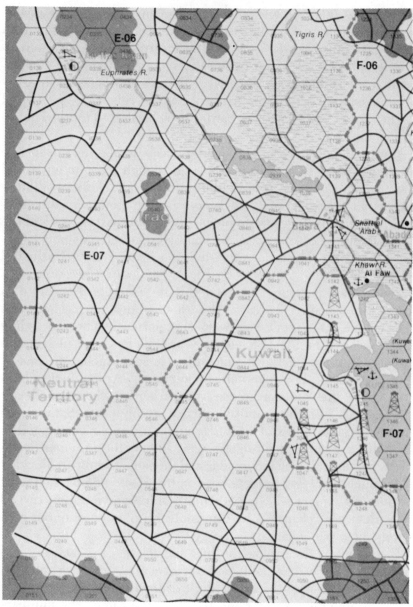

Hexagons form the grid of a map used in a hobbyist game. Such hexagonal maps are also used in official war games. Game play is controlled by complex rules based on movement across numbered hexagons. The terrain symbols indicate geographic reality that controls speed of movement. (Credit: From "Gulf Strike" Strategic Map, Victory Games, Inc.)

"It is impossible in a combat procedure governed by hexagons," he wrote, "to replicate the real-world process of an attacker with sufficient combat power advancing against opposition at a rate realistically reflecting all of the many factors that govern opposed movement. Either there is no advance or there is an unrealistic hex-by-hex movement, with the distance advanced related to the size of the hex. . . . Real-world combat must be related to terrain, not to artificial grid lines."

Many computerized military games are based on the hex for the terminal display's "game board," even though a computer would allow a more realistic representation of terrain. Serious military gamers have also adopted the pieces used in commercial games. SAS, for example, uses these pieces. Both hobbyist pieces and the military pieces are labeled with essentially the same nomenclature and the same abbreviations for such information as unit size and combat strength.

Pieces used in hobbyist games closely resemble those used in secret games played to analyze the use of tactical nuclear weapons in a real NATO-Warsaw Pact war. Those crucial games, including one developed by the U.S. Army War College, were played in a study commissioned by the Defense Nuclear Agency to "assess alternative tactical nuclear warfare doctrines." As in a typical hobbyist game, the cardboard pieces, or markers, in the war college's game represent either combat units or supply units. Symbols on the rectangular pieces identify the unit. A number is assigned to the unit's combat or defense strength, and another number determines movement. The movement and the fate of a unit are decided by formulas involving the numbers on the piece, numbers on what commercial gamers call combat results tables, and numbers turned up by devices of chance—dice for hobbyists, random-number generators for professionals.

The numbers determine the hits and misses, the fatalities and the survivors of warfare. But, in the games that military strategists play, the numbers of the living and the dead are not just numbers for determining the winner and the loser of a game. The numbers from the Army War College games and others like them eventually find their way into serious analyses of a NATO-Warsaw Pact war in which tactical nuclear weapons may wipe out parts of East Germany and West Germany.

According to the scenario of one of these NATO nuclear games, "terrain degradation due to nuclear strikes" appears as a blot over some 830 square miles along the East German-West German border. In this, as in other typical military games (officially, "models"), there is no civilian panic, no provisions for handling vast numbers of casualties, and none of the sophisticated interplay of politics and warfare that is the hallmark of a good hobbyist game.

Most hobbyist games, in fact, avoid the use of either battlefield nuclear weapons or city-killing missile warheads. Nuclear war, as a game designer once remarked to me, "has only two data points—Hiroshima and Nagasaki." I also got the impression that games about nuclear war are not particularly interesting to most hobbyists, whose games are typically anchored in the past or in fantasy. "I call war games the hobby of the overeducated," Dunnigan told me. "A lot of them are technical people. These games appeal to smart people in general, people diligent about history. They demand a disciplined model, not a bullshit model."

Commercial designers have great faith in history as an armature of reality. And it is that concept of history that generates the greatest difference between designing a game for a hobbyist and a game for a general. The hobbyist (who believes with Thomas Hardy that war "makes rattling good history") wants the thrill of participating in the inexorable movement of history and verification of the adage about the past providing lessons for the present. The modern general wants what he thinks modern technology can give him: computer-guaranteed facts.

Facts, however, are not what drive recreational war games. And it is here that the connection between reality and recreation breaks down. Because there is no history of nuclear war—just those two awesome data points—military war gamers, such as the indefatigable NATO scenarists, must create an artificial nuclear war and then try to make it real with numbers plucked from reality: kilometers, artillery ranges, the sizes of Army units, the speed of aircraft carriers.

Unlike most military scenarists, the hobbyist is realistic about the chances of victory. A hobbyist game usually has "victory conditions" or a "victory level." As Dunnigan put it in his book on war games, "Victory is often a fairly vague thing in military history. . . . Certain levels of victory have been established through the years. These levels are, in ascending order, draw, marginal victory, substantial victory, decisive victory, overwhelming victory." A general wants unconditional victory; an amateur will settle for an entertainingly realistic war.

When hobbyist games enter the present or the near future, an eerie kind of realism sometimes does appear, and that realism draws military gamers again and again to the quest for the simulation that will predict the future. Military model-makers know, as an article of faith, that the future cannot be predicted. And yet, without daring to say it aloud, many do want to discover some way to simulate the future. These heretical modelers, like medieval alchemists, are sometimes implicitly encouraged by their orthodox superiors.

Hobbyist games have produced tantalizing examples of a predictable future. An Indochina game that Dunnigan published in 1972 was played in Thailand by a group of American officials who substituted top-secret information for the game's data, which Dunnigan had found in *The New York Times* and *Newsweek*. Dunnigan said he was told by "a guy from the State Department" that there was little difference between playing the game with Dunnigan's information or with official secrets. The 1973 Arab-Israeli War seemed to follow the scenario of Sinai, an earlier Dunnigan game in which the Egyptians manage to cross the Suez Canal into the Sinai—a maneuver that seemed extremely unlikely when the game was developed.

Civilians have also contributed to the prophetic mysticism of war games. A neighbor of mine stopped playing a computerized game, B-1 Nuclear Bomber, with his son because the youngster seemed to be enjoying not the challenging flight through Soviet defenses but the nuking of Soviet cities. Balance of Power, another best-selling computerized war game, pits the United States against the Soviet Union in an ideological struggle, with the object of the game being to keep the bombs and missiles sheathed. If the superpowers fail to avoid nuclear war, the screen blacks out and the player gets this message: *You have ignited a nuclear war. And no, there is no animated display of a mushroom cloud with parts of bodies flying through the air. We do not reward failure.*

SPI's Würzburg so disturbed the real citizens of the real Würzburg (which frequently experiences real, full-scale NATO war games) that NATO, in response to Würzburg queries and protests, officially stated that there were no real plans to explode nuclear weapons near the town. During debate in England over nuclear weapons, two members of Parliament denounced not only the reality of the weapons but also a "disgusting and offensive" American import— Nuclear Escalation, a card game with a six-sided die that glows in the dark and has a mushroom cloud instead of a 1; its 2, 3, 4, 5, or 6 indicates in millions how many people are killed by radioactivity after a nuclear attack.

SAS, the game Dunnigan developed for Andy Marshall, is a further link between the reality of war and the make-believe of games. SAS has passed to Mark Herman, another commercial gamer and New Yorker whose laid-back, fast-talking style makes him as rare a specimen as Dunnigan in the corridors of the Pentagon. "I put into this SAS, right off the bat, right to their faces at the briefing, that there are three words that do not belong in the military lexicon: unsinkable, unbreakable, and indestructible," Herman told me. "I got the Navy people to accept that carriers can be sunk, and they do sink in the game. I told them I wouldn't stand for it otherwise. The Navy has done one thing that is sort of unrealistic in its thinking. They really don't like to think about an attacking nuclear warhead at sea. One hit with a nuclear warhead on a carrier and that's it."

When I met Herman he was working on a paper for officials in the Department of Defense who wanted to know how commercial games were designed. He said he raised several important questions in the paper. He would not describe what he would be writing in the paper, but I could imagine, from our conversation, that one of the topics would be participation. Like many civilian gamesters I talked to (both commercial and military), Herman was critical of high-ranking officers who decline to play war games because they take too much time.

The long gaming tradition in Germany, for example, is for generals to play at the level of generals and for general staff officers to play at their level. "Our higher-level flag officers, not having the time to play the game, are getting the information secondhand," Herman said. These same officers may also be hearing the results of games that reflect doctrine rather than reality.

"The game should not have any doctrine. The players themselves are the ones putting in doctrine. If the game models reality in some way that you're comfortable with, then how you employ the forces is the doctrine. The players have to play faithfully to the reality. What you do get a feeling for are the first-order magnitude problems. You get insights but no conclusions.

"Let's say someone in the game uses strategy that makes aircraft carriers vulnerable to some kind of weapons system. You may or may not have liked the way the game dealt with it, but if you then go back into the reality—no game now—and you say, 'How do we deal with this particular technology breakthrough or problem?'— then you know that in the game you've asked a good question. That's what the game will do for you: raise questions."

Herman began playing war games when he was twelve years old. He pushed lead ships around the floor and marked explosions with upside-down golf tees "because I thought they looked like splashes." He was already thinking about creating games about that time, and, at around the age of twenty-one, he began working for Dunnigan at SPI. "I learned the trade under Jim Dunnigan," he said, looking back from the perspective of his thirty-one years. "He has an unusual style. He sparks." Herman also worked on Department of Defense games at BDM International, Inc., a Washington think tank. Although not as well known to the public as Rand, BDM, like SAI, ranks as one of the inner-circle defense research organizations.

At BDM Herman worked with General William Depuy, an Army intellectual who spent time in the CIA and, as a key adviser to General William C. Westmoreland in Vietnam, was credited with devising the search-and-destroy strategy. "I worked for him for two years. I used to design the games at BDM and he actually played them. And he used to tell me what it was like to be in a real war. He's been in everything since World War II. He told me that war games are played too fast. In real life, people don't fight as quickly as games portray them to fight."

I talked to Herman at the headquarters of Victory Games, on the sixth floor of an old loft building in the garment district on Manhattan's West Side. Proofs of maps and rules and complex tables lay in piles on worktables. Next to Herman, at a desk as cluttered as his, sat another Victory Game designer who was working on a Civil War game. A third was hunched over a computer keyboard, scrolling a set of rules down the screen of a monitor. The loft had the air of an artists' co-op. The uniform of the day was jeans and T-shirts.

Herman was putting the final touches on War in the Pacific, which is what game players call a "monster game" (in contrast to a short and relatively simple "beer and pretzel" game). War in the Pacific begins on December 8, 1941, and ends on July 31, 1945. The game ends just short of when the war actually ends because, Herman explained, "I don't bother with the A-bomb." Monster games usually take at least forty hours to play. Herman estimates that War in the Pacific will last about ninety-five hours, although a player may choose to play only a scenario at a time. Each of the game's twenty-one scenarios, such as the Battle of Midway, can be played separately in around two hours—"like Monopoly," Herman said disdainfully.

Herman's way of designing a game, as he was presumably about to tell the Department of Defense in a paper for restricted circulation, is vastly different from the way most DOD games are designed. He does not make any major distinctions between commercial games and made-for-the-Pentagon games. As he put it, "The techniques I use on the commercial side are certainly the ones I use on the other side. But at Victory Games we do commercial games for a living. I give them"—meaning unnamed officials at the Pentagon—"good ideas and save them money. I keep them"—commercial and military games—"very separate. They do look my stuff over for that," *that* meaning secrets. Herman was uncharacteristically vague and hesitant when he spoke about working the two sides of the war-game street. He was obviously not comfortable with handling secrets.

As he began taking me through his reasoning about the design of the game he was finishing, he became more at ease. "It begins when I create a hypothesis," he said. "Now the hypothesis may evolve along the course of the project, but the gist of it is this: If I'm going to a do a game on the war in the Pacific from 1941 to 1945, at that point I have to decide what the focus has to be—who are the players and what kind of decisions are they going to make? At this point I decided the player is going to be MacArthur or Nimitz. That to me was where it was.

"I didn't want to design a war game that was logistically heavy.

I wanted to give the people ships, give them guys, let them go out and invade things. Basically, let them do military strategy. I centered the design around an operation, such as the Coral Sea operation [a May 1942 carrier battle], and I don't mean just the two days of combat but the whole thing—closure, the combat, to the withdrawal. You can use the Coral Sea in that sense. And that's the part of the game I call the battle cycle.

"The other player is going to be the reactive player. The reason I created that kind of a system was because one of the most important things in the war in the Pacific was the Ultra information—the breaking of Purple, the Magic codes. So I created this whole system around closure based on the intelligence: how much reaction the other guy had was based on how much he knew about code break. This in essence would allow you to go out and ambush him—and do Midway [June 1942], probably the most notable decisive battle in history.

"The other kind of operation was the intercept: Coral Sea. The other guy came out and you knew he was going to do an operation. And we would meet somewhere out in the ocean and have a battle, usually somewhere around an island that was of strategic significance. The game centers on three kinds of operations—surprise attack, intercept, and ambush. Ambush and surprise attacks are obviously the extremes. Intercept was really the norm of the war. That's really the center of the game.

"There are, of course, other things, strategic bombing, submarine attrition, all that kind of stuff. But that is really extra to this main central theme. So once you have developed the central theme, then you have to work in all the details of getting the pieces to physically move on the map, fly to each other, bomb each other. Combat, movement, attrition, which every game has in one form or another.

"Just to characterize that against what DOD does, DOD usually starts with the movement and attrition side of things. They immediately come into it with the numbers and the number crunching. They usually start with the attrition modules." I asked if QJM supplied any of the attrition data. He said that he did not use QJM data but assumed that Dupuy's information was being marketed in the Pentagon, "where marketing is the standard of life and getting projects done is the secondary issue—and getting another project is more important than doing the one you've got." When Herman works on DOD games, he said, he relies on data provided by the armed services, such as data from simulated dogfights between Air Force pilots flying U.S. aircraft and "Red Flag" Air Force pilots trained in Soviet tactics and flying aircraft adapted to perform like Soviet warplanes.

"DOD usually starts with what they can understand. They do what they call war models. And if there is anything I love about military history, which I have been reading since I was twelve years old, it's that there's no model of warfare. It's direct chaos. If you put a hundred people into the field, fifteen will do something. The other eighty-five are going to say, 'Those fifteen guys got killed. I'm out of here!' Or, 'Those fifteen guys are winning. I'm staying here.' It's going to be one or the other. There's really not a whole lot in between there. So to model everybody shooting at everybody else doesn't model warfare. It just models everybody shooting at everybody else. And unless you sensitize that to some kind of reality, you've got nothing of any use."

In DOD games, Herman said, "everyone's moving around like robots. Nobody's tired. Nobody sleeps. Nobody eats. That's why I used closure. Once you get close, there is combat. Things happen more slowly when there's no one around you shooting at you. And the guy on the spot doesn't know what the heck's going on. Nobody directs that. They just keep on feeding more guys where they think things are going well and hope for the best. Feeding those extra guys in at the right moment is more an art form than a science. And half of the information you're getting is wrong."

Some game designers inside the military establishment are following the commercial gamers' lead and are now using history as an ingredient of Department of Defense games. They may have been inspired, Herman said, by the fact that military history is an essential ingredient in Soviet strategic and tactical thought, as revealed in current Soviet military writings. But he questioned most of the models used by the Pentagon. "Numbers are meaningless. It's the relationship of the numbers to each other that's significant. That's why the system design is much more important than the data.

"The system creates the conditions under which combat occurs, but the combat itself is just shooting a missile and seeing something

get blown up. That's really of small significance because you can
always just change the number. Everybody has a different attrition
model. They can bring up test data or Falkland War data or Israeli
war data, and it will substantiate almost any point you want to
make.

"I call it the system of Kentucky windage. You know, 'Well,
there need to be a *one* there,'" he said, in what he assumed to be
the accent of a frontiersman aiming a muzzle-loading eighteenth-
century Kentucky rifle. "You modify it as you go along.

"In what I call modern fiction—contemporary war games—you
have to start suppositioning. For instance, if you have a division
attacking somebody, not everybody in the division's on the front
lines. In fact, very few of them are on the front lines. Especially
if they're marching to the place. They're strung out in long formations. So the strength of the division is no more than one battalion's worth of the whole division. A defending battalion can
repulse a division, make them retreat."

As an example of modern fiction, Herman told of an Israeli test
of a U.S. Army combat war-game model that was adapted to run
a battle between an Israeli tank platoon and a Jordanian tank battalion. The model showed that the outnumbered, outgunned Israelis lost. In an actual battle the Israelis had won by forcing the
Jordanians to retreat.

"What I've learned," Herman said, "is that troop quality, troop
training, and unit cohesion matter far more than how many guys
you have. Bean counting is probably the least accurate measure of
military strength that there can be. In the Pacific war one third as
many Japanese beat up three times as many British in Malaya, and
in the Falklands one third as many British troops beat three times
as many Argentineans. Some British commandos were outnumbered eight to one. Combat ratios are almost meaningless to battle
outcome. Alexander conquered the world with a lot fewer guys
than those who opposed him."

Rand analyst Davis, in his critique of combat models, also singled
out bean counts, or "static measures," as they are more formally
known. "Over the last two decades," Davis wrote, "the effort to
count weapons and divisions and to deal in other miscellaneous
static measures has become somewhat of a cottage industry." The
bean counts, he said, "have had a profound effect" in convincing
people that "more money should be spent on defense."

The "cohesion" Herman mentioned is the shorthand word for the elusive human elements that drive soldiers to stand shoulder-to-shoulder and fight. An analysis of cohesion, published by the National Defense University, cited Dupuy's QJM and noted that although Dupuy considered that morale and leadership "probably have more influence on the outcome of a battle than any of the other qualitative variables of combat," there seemed to be no way to measure them.

In developing the SAS game for the Department of Defense, Herman combined the reality of modern war with the concepts of commercial gaming. As an illustration, he talked about the Argentine submarine that stalked the British Fleet during the Falklands War with no success. Herman's game system would bring the submarine to a position where it could attempt an attack on a British ship. A Combat Results Table would then determine whether the attack succeeded. "Getting the submarine to the point where it has a firing solution," he said, "where he detected the ship, got into the right position—all that is in the conditions of the game system. But the point where he pushed the button and the torpedo blew, that's the Combat Results Table, anywhere from 'I missed' to 'I just hit you.' And if you get a critical hit that would have sunk the ship. A magazine. The kind of thing that happens.

"The military starts with 'I push the button and the torpedo goes away.' I start with 'the submarine's got to find the ship and shoot at it.' I don't really care if the torpedo hits or not. That's the luck of the war. In our games, you have to roll dice for that. The impression that our war games are luck-generated is in that one fallacy: that we roll dice."

The only times players do roll dice is to determine something that is greatly dependent upon chance: One player's searching ship is near another. The pursuer determines whether it has found its target by rolling a die; certain numbers, looked up on the Combat Results Table, call for certain results (much as the luck of the roll gets a Monopoly player out of jail). If the target is found, another roll and another check of the table decides a hit or a miss; a hit engenders another roll, which the table translates into a description of the damage.

"People in DOD aren't stupid. They're really not. There's a lot of politics, but down deep, if the model stinks, the people know it and they don't put any credence in its results. And if the model's

very good, they'll still be skeptical. And they should be. The first
rule of operations research, which is the mathematical side of this,
and has the best people, is 'A model is not predictive.' Period."

In 1983, a year before I played the Persian Gulf game at the
Naval War College, Victory Games published Gulf Strike, a monster game that Herman designed. Rereading the scenario for the
war-college game, and comparing it to Herman's complex and stimulating game, I found it hard to believe that it was his game that
was supposed to intrigue hobbyists—and the war college's game
was the one that was supposed to inspire informed thought about
U.S. strategy in the Middle East.

In the introduction to Gulf Strike, Herman describes the game
as an examination of the ramifications of the U.S. decision to create
a Rapid Deployment Force for possible use in the Middle East "in
the face of logistical constraints, established and hostile military
forces, and an uncertain mosaic of shifting political animosities."
The game, through a series of scenarios, offers an "understanding
of the operational alternatives available in the Persian Gulf in the
event of widespread conflict."

There are 910 playing pieces representing fourteen nationalities.
There are sixteen tables, charts, and other aids in the sixteen-page
auxiliary booklet that accompanies Gulf Strike's sixty-page "Rules
of Play" and three large hexagon-gridded maps. There are five
scenarios, each one looking at the Persian Gulf from a different
perspective. Herman estimates that the length of time needed to
play out the scenarios ranges from forty-five hours for a Soviet
invasion of Iran to two hours for a "showdown in the Horn of Africa."

A young Washington man who has played Gulf Strike said that
it would take him about ninety hours to fully explore the richness
of the game. "If it took you all weekend to play five turns, you
would have done ten days' worth of the game for two days of your
own time. And I can speak from experience as an Iranian air commander."

The game is based on reality. Players are assumed to be strategists, not just people killing time at an advanced form of tic-tac-
toe. I found this out when I asked two veteran gamers to walk me,
aided by a couple of other absolute amateurs, through an hour or
so of Gulf Strike. The gamers set up a small demonstration area

with pieces indicating planes, ships, and bases, and asked, "Does someone want to do an American air strike on an Iranian base?"

One of the first-time players, taking the American side, saw all the resources available to him and announced, "I've always been bugged by the Iranians. I want to take out Khark Island."

"There's nothing *on* Khark Island," a gamer said. "Why not go after something worthwhile?"

"But there are all those great oil fields on Khark."

"This is a war. You're protecting the Kuwaitis. If you're a Kuwaiti and your land is being invaded, you're going to attack the incoming Iranian tanks, not some oil field."

Elaborating on this later, the gamer said, "You have to have a goal to achieve. Otherwise, there's no point in playing a game, even if it's Monopoly. That's why the Iran-Iraq war is not a game. There's no point to it. When you put a point to something, then you make it a game."

The creation of reality for a game can be extensive. In War in the Pacific, the battlefront player makes such homefront decisions as the allocation of time and money for weapons production and the length of time necessary to train aircraft carrier pilots. Games are also tested—much more rigorously, Herman and others have said, than Pentagon games.

In the real Pacific war, according to Herman's research, the Japanese still had ships capable of carrying 2.03 million tons at the end of the war. In a validation test of the game—a playing of a handmade, preproduction version with drafts of Herman's rules but no Herman ("I don't come in the box")—the tester ended the game with 2.09 million tons of shipping. Herman seized on this bit of information as a validation indicator. He also was glad to see that the test game had produced a battle that closely resembled the Battle of Midway; it did not matter that the battle was fought on somebody's dining-room table.

Victory Games, which expects to sell about fifteen to twenty thousand copies of a game like Gulf Strike or War in the Pacific, is a subsidiary of Avalon Hill, the pioneer American war-game producer and the company that published Dunnigan's first game. Avalon Hill traces its origins to 1954 with the publication of Tactics by Charles S. Roberts, a young Baltimore advertising man who had once planned on an Army career. Tactics, a game about a war between two hypothetical countries, was brought out by Stackpole, a publisher that specialized in military books. Roberts also unsuccessfully tried to sell a game called Game/Train to the Infantry School at Fort Benning.

Roberts named his company after the Avalon, Maryland, neighborhood he lived in, set up a mail-order business in his garage, and started selling such games as D-Day, U-Boat, and Gettysburg to fans who were at first bewildered by this new kind of game. "It was revolutionary," Roberts recalled, "to say that you could move up to *all* of your pieces on a turn, that movement up to certain limits was at the player's option, and that the resolution of combat was at the throw of a die compared to a table of varying results. As simple as this sounds now, the new player had to push aside his chess-and-checkers mind-set and learn to walk again. After he learned to walk, he had to master the intellectual challenge of the game itself, usually without the benefit of an experienced opponent."

Roberts lost Avalon Hill to creditors in 1963, but by then he had established many of the enduring elements of commercial games. One of the new owners' first games was Midway, with historical notes supplied by Rear Admiral C. Wade McClusky, Jr., air group commander aboard the aircraft carrier *Enterprise* during the Battle of Midway.

Avalon Hill published Dunnigan's first effort, Jutland, in 1967. (It did *not* publish a game he designed about the student riots at Columbia University in 1968, when he was a student there. The name of the game—Up Against the Wall, Motherfucker!—preserved two expressions made famous by the riots. The game was published in 1969 by *The Spectator*, the Columbia student newspaper.)

In 1969, a year before he graduated from Columbia, Dunnigan launched SPI "in a rather haphazard fashion, much like the old 1930s movie in which a group of bright young kids gather around and say, 'Hey, gang, let's put on a Broadway musical in Dad's garage!'" At around the same time he took over a fledgling magazine, *Strategy and Tactics*, which became a gamers' source of lore and criticism. According to Dunnigan, by 1972 more than 200,000 games were being published by SPI and Avalon Hill.

Role-playing fantasy and science-fiction games began appearing in the war-game market in the 1970s, following the spectacular success of Dungeons and Dragons, developed by E. Gary Gygax and Dave Arneson. It was probably the emergence of fantasy gaming—and reaction to the movie *WarGames*—that caused Pentagon war gamers to decline to talk about the connection between hobbyist games and real war games.

"This is not Dungeons and Dragons we're doing here," a Pentagon officer indignantly told me in a discussion of what he called "serious modeling and simulation." Like many military officers and Defense officials I talked to, he was embarrassed by the hobby-Pentagon connection. One official, a civilian, visibly winced when I referred to "toy-store games." He had three games on the bookshelf behind his desk. He, like Dunnigan and Herman, seemed to believe that *hobby* was a more sober-sounding word than *toy*.

Dunnigan, a representative of the hobby-reality connection, was not universally admired, especially by senior Army officers who believed there was something terribly wrong about a man like Dunnigan careening around military territory. Some of that enmity was inspired by his irreverent attitude.

"Seventy-five percent of the Army's high command is incompetent," he told me. "We don't really need the Army. There's no such thing as Attila the Eskimo. We have no natural enemies. The Navy is sexy, effective. But the Army? It's something we are going to sacrifice for our European allies. I've done several global war games, and overseas armies disappear to be replaced a couple of years later by the mobilized citizenry, who are a sharper article."

I was sure he had said all this many times before, in Pentagon and Army War College offices. Dunnigan never said what the Army wanted him to say, and, perhaps understandably, he has shifted careers from gaming to writing books and developing financial software. He says he turned down an offer to be the first director of the National Defense University's War Gaming and Simulation Center. He suggested a commercial game designer, Sterling Hart, who became the first director—and was quickly succeeded by a military officer.

Many officers and Pentagon civilians play games as a hobby. They include a Marine Commandant, General Paul X. Kelley, Jr., and an Army Chief of Staff, General Edward C. Meyer, who played Missile Command on an Atari for relaxation. Dr. Peter P. Perla, a wargaming expert at the Center for Naval Analyses, is one of a number of researchers and analysts who play games both for recreation and for a living. "A lot of us began playing as kids and then moved into that kind of work," he told me.

Young officers—members of the video-arcade generation—play commercial games and consider the time at games well spent. The *Proceedings*, the prestigious professional journal published by the U.S. Naval Institute, in a review of two computerized commercial games, suggested that they were "worthwhile and educational" for naval officers, especially at a time when budgetary considerations limit days at sea and tactical exercises. "One of the first lessons both games teach," the reviewers wrote, "is that ships, with the exception, of course, of the *Iowa* [battleship] class, are easy to sink."

The original publisher of Dungeons and Dragons, TSR (Tactical Studies Rules), previously had concentrated on distributing rules for games using miniatures. This branch of gaming involves combat between scale-model figures on large, realistic terrain platforms that resemble elaborate model-train layouts.

In war games played with miniatures the emphasis is on the accuracy of the weapons and accoutrements of the exquisitely made figurines: Roman legionnaires fighting Goths; mounted knights charging a pike-wielding peasant army; and, in a classroom of the Command and General Staff College at Fort Leavenworth, Kansas, $\frac{1}{285}$-scale models of NATO and Warsaw Pact tanks on a large table bearing the valleys, roads, and villages of a typical German countryside.

The history of civilian wargaming begins with such figurines. And if history is the key to understanding war, history is also the key to trying to understand how games created for hobbyists can influence games created for generals. The history of war games has three branches. One is old and gnarled with military tradition. The second, bearing the creations of designers like Dunnigan and Herman, is new and vibrant with civilian ideas. The third branch, bearing military-political games, is entangled with the ivy of academe. Military war games, the eldest games of all, are rooted in the origins of war itself. Commercial war games emerged, playfully, from the modern world. Academic games were inspired by the professorial belief that gaming was too important to leave to the generals.

To begin the march through all these aspects of wargaming, the next move is to the past.

CHAPTER 6

Playing Pearl Harbor and Other Games

One of the last of the Anglo-Saxon poems of heroes celebrates the Battle of Maldon, in A.D. 991. The poem tells how the Vikings, about to attack, asked their foe to allow them safe passage as they crossed a ford. Birhtnoth, the English commander, says, "The ground is cleared for you—come quickly to us," and his troops— "those stouthearted warriors at the war play"—let the Vikings make it to firm ground to launch their attack. Thanks to being granted this first move in the war play, the Vikings won. The sporting Birhtnoth fell in the battle, but he enshrined in Western tradition the view of war as a kind of game.

The envisioning of battle through games—as in chess, the Hindu *chaturanga*, and the oriental *go*—is an idea so old and universal that its origins are lost in time. But the idea that real war itself could be waged like a game, first recorded at Maldon, was well chronicled through the age of chivalry. As chess became a European game, the king, queen, castle, knight, bishop, and peasant pawn mirrored the world and the warfare of the era. By the seventeenth century the tradition had manifested itself in other forms: military chess and toy soldiers.

In military chess, a German invention, traditional chess pieces evolved into specific battlefield figures, from marshal and colonel to captain and private. While officers learned war at this militarized chessboard, kings and princes played games with tiny soldiers made of silver and gold—and sent real soldiers off to highly stylized battles. In the real wars of the seventeenth century, armies fought in tight formations, following tactics as formal as the moves in chess. War was seen as a phenomenon subject to rules of logic and mathematics.

"Battles were no longer fought from motives of patriotism, but for art's sake," a general later wrote, "and it was deemed preferable to forgo victory rather than to achieve it by unscientific methods." An eighteenth-century German tactician who wrote *A Mathematical System of Applied Tactics and the Science of War Proper* also wrote *Rules of a New War Game for the Use of Military Schools*. This game, known as the New Kriegsspiel, was played on a board whose 3,600 squares represented the terrain along the Franco-Belgian border. A sixty-page rule book governed the game, which was soon eclipsed by another Kriegsspiel, this one played by Prussian Army officers on realistic maps. In 1824, when the chief of the German General Staff saw Kriegsspiel being played, he exclaimed, "It is not a game at all! It's a training for war!"

Every Prussian regiment was issued Kriegsspiel and ordered to play it regularly. Kriegsspiel became endlessly complex as data from nineteenth-century wars were incorporated into more and more charts and tables. The game became known as "rigid" Kriegsspiel to distinguish it from the evolving "free" Kriegsspiel, which was governed by the verdicts of umpires rather than voluminous rule books. (Game designer Mark Herman used the old terms "rigid Kriegsspiel" and "free Kriegsspiel" in distinguishing between gaming with and without computers. Rigid Kriegsspiel, he said, "is what a computer model would be because you program it. Free Kriegsspiel allows for clever thought during play.")

As Kriegsspiel became popular, German artisans began producing *Zinnfiguren* ("tin figures"), flat, inexpensive soldiers that transformed a royal diversion into a commoners' pastime. Toy soldiers marched through the lives of Goethe and Churchill, of Robert Louis Stevenson and Anatole France, of G. K. Chesterton and H. G. Wells. Playing with soldiers had nothing to do with real war. But the toy soldiers that belonged to Wells, a pacifist, were to launch the enduring relationship between hobbyist games and military games. Elements in Wells's toy-soldier games can be found in today's Pentagon games and in the games sold in hobby shops.

Trevor Dupuy, like many officers schooled between World War I and World War II, discovered wargaming in the pages of Wells's *Little Wars*, which was published in 1913, the year before the First World War began. Dupuy's father, an Army officer, gave his son a copy, starting Trevor Dupuy off as a second-generation student of war and writing.*

" 'Little Wars'," Wells wrote, "is the game of kings—for players in an inferior social position. It can be played by boys of every age from twelve to one hundred and fifty—and even later if the limbs remain sufficiently supple—and by girls of the better sort, and by a few and gifted women."

In Wells's first little war against little soldiers, he fired a spring-powered breech-loader gun, "a priceless gift to boyhood," a gun "capable of hitting a toy soldier nine times out of ten at a distance of nine yards." It fired a wooden bullet about an inch long. His first battlefield was a room littered with volumes of the *Encyclopaedia Britannica* and other hefty books forming barricades for lead solders to hide behind while players fired at them.

Reading about this first game, I remembered setting up lead soldiers in the driveway of my house and, crouched at an open living-room window, picking them off with a Red Ryder BB gun. I lost quite a few lead soldiers that way and soon changed my solitary game to a kind of maneuver warfare in which I moved my troops but did not slay them. Wells, too, changed his game, creating what he called simply Country—paper houses, a castle of cardboard, a chalk-line river, forests of twigs. Rules became more elaborate—"as many men are dead as a shot knocks over or causes to fall or to lean so that they would fall if unsupported."

Then as now, war gamers felt compelled to call in experts. Wells's consultant was a British Army captain back from the Boer War. The captain speeded up the battle action, but the war itself became more complex with strict rules about the size and disposition of troops. The new rules led to the question of taking prisoners. "Now,

*Father and son wrote *Military Heritage of America*, which became widely used as a Reserve Officers Training Corps textbook.

in actual civilized warfare," Wells wrote, "small detached bodies do not sell their lives dearly; a considerably larger force is able to make them prisoners without difficulty. Accordingly we decided that if a blue force, for example, has one or more men isolated, and a red force of at least double the strength of this isolated detachment moves up to contact with it, the blue men will be considered to be prisoners." *Red* and *Blue* would continue to be the names of opposing game forces.

"You have only to play at Little Wars three or four times to realise just what a blundering thing Great War must be," Wells wrote. "Great war is at present, I am convinced, not only the most expensive game in the universe, but it is a game out of all proportion. Not only are the masses of men and material and suffering and inconvenience too monstrously big for reason, but—the available heads we have for it are too small."

During the Great War that came, the idea of war as a game lived only in the sky, where pilots chivalrously called themselves knights of the air. The carnage of trench warfare inspired a poetry of death. ("What passing-bells for these who die as cattle?" Wilfred Owen wrote before he himself fell.) War had come a long way from Maldon, but the idea of battle as a contest, a game of chess, endured. The term *war games* remained in the military lexicon.

The knights and bishops and medieval chessboard were replaced by tanks and cannon and a sand table with contoured terrain. Military war gamers, inspired by Wells's miniaturization of battle, made sand tables more realistic. Cadets in military academies gathered around the tables and deployed lead soldiers and model tanks, though often the tanks were arrayed alongside lead generals who still rode lead horses.

Kriegsspiel—war play—became a way to go to war. Gradually the tactical game evolved into a strategic one. In 1848 the Prussians played the game to simulate a war between Prussia and Austria. This and subsequent gaming laid the strategic foundation for the decisive Prussian victory over Austria at Sadowa in 1866 in what became known as the six-week war. The Austrians learned a lesson and began playing war games themselves, as did the French, English, Russians, and Italians, especially after the Prussian success in the Franco-Prussian War of 1870. About that time Japan also took up gaming—with, by some judgments, victory in the Russo-Japanese War of 1904 the result.

Although gaming had been practiced spasmodically in U.S. military establishments since early in the nineteenth century, not until the 1880s did a specific game receive serious attention. The game was brought forth in the book *American Kriegsspiel* by William R. Livermore, a major in the Army's Corps of Engineers, with assistance from Hugh G. Brown, an infantry captain and Civil War veteran. The game ushered into U.S. military thinking many ideas that would survive in modern gaming, both hobbyist and military.

The game, played up to a level of sixteen companies, used topographical maps and colored blocks for troops (red-and-green versus blue-and-orange). Metal pointers of various sizes indicated the marches and firing patterns of the troops. A pegs-and-holes "firing board" startlingly resembled a data-processing punch card of early computer days. Tables that anticipated Mark Herman's Combat Results Tables handled questions of firepower and troop movement. As in games today, a roll of dice decided issues that in real war would be decided at least somewhat by chance.

The Possibility of Firing Table looks like a QJM ancestor, with considerations of firing conditions ("No distant fire in dense forest.") and weather ("Rain: Can see only as far as umpire permits. [Throw die.]"). Attrition formulas were based on data from the Civil War and nineteenth-century European wars, with calculations that foreshadow those used by the QJM: A casualty rate of one man per minute at a range of five hundred yards, three per minute at one hundred yards; encounters were called "melees," a word that Wells had used in dispatches from his Little Wars.

The complexity of the American Kriegsspiel was criticized by an Army officer in words that some modern critics would recognize: The game "cannot be readily and intelligently used by anyone who is not a mathematician, and it requires, in order to be able to use it readily, an amount of special instruction, study, and practice about equivalent to that necessary to acquire a speaking knowledge of a foreign language." The officer who said that invented his own war game, for gaming seemed to be in the military wave of the future. Kriegsspiel had reached the United States and its stay—particularly for the Navy—would be permanent and pervasive.

In 1889 Major Livermore brought his American Kriegsspiel to the five-year-old Naval War College in Newport, where the battlefield war game quickly went to sea. The champion of gaming at the new college was William McCarty Little, retired from the Navy as a lieutenant in 1884 because he had lost the sight of one eye in a gun accident ashore. Rear Admiral Stephen B. Luce, founding president of the college, had invited his friend McCarty Little to join the faculty, and in 1886, when Rear Admiral Alfred Thayer Mahan became president and lecturer on naval history and strategy, he encouraged McCarty Little's interest in gaming.

Mahan and McCarty Little, as he was inevitably called, demonstrated tactics by moving cardboard ships around on sheets of drawing paper. By the time Major Livermore arrived in 1889, McCarty Little had been lecturing on gaming for two years and Naval Kriegsspiel was permanently part of the college's curriculum.

McCarty Little linked gaming to "the annual problem," an imagined sea battle or war issue presented to the students for a solution. He likened this to "a play at a theater with full explanations of the reasons for each move." In 1895 this was the annual problem: Great Britain, after declaring war in October, dispatches a large fleet to New York Harbor. "Enemy also assembles in Halifax a force prepared to descend upon our coast between Boston and New York. This force will sail from Halifax November 10; consists of 10 [ships of the] line, 20 heavy cruisers, 20 torpedo boats, and 10 destroyers and scout class, also a corps of 30,000 men of all classes in 100 transports." The U.S. Fleet, greatly outnumbered, deploys to Gardiners Bay, on the eastern end of Long Island, and to New London, Connecticut. The Halifax fleet lands its troops at Narragansett, Rhode Island. The students had to devise tactics to counter the invasion.

Two years after this theoretical problem, Assistant Secretary of the Navy Theodore Roosevelt gave the Navy War College students a "special problem." His scenario: "Japan makes demands on Hawaiian Islands. This country intervenes. What force will be required to uphold intervention? Keep in mind possible complications with another power in the Alantic."

The other power and its Royal Navy became an obsession through the early years of the twentieth century, when the college inaugurated the Strategic Naval War Game and established many of the basic rules of American wargaming. Game fleets were assigned colors. Blue became the permanent color of the U.S. Navy. The Royal Navy was Red; the Canadian, Crimson; the German, Black; the Japanese, Orange. The Naval War College at that time selected America's probable future enemies and plotted the Navy's likely battles through exercises at sea and games in the war college.

One war plan, developed at the college, called for U.S. Marines to sail to Halifax and seize that port, along with the fortress Louisbourg, in six days. Meanwhile, Bermuda would be taken in seven days, and Jamaica in seventeen. But in playing out the plans, the Blue Fleet was always beaten by the Red. The Blue Fleet, threatened by the Red Fleet's long-range guns, could not get close enough to fire effectively, and if the Blue Fleet did get close enough, it would be in a "fatal zone." At least partially as a result of these war-game battle losses, U.S. warships got steel deck plates, guns were given higher elevations, and long-range gunnery training was stepped up. By 1938 the Blue Fleet's guns outranged the Red Fleet's by 10,000 yards, and in Newport war games Blue began winning against Red.

The Royal Navy was not the only potential foe being fought at the college. In 1900 students had worked on the problem of a Japanese invasion of California. In 1911 Newport produced a "Strategic Plan of Campaign Against Orange," which envisioned a war ending with a Navy blockade of Japan. Much of the plan, which foresaw the fall of the Philippines, was still accepted doctrine on December 7, 1941.

Of 136 strategic games played at the college between 1919 and 1941, a total of 127 were played against Orange. In a 1939 game the scenario begins in the third year of a Blue-Orange war, with 400,000 Americans in New Guinea on the eve of an amphibious invasion of the Japanese-held Philippines. Games in 1934 and 1937 had Blue and Purple (then the Soviet Union) allied in attacks against Orange in Manchuria.

In Europe Kriegsspiel became the foreplay for World War I. British Army officers in 1905 prophetically gamed a German invasion of Belgium—and, at times consulting a German bicyclers' map, discovered some mobilization and logistical problems that were solved in time for the real mobilization for war. German players at the war school in Engers, preferring an elegant realism over the traditional game blocks and pieces, deployed troops of figurines similar to those Wells was shooting at with his toy cannon. Kriegsspiel once again was a royal sport. The Kaiser himself is said to have appeared at some games in full uniform, complete from helmet to spurs, eager to play and imperially reluctant to lose.

From 1881 to 1906, during this heyday of Kriegsspiel, Alfred Graf von Schlieffen, as chief of the German general staff, developed the war plan that was to bear his name: a powerful right wing of the German Army would swing through the Low Countries, seize the Channel ports, and sweep down on Paris while the much weaker left wing engaged French troops, preventing them from reaching the major German force. But Schlieffen and the German gamers who followed him were being drawn to the war game's seductive and disastrous fallacy: A win in a game was a guarantee of victory in the real world. In Schlieffen's gaming he had not counted on fierce Belgian resistance—or on the military wisdom British officers had extracted from one of their own rare war games and shared with the French and Belgians.

"Such lack of imagination," a British military authority wrote in a modern assessment of gaming, "was common in the German army; and the constant playing of war games [against] a mental replica of itself almost certainly made it worse. In the case of 'free' war games the problem was aggravated by uniform solutions imposed by General Staff umpires, trained on the principle that the army should 'think with one mind.' 'Rigid' war games produced their own brand of mental rigidity."

German faith in gaming did not wane through the war or even after defeat. Wargaming became the secret weapon of a Germany whose war machine had been dismantled by the Treaty of Versailles. Confronted by the restrictions that the Allies had imposed on the German Army, senior staff officers used war games, sometimes blandly called map maneuvers, to train a new generation of officers. Because there were no real armies to put through real field exercises, officers conducted "theoretical exercises," issuing orders to phantom troops.

In 1929 German Army officers staged what may be the first political-military game, one involving both the army and the Foreign Ministry. The ironic scenario called for a possible Polish invasion of German territory. Members of the Foreign Ministry played the president of the League of Nations and German and Polish foreign ministers; army officers played military roles. The German playing the Polish minister explained that the Poles had to invade because of hostile German actions. "His ability to invent German provocations left his adversary speechless," a participant would later note.

German-style wargaming was not widely adopted by the U.S. Army. Although gaming became a major enterprise at the Naval War College between world wars, the U.S. Army during that same period did not do much gaming. "Map maneuvers" and "Command Post Exercises" gave young officers training in decisionmaking and maneuver warfare, but in such simulated warfare there was no sense of enemy, no game against even a simulated opponent.

Games, however, easily simulated the simple purity of the Navy's theater of battle—the vast sea, the precise placement of ships. When battle was joined at sea, ships wheeled about and performed as colossal machines, obeying such coolly uttered orders as "You may fire when you are ready, Gridley." But on land, battle was often decided by how small units performed and by how individual soldiers did their jobs. A battalion did not have a rudder. An army battlefield was messy. Battle's carnage spread beyond the battlefield, to shell-torn towns and to roads clogged with refugees. On land, as General Sherman said, war is hell.

The U.S. Marines, who sang of fighting on the land and on the sea, developed their own war-game tradition. They began gaming amphibious warfare tactics soon after the Spanish-American War, picking likely sites and scenarios for future landings. There is evidence that between the world wars Marines gamed a landing at the Caribbean island of Grenada—a scenario made real in 1983. Marines studied historical battles and sometimes refought them. In the 1920s Marines reenacted Pickett's charge at Gettysburg and the Battle of Newmarket (with cadets from the Virginia Military Institute). In 1924 Marines fought at Antietam, using modern battle tactics. They looked to the future as well. A 1921 Marine war plan, which included landing in the Marshall Islands, anticipated the U.S. island-hopping campaign in World War II.

American wargaming for entertainment began in the 1930s. Adults not only played Wells-inspired little wars with toy guns and miniature soldiers but also crawled about the floor to deploy lead ships

in a made-in-the-U.S.A. game that took toy war to sea. What became known as "Fletcher Pratt's Naval War Game" was the brain-
child of a writer who, along with his friends, "became a trifle bored
with the poker games that were the staple amusement of their
gatherings." So he started playing an impromptu naval war game
that marched fatefully along the same path that Wells's game had
taken: Conceived in a jittery prewar era, played at first only for
the enjoyment of its creator and his friends, it was published on
the eve of a world war—and all but vanished in the horror of real
war.

Pratt, whom someone once called a one-man war college, was a
prolific writer on Napoleon, the U.S. Navy, the War of 1812, and
the Civil War. He was also a founder of the American Rocket
Society, a science-fiction writer, and a journalist who in World War
II became a war correspondent. In the 1930s he welcomed to his
home in New York City fellow writers and visiting Army and Navy
officers who joined him in his naval game. One of the players was
a brand-new graduate of West Point who was dating a young woman
in New York. He was Second Lieutenant Trevor Dupuy.

At first players aimed flashlights at ships and shot beams of light.
Then came toy cannon that fired wooden plugs. Then came "air
pistols, with which the contestants retired to another room to shoot
at pictured ships on a wall," Pratt wrote. But all the weapons "had
the fatal objection of turning the game into one of chance, or the
dangerous one of introducing a violent sense of unreality." Game
nights originally were stag affairs, like the poker nights. Soon,
though, the naval game attracted "nearly as many players of one
sex as of the other; and one of the feminine delegation has been
praised by a naval officer player as the most competent tactician
in the group."

Pratt finally developed a complex formula involving the thickness
of a ship's armor, her speed, and the caliber and number of guns
on each of the ships (transformed from pictures on a wall to lead
waterline models on the floor). "The formula can, of course, be
criticized as bringing into mathematical relation widely different
elements among which there is no genuine mathematical relation,"
Pratt wrote. "It is extremely arbitrary; but the only answer is that
it works out in practice rather better than some more common
methods of computation."

As proof of his method Pratt used the real-world fate of the *Admiral Graf Spee*. The German "pocket battleship" went to sea on August 20, 1939. When World War II began on September 1, she began hunting down British merchant ships. She had sunk or captured eleven of them by the time three ships of the Royal Navy—the heavy cruiser *Exeter* and the light cruisers *Ajax* and *Achilles*—found her off Montevideo, Uruguay, on December 13.

In the battle that followed the *Graf Spee* took more than fifty hits. She limped into Montevideo, where her captain scuttled her, on orders from Germany, rather than risk being sunk at sea by the British warships waiting outside the harbor. "Rated on gun-power and armor," Pratt wrote, the *Graf Spee* "should have been more than a match for the three British cruisers; but by the formula here [in his book] they should have beaten her. They did."

The complex formula, by which Lieutenant Dupuy played, fore-shadowed those equally complex formulas that Colonel Dupuy would develop. In fact, Dupuy mentioned the *Graf Spee* game-and-reality phenomenon when I interviewed him. I have also heard it cited by two other professional gamers. Fletcher Pratt's naval war game has become part of the lore of both commercial and military war-gaming.

On September 2, 1941, about a year after Fletcher Pratt's naval war-game book was published, a naval game was begun at the Japanese Naval War College in Tokyo. For eleven days officers of the Imperial Japanese Navy played a game that climaxed with a massive surprise attack on Pearl Harbor. Vice Admiral Chuichi Nagumo, who would lead the real attack, played himself in the game.

In assessing the results the Japanese estimated that two thirds of the principal ships of the U.S. Pacific Fleet at Pearl Harbor would be sunk, but at great loss to the attackers. Six carriers carried out the raid in the game, which indicated that U.S. patrols would discover the attack force and sink or put out of action two or three carriers. Nagumo, convinced by the game, argued against imper-iling the entire Japanese force of large carriers. But he was over-ruled by Admiral Isoroku Yamamoto, commander in chief of the Navy—and chief umpire of the game.

A month before the Pearl Harbor game officials of the Total War Research Institute had gathered for another game in Tokyo, this one a global political-military game designed to examine a long-term war. The time span of the game was from the middle of August 1941 through 1943. Players represented the German-Italian Axis, the Soviet Union, the United States, England, Thailand, the Netherlands, the East Indies, China, Korea, Manchuria (then under Japanese control), and French Indochina.

Japan was played by several players rather than one. Reflecting the reality of the nation itself, the game's Japan was a shifting coalition of often competing interests—the Army and the Navy, other cabinet members, and industry. "Measures to be taken within Japan were gamed in detail and included economic, educational, financial, and psychological factors," said an account of the game based on a postwar examination of institute records. "The game even included plans for the control of consumer-goods, plans, incidentally, which were identical with those actually put into effect on December 8, 1941."

In the first phase of the game, war began on December 15, 1941, when Japan, allied with Germany and Italy, launched surprise attacks on the Philippine Islands and several nations in Southeast Asia. (There was no attack on Pearl Harbor, probably because high-ranking officers in the Imperial Navy were still arguing about the Pearl Harbor plan.) "The calculations of the first phase of the war were detailed and daring," says the account. "They were enormously complex in terms of timing, geographic spread, and close matching of Japanese and U.S. forces-in-being."

But, just as German war gamers let games eclipse reality, the Japanese players, projecting their game into the future, let wishes conquer sense. In the game, while the Soviet Union is fighting a losing battle against Germany, Japan invades and seizes Siberia. Germany wins over the Soviet Union and England. The United States is not treated as an important force. There is no real assessment of risk or of Japanese war objectives. But what happened in this unrealistic game was essentially what Japanese military officers, during Japanese debates over going to war, insisted would happen in reality. The results of the game, presented by the officers as convincing arguments to political authorities, led directly to Japan's decision to go to war.

The U.S. Navy was also run by officers trained through gaming. While they learned the fine points of their profession on the gridded tile floors in the game room of the Naval War College, their families had a chance to enjoy life in Newport, the genteel resort of American aristocracy. Sometimes officers' children were allowed into the observation balcony overlooking the gaming so they could watch their fathers.

One of those children in the 1930s was Isaac C. Kidd, Jr., son of a man who would become an admiral and Commander of Battleship Division One. The son, too, would become an admiral and, as Commander in Chief of the Atlantic Fleet, would pioneer the use of computers in the war-college games. He would be credited, in fact, for restoring significant gaming to the college. In those gaming sessions he told his junior officers he was training them hard because "we don't want any more Pearl Harbors." It was at Pearl Harbor, on December 7, 1941, that Ike Kidd's father went down with his flagship, the *Arizona*.

When World War II began, the Naval War College began looking into tactics for carrier-based aircraft and potential American participation in two-ocean warfare. The war college's two-ocean study was one of the ingredients in what became the basic U.S. war plan.

All but one wartime admiral and all the Navy's principal leaders—Halsey, King, Spruance, Nimitz—had played games at the college and, after the war, Chester Nimitz summed up the results: "The war with Japan had been reenacted in the game rooms at the Naval War College by so many people and in so many different ways, that nothing that happened during the war was a surprise . . . absolutely nothing except the kamikaze tactics toward the end of the war; we had not visualized these."

Nimitz's kamikaze exception could be a symbol for what can happen from too much reliance upon war games as a forecast of reality. Playing by the rules, seeing war as a game of chess, participants in war games sometimes overlook the human element that brings the unexpected hurtling into battle. What Nimitz failed to say was that the kamikaze problem was solved, both by a new and special kind of mathematical gaming and by the traditional skill of experienced commanders—intuitive changes in tactics. And what the wargaming enthusiasts then and now often fail to say is that the losers of World War II took gaming far more seriously than the winners did.

The German campaigns against France and the Soviet Union, according to Generaloberst Heinz Guderian, the inventor of the German armored blitzkrieg, "were so thoroughly rehearsed in advance in war games and map exercises that actually every single commander down to the company level was thoroughly familiar with his initial missions, with the difficulties which were presumably about to confront him, and with the enemy forces situated opposite him." When Guderian planned the invasion of France in 1940, he merely changed the dates and time on the orders used when the campaigns had been gamed. A game helped convince German planners that an invasion of England was not feasible; a game called Otto Map Exercise paved the way for the invasion of the Soviet Union.

Gaming on the grand scale, Japan captured Ceylon, linked up with German troops who had conquered Egypt, then triumphantly marched on toward the Orient and took the Aleutians. This bold and satisfying Aleutian move came in a game played in May 1942 on board the battleship *Yamato*, flagship of the Combined Fleet. The game began with an invasion of Midway Island, apparently as a way to exorcise military embarrassment, for, by some accounts, several Japanese officials believed Midway to have been the base from which American bombers had taken off for the stunning raid on Tokyo the month before.*

In the game U.S. planes from Midway bombed Japanese aircraft carriers after they had launched their planes against Midway. The rules of the game at this point called for a roll of the dice to determine hits. The umpire rolled and declared nine hits, with two carriers sunk. The presiding officer, however, overrruled the umpire, a decision that "arbitrarily resulted in *Kaga*'s still being sunk but *Akagi* only slightly damaged," according to a Japanese naval officer at the game. "To [the umpire's] surprise, even this revised ruling was subsequently cancelled and *Kaga* reappeared as a participant in the next part of the game. . . . The verdicts of the umpires regarding the results of air fighting were similarly juggled, always in favor of the Japanese forces."

A month after the game the real Battle of Midway was fought.

---

*The bombers, led by Lieutenant Colonel Jimmy Doolittle, flew from the aircraft carrier *Hornet*.

The Japanese did attack Midway, and land-based U.S. planes did attack the Japanese force, but they had little effect. Planes from U.S. carriers proved to be the real ship killers. These planes, which had no role in the juggled game, sank the *Kaga*, the *Akagi*, and the two other large Japanese carriers. The United States lost one carrier and a destroyer. As the Japanese game had predicted, the battle would be decisive. But the game had been rigged to pick the wrong winner.

The Germans played their most realistic game of the war in the Ardennes in November 1944, when the staff of the Fifth Panzer Army was meeting to game defensive moves against a possible American attack. "The map exercise had hardly begun," an account of the game says, "when a report was received that according to all appearances a fairly strong American attack had been launched in the Hürtgen-Gemeter area," exactly where the game battle was being played. Generalfeldmarschal Walther Model, whom Hitler had frequently called his best field marshal, ordered the players— except the commanders of units under attack or directly threatened—to continue playing, using messages from the front as the basis for game moves.

For the next few hours the game reflected the real threat the Germans faced. At the game table players decided that a reserve unit, the 116th Panzer Division, had to be thrown into battle. The commander of the 116th, who was playing the game, turned to his operations officers and couriers and issued actual orders that paralleled his gaming orders. "The alerted division," according to German General Friedrich J. Fangor, "was thereby set in movement in the shortest conceivable time. Chance had transformed a simple map exercise into stern reality." The American forces were repulsed in what was a prelude to the Battle of the Bulge, itself an operation that would be frequently gamed, both by hobbyists and military officers. (In 1985 the battle was replayed at the U.S. Army Command and Staff School with up-to-date weapons and the German troops replaced by Soviets. "One time the good guys won and one time the bad guys won," I was told.)

Things German won little admiration in Great Britain on the eve of World War II, and perhaps for that reason Kriegsspiel lost favor in British war councils. But a modern, scientifically based version of wargaming, which emerged early in the war, would have a profound and lasting influence on gaming in the United States. At first glance this new way of gaming did not resemble classic war-gaming. It was called operational research, a single label for the numerous jobs being done by the assorted academicians, scientists, and eccentrics rounded up to contribute, somehow, to the British war effort.

The term *operational research* traces back to 1937 when some workers at the Air Ministry Research Station were given the task of figuring out how to put to use a promising new radio-echo detection system that later would be dubbed radar. The word *operational* was inserted into the group's research title to set off what they were doing from traditional research and development activities. Thus was born operational research.

Among the advocates of operational research was Professor Solly Zuckerman, a zoologist (monkeys were his special interest). He applied scientific techniques from biology laboratories to his wartime work, pondered war as a scientist, and decreed that a military operation was "an experiment of a very crude kind." This became a kind of anthem for operational research, which became known by the initials OR. These were the firstborn of a new breed—partial civilians looking at war in a way that full-blooded military people never did.

The operational researchers saw war not as an art but as a science. They discovered, for example, that the mathematical concept of "constant effectiveness ratio" could be applied to warfare; results obtained in one theater of war or in a series of battles could be considered valid—"constant"—elsewhere. One constant effectiveness ratio study showed that whether British aircraft laid mines in German sea lanes or German planes dropped mines in British ports, the effectiveness ratio remained constant: For every sixty mines laid, one ship was sunk.

OR analysts pored over reports of the bombing of railways in Italy and then translated that data into plans for the massive bombing of German railroads. Knocking out a railroad system was merely an exercise in network mathematics. OR analyzed the convoy system and found that the percentage of losses in large convoys was much smaller than losses in small ones. A mathematical analysis showed that seven escorts could protect eighty ships as well as six escorts could protect forty. Such convoy studies still go on in an era of nuclear-powered submarines and seagoing nuclear-tipped missiles.

An OR scientist looked into how antisubmarine warfare planes fought their war and saw what they did as an operational system. When one of the planes spotted a U-boat on the surface charging its batteries, the pilot made a depth-charge run. The U-boat had about two minutes to try to escape before the attack. The depth charges were set to explode when they were one hundred feet below the surface.

The OR scientist, from his operational-system viewpoint, figured out that the U-boats were escaping because they were nowhere near one hundred feet down by the time the depth charge exploded. He suggested that the depth charges be set to explode at twenty-five feet. So devastating was the effect of the change that the Germans believed the British had developed a powerful new explosive.

U.S. military officers discovered operational research (which became *operations* research when it crossed the Atlantic) soon after they began joint warfare with the British. One of the first U.S. organizations devoted to OR was the Naval Research Laboratory's Operations Research Group. The ORG studied several naval problems, including what to do about kamikaze attacks.

The suicide pilots began making their death dives on U.S. ships in October 1944 during the battle for Leyte Gulf. A special ORG research section found that the kamikazes were far more effective than dive bombers or torpedo planes: of the kamikazes that managed to aim themselves at ships, 5 percent were getting through intense antiaircraft fire and making hits that sunk ships.

The ORG looked at antiaircraft guns' angles of aiming and rates of fire, at the dive angles of the kamikazes, and at the varying ability of large and small ships to take evasive action. Smaller ships rolled and pitched so violently during sharp turns that antiaircraft fire became erratic. The conclusion: A ship under attack should base its evasive tactics not on eluding the kamikaze but on bringing its best concentration of antiaircraft fire to bear on the attacker; a ship should turn beam toward a kamikaze coming in high and turn beam away from a plane coming in low.

The Washington-based Operations Research Group was also asked to find out why there were so few successes in what had seemed to be a sensible Navy plan for hunting U-boats. The plan took advantage of the U-boat's need to surface frequently to recharge the batteries that powered it underwater. If patrol planes could cover a large enough area in sea lanes where U-boats were known to patrol, at some time in, say, forty-eight hours, a submarine would surface and would be detected. But the plan was not producing very many sightings.

The researchers played an OR-style war game with planes and submarines. They decided that a submerged submarine, through its periscope, was able to see the hunting planes and figure out their courses. Borrowing from the old war game called chess, the researchers concluded that what was needed was a "gambit," a calculated risk that would induce an opponent to make a careless move. The new plan had the patrol plane fly *outside* the area, presenting the submarine with an inducement to surface and head along the safest course—that of the disappearing plane. If the gambit worked, the submarine's move would produce a counter-move that had the plane return, and, with the aid of surface ships, attack. The OR gambit was tested in Atlantic patrols, and it worked.

Such OR achievements began to convince military officers that the civilian thinkers had something to offer. After the war, OR, which became known as MOR (for military OR), was not mustered out.

"By the time the war ended," Andrew Wilson, defense correspondent of the London *Observer*, wrote, "there was scarcely a field of military activity, on the Allied side, that had not been profoundly affected by operational research. Its impact ranged from improvements in tank gunnery and field engineering to the complete recasting of aid and naval procurement programmes. If its findings were sometimes over-optimistic (as in the case of the damage inflicted by strategic bombing on German war production), at least they were much nearer the mark than the intuitive estimates of military commanders; and they demonstrably saved many lives, and vast resources, that would otherwise have been wasted on unprofitable activities. This made it inevitable that operational research methods developed in war should be retained as an aid to military planning in an uneasy peace."

In that uneasy peace the United States exploded the first hydrogen bomb and began a series of test explosions that made atolls in the Pacific laboratories for the study of a new kind of war. Scientists and strategists began to, in the words of one of them, think the unthinkable. ENIAC (*E*lectronic *N*umerical *I*ntegrator *a*nd *C*omputer), a pioneering computer set up at the University of Pennsylvania in 1946, was given a job: calculating bombing data for nuclear war. Computers, even the early crude ones, could extrapolate "the two data points" of Hiroshima and Nagasaki in a search for the constant effectiveness ratios of nuclear war.

In 1947 the Navy's Operations Research Group became the Operations Evaluation Group, with a Massachusetts Institute of Technology contract. In 1948 the Operations Research Office was established under contract with Johns Hopkins University in Baltimore. Researchers there played Tin Soldier, a kind of checkers match with opposing tanks on a hexagonal battlefield. The game was the creation of Dr. George A. Gamow, a mathematician who had worked on the first atomic bomb and in the 1950s wrote popular books on science. Tin Soldier in 1954 evolved into the Maximum Complexity Computer Battle. In the annals of MOR, this is credited as the first computerized analytical war game for research studies.

ORO's simulations, such as Tacspiel and Theaterspiel, were strictly Army affairs played in starkly military fashion. Twenty or so Red, Blue, and Control players, almost invariably retired officers, gathered around map tables in large rooms equipped with little more than chairs, clocks set to game time, and whirring machines that produced random numbers. Players looked only at the military aspects of potential conflicts in Western Europe, the Middle East, and Southeast Asia, and examined such old problems as logistics and inventory management and such new topics as the radiation exposure of tank crews on nuclear battlefields and tactics for newly developed antitank weapons. In the Quick Game, players squeezed two days of combat into a day of play; the Super-Quick Game packed as many as sixty days of combat into half a day. But researchers were already discovering one of the realities that still haunts gaming: It takes a great deal of time.

The computer eased that problem by quickening the pace of games. The computer also began changing the nature of games, pushing them off the tabletop and into the theoretical realm of modeling. To the ranks of military gamers were added modelers, who would rather analyze phenomena than launch attacks, who would play at war not in quest of victory but in a mathematical search for a weapon's "measure of effectiveness." Gaming was becoming less the avocation of retired military officers and more the vocation of scientific modelers. The soldier and the sailor and the Marine still had their practical, traditional view of wargaming, but they were no longer alone. A new and powerful player was discovering war games.

In the most important espousal of operational research, under a wartime contract with the U.S. Army Air Force,* the Douglas Aircraft Company, which had been building planes in Santa Monica, California, assumed postwar management of the wartime Project RAND (an acronym for "research and development").

Douglas, at the same time, looked to future war-business prospects with the aid of the Douglas Threat Analysis Model, which examined international crises from 1945 to 1967 and ranked nations from "Dominant World Powers or Actors" (the United States, No. 1; the Soviet Union, No. 2) to "Local Actors" (Lebanon No. 71, Nicaragua No. 89). The model also examined 3,000 incidents between No. 1 and No. 2 and graphed the encounters in terms of "tension-level units."

Rand, spawned and nurtured by the Cold War, became the nation's major powerhouse for creating strategy in a nuclear world, a world dominated by two nuclear-armed opponents, a world of Blue and Red. The new strategy was built upon operations research but ranged beyond it. Edwin Paxson, a Rand mathematician, called what he did "systems analysis"—an attempt to quantify everything possible in complex operations, such as a Strategic Air Command strike against the Soviet Union. The analyses often took the form of simulation, involving not just pilots but also policymakers. And this evolved into a Rand specialty that quickly spread to the academic community—the political-military war game.

By the time of the Korean War, gaming occasionally blurred the edge between theory and reality. Marine Colonel Victor H. Krulak, Chief of Staff of the First Marine Division and one of the planners for what would be a flawless amphibious assault at Inchon, one day was in the midst of working on real matters—such as "the effects

---

*It became the U.S. Air Force in 1947. Douglas, now McDonnell Douglas, was the leading U.S. defense contractor in 1984. It builds the Navy-Marine F/A-18 Hornet, the Marine AV-8B Harrier, and the Air Force F-15 Eagle.

of tides and currents, the consistency of the Inchon mud flats, height and character of the Inchon seawall." At that point Lieutenant General Edward M. Almond, General Douglas MacArthur's chief of staff, decided he wanted a war game of the Inchon operation. A liaison officer on Almond's staff delivered the war-game directive to a Marine colonel who, Krulak later wrote, "took the directive, folded it several times, tucked it into the liaison officer's pocket and told him to take it back to GHQ."

In 1961 the Marines established a permanent war-game organization in the Landing Force Development Center at the Marine base in Quantico, Virginia. There Marines still play a continually updated Marine Corps Landing Force War Game, which can call up scenarios in Latin America, Southeast Asia, and the Middle East. Players can change the weather, call in naval gunfire, and order tactical air cover. Wars range from skirmishes with guerrillas to full-scale battles against enemies using chemical and nuclear weapons. And, since the 1960s, the game frequently has been a meticulously planned invasion of Cuba.

Meanwhile, in the Army and the Navy, the computer was changing gaming's relationship to actual warfare. Sand tables and game rooms were obviously only icons of the real world. But computers and electronics were often performing in game rooms the way they would in battlefield command posts and in warships' combat information centers.

In 1958 as, coincidentally, nuclear power was revolutionizing the Navy, the Naval War College retired its wargaming room, with the gleaming tile-floor ocean and the sliding curtain that had played the role of horizon for the tiny fleets of so many little wars. In its place came the Navy Electronic Warfare System, which had been under development since 1945, occupied a three-story building, and cost upwards of $10 million, an astonishingly high price in 1958 dollars.

NEWS, as the system was called, had twenty realistic command centers, complete with simulated radar and communications equipment. A fifteen-foot screen glittered with lights signifying ships, planes, submarines, and missiles. The movements of ships were controlled by computers fed navigational information.

NEWS computers were supposed to determine whether the target was hit and, if so, what damage had been done. The navigational computers would then slow down the ship that was hit and eliminate damaged weapons from its armament repertoire. But, as it often turned out, the human umpires did the reacting, making speed-and-damage decisions much as human players had in Fletcher Pratt's living-room games.

Lincoln Bloomfield, an MIT professor, was at the beginning of his political-military gaming career when NEWS arrived at the war college. "It was full of vacuum tubes," he recalled. "But it couldn't indicate anything going more than five hundred knots, and already there were missiles at twice that speed. The officers bypassed the electronics when that problem was discovered."

Nuclear grand strategy and even tactical nuclear weapons had been hovering around the game tables since World War II. Nuclear weapons—for deterrence or for use—inspired their own strategic games, which often were played with military tactics on the sidelines. As Bloomfield put it, "In the course of the political exercise we would bring in the military with their grease pencils and overlays over the map and they would give military counsel about political movements—a kind of war military game tucked into a political game."

Systems analysis was no overlay. From the time Secretary of Defense McNamara introduced it to the armed forces in 1961 until today,* systems analysis has been an integral part of Pentagon-directed wargaming. Operations research primarily had dealt with the real weapons and real events of a real war. Systems analysis dealt with the concepts and the costs of potential war. Since these were the subjects of Cold Wargaming, it was only natural that systems analysis would be incorporated into the games.

A game with systems analysis in the scenario used numbers not for keeping score but for measuring aspects of the game's artificial reality. For the most part, only the analysts understood the inner workings. Computer simulations, wrote two chroniclers of Pentagon gaming, took on "quasi-religious overtones. Offerings are put into the black box by acolytes who are never sure what is going to come out; those who come to worship are often not sure what has happened either." Players and policymakers, unable to see this

---

*McNamara's Office of Systems Analysis lives on under another name, the Office of Program Analysis and Evaluation.

peculiar kind of reality, talked, at first nervously and then trust-
ingly, about the black boxes and the "model opacity"—the computer age's cave of the oracle, the analyst's screen around the
wizard.

One of McNamara's wizards was Alain C. Enthoven, who, like
many other wizards (more often called whiz kids), had arrived at
the Pentagon from Rand. Enthoven was critical of large, computerized war games because, although they "tell you the outcome of
the war," most of them "are not constructed to tell you why the
outcome came out the way it did and not some other way."

As the first head of the Pentagon's Systems Analysis Division,
Enthoven was both manager and missionary. He rode the military
circuit preaching cost effectiveness. "Numbers are a part of our
language," he told a Naval War College audience. "Where a quantitative matter is being discussed, the greatest clarity of thought is
achieved by using numbers instead of avoiding them, *even when
uncertainties are present.* [His emphasis] . . . Systems analysis
takes problems that are not defined and attempts to define them. . . .
Rather than trying to select a precise maximum or minimum, a
motto of the Systems Analysis Office in the Office of the Secretary
of Defense is, 'It is better to be roughly right than exactly wrong.' "

Although political-military gaming continued in the Pentagon
basement, elsewhere in the defense establishment black-box computer models handled the puzzles of nuclear war, from missile
defense techniques to civil defense, from the use of tactical nuclear
weapons to concepts of strategic arms control. An analyst who
worked on nuclear strategy in the Office of the Secretary of Defense
said that in the 1960s models of U.S.-Soviet missile exchanges
greatly influenced strategic thinking.

"The 1960s world of systems analysis and computer simulation
models had departed—with gradually increasing speed—from the
1940s world of wartime combat operations research," wrote Clayton
J. Thomas, an operations research veteran of both worlds. At first,
he said, models were looked upon as experimental. But as "computer programs grew larger and larger, it became more and more
difficult to resist the temptation to transfer large models from an
experimental role to an institutionalized productive role."

By 1966 a Joint Chiefs of Staff study listed 103 different war
games and simulations run by fifty organizations, of which only

thirty were military. Academic researchers seeking financial aid for gaming and analytic models found a welcome in McNamara's rational Pentagon. On many campuses the political scientists worried not about publish-or-perish but about play-or-perish.

"The community of 'simulators,' particularly those who had played in serious policy games, seemed to have reached a consensus that for purposes of policy analysis the PE [political exercise] was, at best, a form of organized mind-blowing, with serendipity the chief objective," Bloomfield later wrote. "The players," he recalled, "were most often undergraduate students (and often high school-level youngsters in boot camp). This clearly threw in doubt the validity of their decision behavior as a reasonable analogue to governmental policy."

There was a passionate belief in gaming, in numbers, in simulating, in manipulating models. To the ultimate believers, if something could be modeled, it could be done. Fate was easily adjusted through mathematics and analysis. American policy mandates could be analyzed and modeled in Washington and then, in the real world, operators could shape that policy by working the humble clay of developing countries.

At the Special Operations Research Office at the American University in Washington, academicians working under a Pentagon contract in 1964 launched Project Camelot, a secret operation aimed at finding a way to predict revolutions. Camelot's designers hoped to develop "a general social systems model which would make it possible to predict and influence politically significant aspects of social change in the developing nations of the world" and "assist the Army in planning for appropriate advisory and assistance operations in developing nations. . . ."

To test the model, Camelot's designers needed a data base from a specified country. Chile (No. 70 on the Douglas power-ranking chart) was chosen, and Rex D. Hopper, director of Project Camelot, wrote to a social scientist there and asked him to participate, for a fee of $2,000, in helping to design a study of "internal war potential and the effects of government action on such potential." The result of this letter was a Chilean newspaper story headlined "Yankees Study Invasion of Chile" and a diplomatic flap that produced U.S. headlines and the quiet death of Project Camelot.

Such analysis-spawned creatures as Camelot bewildered and infuriated military officers trained in old and battle-tested ways. Officers scorned McNamara and his whiz kids, if not very often at Pentagon meetings, at least over drinks at the Army and Navy Club. Rarely was criticism heard in the public arenas of Washington. One of the few officers to openly criticize the rising influence of systems analysis was the Navy's Hyman G. Rickover, who rarely suppressed his contempt toward superiors.

Testifying in Congress in 1966, at the height of his power as a vice admiral in charge of the Navy's nuclear reactor program, Rickover made a typical historical wisecrack: "On a cost effectiveness basis the colonists would not have revolted against King George III, nor would John Paul Jones have engaged the *Serapis* with the *Bonhomme Richard*, an inferior ship. . . . Computer logic would have advised the British to make terms with Hitler in 1940." He also gravely spoke to war itself—and to the war then going on: "A war, small or large, does not follow a prescribed 'scenario' laid out in advance. If we could predict the sequence of events accurately, we could probably avoid war in the first place. The elder Moltke* said: 'no plan survives contact with the enemy.' Are we not re-learning that bitter lesson every day in Vietnam. . .?"

What the critics like Rickover did not realize was that the Vietnam War, so different from all other wars, was a massive version of a Project Camelot, a model become real, a model designed to show it was possible to fight and win not only Vietnam's internal war but also an American limited war. "The greatest contribution Vietnam is making—right or wrong is beside the point—is that it is developing an ability in the United States to fight a limited war, to go to war without the necessity of arousing the public ire," McNamara was quoted as saying in explanation of the war.

Vietnam, by that view, was playing out the scenario of limited war—a measurable war, a war that could be logical and reasonable. "As we go back and read the writings of political scientists and systems analysts on limited war," wrote Army Colonel Harry G. Summers, Jr., "they are noteworthy for their lack of passion. The

---

*Count Helmuth von Moltke, chief of the Prussian General Staff (1858-88), was an advocate of wargaming.

horror, the bloodshed and the destruction of the battlefield are remarkably absent." It was Summers who circulated a "bitter little story" told during the final days of the Vietnam War:

"When the Nixon Administration took over in 1969 all the data on North Vietnam and the United States was fed into a Pentagon computer—population, gross national product, manufacturing capability, number of tanks, ships, and aircraft, size of the armed forces, and the like.

"The computer was then asked, *'When will we win?'*

"It took only a moment to give the answer: *'You won in 1964!'*"

CHAPTER 7

Gaming Under Analysis: The Political-Military Games

Soon after World War II, wargaming, like nuclear weapons, passed out of military control. If nuclear weapons were too dangerous to leave to the generals, then so was nuclear strategy. And as wargaming flourished in the Cold War world of militarized civilians so, coincidentally, did a different kind of gaming called game theory. Unlike wargaming, with a heritage as old as war itself, game theory was new and unconcerned with battle. But just as atomic theory had been quickly conscripted into weaponry, game theory soon became the foundation of a battle plan for nuclear war.

Much of that planning was done near the beaches of Santa Monica, in the headquarters of Rand, which in 1948 became a private, nonprofit corporation. The Air Force was Rand's only client, but Rand was looking beyond Air Force matters to questions of national strategy in a nuclear world. Playing old, military-type war games, Rand's nuclear strategists learned little. Hasty war games—"scratch-pad wars"—began with the Soviets invading Europe and ended with Air Force planes dropping hundreds of nuclear bombs on the invaders, incidentally causing the destruction of Western Europe. There had to be a more rational way to deal with nuclear war. Game theory seemed like a way.

Game theory was a mathematical concept, at first studied in such games as poker and bridge and then found as deeply embedded in the human condition. Robinson Crusoe was an isolated economic unit competing with nature. Othello and Desdemona were playing a game of trust and jealousy. Chess players were engaged in a direct conflict of interests. Two department stores competing for one group of customers were in a plural game. And the United States and the Soviet Union were playing the nuclear war game.

John von Neumann, a mathematician who is said to have first thought of game theory during a poker game, was well known at Rand for his work on the hydrogen bomb. He and his game theory appealed to Rand's mathematically minded nuclear strategists, who were developing the concept of nuclear deterrence. To them game theory looked like a rational, dispassionate method of examining the strategic behavior of nations, especially the United States and the Soviet Union. The theory envisioned each country as a "rational actor" who consistently followed his best strategy. A political game, or simulation, as some preferred to call it, gave game theorists a chance to look through the real-world aspects of conflict and peer into what they perceived as the system that actually governed the conflict.

Game theory's calculations and tables frighten off the non-mathematician, but a mathematician and game theorist like Martin Shubik can use words as well as numbers, as when he explains the game in which two people bargain in a transaction that can either leave both of them broke or give them both some part of what they are bargaining over. In Shubik's terms this is "the simplest possible case, in which two persons with transferable utilities and a fixed disagreement point bargain over the division of a specified gain." In such a case, he says, "split the difference" is the "only reasonable solution." But as this simple case broadens, complications develop: More players mean the possibility of coalitions; more strategies may mean the introduction of threats. A coalition drastically changes the game, and the forming of an alliance may be more costly than a player realizes. Shubik gives this trenchant example: "If a wealthy woman marries a man who is subject to blackmail, both may suffer."

We do make judgments that game theorists call utility functions when we say "He is the lesser of two evils" or "He has gone up in my estimation." When we say, usually as a prelude to parental punishment, "This is going to hurt me more than it will hurt you," we are demonstrating what Shubik calls relative welfare. As an example of total welfare, he offers "Do me a favor. It would trouble you only a little and would help me a lot."

In a two-person, zero-sum game, such as chess or a duel, the sum one player loses is equal to the sum the other one wins. In a nonzero-sum game both players may win something or both may lose something by cooperating or by opposing each other, or by a combination of both. This was the mathematical foundation for nuclear deterrence. The classic illustration is the Prisoners' Dilemma.

Sam and Ivan are arrested and put in separate cells. The prosecutor, convinced that the prisoners have been partners in a crime, decides to question them separately. He makes the same speech to each prisoner: "We have the goods on you and your pal. You have two choices—confess or stay silent. If neither you nor your pal confesses, both of you will be put away for parole violations anyway. If you both confess, it'll save the state the cost of a trial and I'll see to it that you get short sentences. But if one of you confesses and the other does not, then the one who confesses gets off with a suspended sentence and I throw the book at the other one. Well, what are you going to do?"

The prosecutor has presented each prisoner with a dilemma. Sam (or Ivan) gains the most if he confesses and his pal does not. But he cannot know what his pal will do. If neither he nor his pal confesses, they both go to jail, though not for long. To the game theorist turned war strategist, the United States and the Soviet Union are in separate cells called nations. One may be planning a first-strike nuclear missile launching on the other (noncooperation strategy). Or one may be seeing the advantages of compromise with the other (cooperation strategy). One cannot know with certainty what the other will do or is planning to do.

A variation on the Prisoners' Dilemma is a game called Chicken. The name comes from the teenage drivers' road game: Two cars speed toward each other. The driver who swerves away is chicken. If no one turns away, the game ends with two total losers, and, to the mathematician and the non-mathematician alike, that possible conclusion makes Chicken—or nuclear war—a game much different from poker or chess or the Prisoners' Dilemma. Daniel Ells-

berg, who joined Rand as an economist and became an analyst and
an authority on the Vietnam War (he later leaked the classified
"Pentagon Papers"), is generally credited with developing the
Chicken game as an analogue of nuclear deterrence. He set up a
game to reduce the balance of terror to lines as stark and dispassionate as an electrocardiograph. He called his creation "a machine
for asking useful questions and for preliminary testing of alleged
answers."

The game theorist's machine is a matrix, a rectangular mathematical array containing numbers giving value to the moves each
player can make. The Cuban missile crisis, seen by a game theorist
as the Chicken game in a matrix, looks like this:

Soviet Union

		Withdraw Missiles [W]	Keep Missiles in Cuba [M]
United States	Blockade Cuba [B]	Compromise (3,3)	Soviet victory, U.S. defeat (2,4)
	Wipe Out Missiles with Air Strike [A]	U.S. Victory Soviet defeat (4,2)	Nuclear war (1,1)

The numbers assigned: 4 = best, 3 = next best, 2 = next worst, 1 = worst.

In game-theory terms, *BW*—a U.S. blockade [B] and a Soviet
withdrawal of the missiles [W] offers, for both, the "next best"
solution to the game and results in a compromise, expressed as a
kind of mathematical tie (3,3). But the Soviet player's "best" move—
keeping the missiles in Cuba—would probably result in nuclear

war, and the Soviet player's winning 4 does not do much good when the U.S. player's losing 1 is nuclear war.

A Soviet version of a game matrix looks like this:

	B_1	B_2	B_3	B_4	B_5
A_1	3	4	5	2	3
A_2	1	8	4	3	4
A_3	10	3	1	7	6
A_4	4	5	3	4	8

The Soviet commentary accompanying this matrix explains a winsome-lose-some aspect of the game, a von Neumannn idea called minimax, which has been carried over to U.S. arms control strategy. In the Soviet matrix the game is kept purely mathematical. But in the commentary a philosophical note surprisingly appears. "Let us reflect for a moment on what strategy we (Player A) are to follow," the author of this Soviet game wrote. We (A) "feel drawn to select strategy A_3," but "the opponent is not exactly a fool! Should we chose A_3, he, out of spite, will chose B_3 that will land us a miserable '1.' No, A_3 will not do! What then? Obviously, from the cautiousness principle (central to game theory) we have to seek a strategy such that our minimal payoff be maximal. This is the so-called 'minimax' principle: make the best of the worst behavior of the opponent."

With the advent of game theory, military wargaming wandered into a new and strange environment. Gaming no longer belonged only to the generals and the admirals. Wargaming had become political-military gaming, and the players were often academicians trying to apply campus-bred theory to the bloody reality of the battlefield. The civilian strategists talked of *escalation* and *limited war* and *signaling*. "Escalation was seen as a multimove game with certain allowed rules and signals between the combatants," writes Paul Bracken, a defense analyst and authority on crisis communications. The mathematical models that were built "were based on an underlying blast damage calculus that had a powerful and pervasive influence on U.S. decisionmakers and their view of the entire U.S.-Soviet nuclear relationship."

One such model that came from Rand was STROP (for Strategic Operational Planning), which could "evaluate the outcome of a nuclear exchange in about ⅟₅₀ of a second," making it easy for someone to quickly "examine a large sample of potential conflicts." In STROP "cities" are "value target units," measured in terms of population, industrial output, "or any other value parameter." (Moscow tops out at twenty-five units.)

"The knottiest problem in analyzing a central nuclear war," the Rand report said, "is the payoff and criterion. Central nuclear war is not only highly nonzero sum but also the outcomes can include cases that are catastrophic to one or both sides. The theory of nonzero sum games for noncooperative situations is in an unsatisfactory state, and methods of dealing with catastrophic payoffs are extremely elementary."

But Rand had found a way to measure the unmeasurable. The "primary function of strategic aerospace weapons is target destruction. Other possible functions—prewar deterrence, intrawar deterrence, backup to threats, etc.—are derivative. Hence, the basic measure of the outcome of a nuclear exchange is target destruction on each side. Similarly, the measure of the value of aerospace weapons in a counterforce role is the measure of target damage saved. Both of these measures can refer either to actual destruction or to *potential destruction*. Thus, the value of withheld weapons can be measured in terms of the potential damage they can achieve."

Accompanying STROP was STROP II, which figured out budgets. "For each force structure paid (one for Red and one for Blue), the routine plays the target allocation game and selects preferred allocations for each side. . . . Different budget levels can be input for each side as well as different force mix selection rules. Cost curves are input for each force element," including bombers, missiles, bomber bases, fighter bases, civil defense, and research costs for new missile systems.

"From the attacker's point of view," the Rand report said, "fallout effects are not only a function of the budget level (e.g., they are dependent on warhead size) but also of policy (airburst vs. groundburst, city avoidance, etc.)." The fallout table in STROP II "does double duty. It acts as a policy formulation and also fixes the fallout effects of a nominal warhead for ground bursts. The fallout effects are modified by a factor which is a function of the budget level. Separate factors are tabled for bombers and missiles."

Rand's Systems Research Laboratory studied the interplay between people (almost invariably men) and machines. The laboratory put a staff in a simulated air-defense center that defended an area of about 100,000 square miles. The center handled some 10,000 enemy and friendly simulated flights in experiments that lasted as long as six weeks. Members of one crew broke down, so real was the stress of simulation. The experiments were called "one-to-one" simulations because they came close to matching what they imitated. From the experiments came an early concept of the simulation spectrum, which the laboratory perceived as demonstrating an increasing degree of abstraction:

Real World	Observations from Real World	One-to-one Simulations	Game-type Simulations	All-computer Simulations

Always there was a search through the abstract for numbers. And the numbers could be found—for fallout, for destroyed cities, for bodies, for budgets, it did not matter for what. Then the numbers could be linked to games and used to buttress nuclear strategy. The numbers and theories were transforming the art of warfare into a science of the nuclear age.

Some of the new theorizing looked back. A form of gaming called Crisiscom tried to simulate the way decisionmakers—such as the German Kaiser and the Russian Czar during July 1914—processed information in the midst of international crises. Gaming political scientists tested players to see if they would follow the moves that led up to the Mexican-American War and the decision by Texans to declare independence and then join the United States. Another one rated nearly one thousand events in European history between 1870 and 1881 on "a conflict-cooperation scale."

Most gaming focused on the nuclear future—and upon the government contracts that were being awarded to bright, competitive political scientists. Under one contract 325 Navy petty officers at the Great Lakes Naval Training Center, aided by a staff of twenty-three students from the international relations program at North-western University, spent eleven weeks playing Inter-Nation Simulation, a game developed by social psychologist Harold Guetzkow.

In the make-believe world of INS, heads of government called Central Decision Makers ruled nations called Algo, Bega, Colo, Dacia, Erga, Ingo, Omne, or Utro. Each nation had a Basic Capability, a kind of GNP that indicated the amount of goods and services available to the country's inhabitants. Each country also had nuclear and conventional weapons (Force Capabilities).

The Central Decision Maker's power was dependent upon pleasing the Validators, "a symbolic representation of politically effective groups in the society, such as military juntas, political parties, or pressure groups." The Validators decided their support on the basis of how well the Central Decision Maker maintained national security and the nation's standard of living. The numbers for this were based on information gleaned from interviews of State and Defense officials, academic experts, and newspaper correspondents, who were asked to rate aspects of national power as viewed by members of the national elite.

Although the Central Decision Makers and their imaginary countries lived on for a while on campuses, such abstract gaming had little impact in the world of real decisionmakers. As one political scientist said, scholars were approaching "serious, even crucial, problems by creating artificial worlds in a manner not entirely dissimilar to that of children playing house or building a space-ship out of cardboard cartons and chairs." Political and military leaders remained suspicious of gaming until it received a seal of approval that they trusted. That seal was Rand's.

Early in Rand's history physicists and engineers were augmented by economists, among them Andrew Marshall, and social scientists, among them Herbert Goldhamer, under whom Marshall had studied at the University of Chicago. In 1954 Goldhamer proposed a Cold War Game, a nuclear-age war game that would have permanent and far-reaching effects. Goldhamer's political-military war game would replace the classical war game, would lead to the creation of a permanent gaming facility in the Pentagon—and would bring into gaming Andy Marshall, who was destined to be one of the most influential figures in the new, nuclear Kriegsspiel.

Marshall would appear in many roles in the annals of gaming. As a player in the first game, he was present at the creation, and he has been present ever since. Four years after the Goldhamer game, Marshall, obsessed by a possible Soviet "Pearl Harbor" attack on the bombers and nuclear weapons of the U.S. Strategic Air Command, produced a dazzling Rand report in which he pioneered the use of game theory and the simulation of missile launchings and bombing runs. His obsession inspired a classified Rand study of the Pearl Harbor attack itself. This report unearthed the Japanese Pearl Harbor Game and, in the form of a prize-winning book, made Pearl Harbor, 1941, into a cautionary tale for Cold War, 1962.

It would be Marshall, as head of the Pentagon's Office of Net Assessment, who would become the watchdog of gaming. Dissatisfied with the stagnation of gaming, in the 1970s he would call into question the worth of the entire U.S. game arsenal. In the 1980s he would dare to commission a strategic game from the hobbyist game designer Jim Dunnigan. And later Marshall would preside over the creation of the ultimate game: Blue computer versus Red computer.

Goldhamer proposed that the government of each country in the game be represented by a player or group of players familiar with the country's political system. One or more players would represent "Nature," whose role was described as providing "for events of the type that happen in the real world but are not under the control of any government: certain technological developments, the death of important people, non-governmental political action, famines, popular disturbances, etc." The American side could follow whatever strategy desired; foreign players had to follow the national policies prevailing at the time of the game. The idea was to give the American team enough flexibility to explore a broad strategic spectrum.

The first Goldhamer games, each one lasting a few days, were played early in 1955, with Marshall and three other Rand workers— Harvey DeWeerd, Paul Kecskemeti, and Soviet specialist Nathan Leites—among the first players. Goldhamer next ran a four-week game in the summer of 1955, followed in April 1956 by a game that had thirteen players, including three senior State Department Foreign Service officers. In that game Goldhamer introduced what would become a permanent feature of political-military games: a projection into the future. The game's scenario, written in March 1956, sketched a world of January 1957.

Goldhamer's games, like nearly everything that went on at Rand, were secretive. He and Hans Speier, director of Rand's social science division, set down "observations" about political gaming in a professional journal in 1959 but told nothing of the April 1956 game scenario, saying only, "The fourth game was focused on the activities of the United States and the Soviet Union with respect to each other and to Western Europe." Goldhamer realistically made secrecy an issue in the game itself. "Nature," which included the game's referees, could leak accurate or distorted versions of classified documents to the press (except for the Communist press, which did not leak because it was government-controlled). Country players would find their moves affected by the fact that the information had leaked. Thus was a realistic distrust of the media built into decisionmaking simulation.

News of Rand's Cold War gaming quickly did leak to academe, and before the fourth game was over political scientists from Stanford, Yale, Harvard, Princeton, Northwestern, and the Massachusetts Institute of Technology, among others, were asking to hear more. Much of the interest centered at MIT, where W. Philip Davison, a Rand colleague of Marshall and Speier, was a visiting professor during the academic year 1957–58. Davison played a scaled-down version of the Cold War Game with MIT students around a seminar table.

Other MIT professors also tried the gaming technique and reported that it helped students become more sophisticated about the intricacies of international relations. But not until September 1958, when Paul Kecskemeti, veteran of Goldhamer's earliest games, went to MIT to play in a game arranged by Professor Lincoln P. Bloomfield, did political-military gaming become an important force outside of Rand.

Bloomfield had joined the MIT Center for International Studies the year before, after eleven years at the Department of State. Bloomfield decided to depart from the Rand and INS style of games and develop his own. Well connected in Washington and in the Cambridge intellectual community, he knew he could find players who would make the games more realistic and more useful to policy planners. Instead of students and Navy petty officers, he turned to government officials, scholars, and specialists on areas or subjects covered in the games.

Getting U.S. officials into the games involved some quiet diplomacy. Bloomfield, with Max F. Millikan, director of MIT's Center for International Studies, and Walt W. Rostow, who, like Bloomfield, shuttled between Cambridge campuses and Washington corridors of power, presented the idea to Acting Secretary of State Christian Herter and Allen Dulles, Director of Central Intelligence. Herter, as Bloomfield later recalled, "enthusiastically [offered] the services of Policy Planning Staff members for all subsequent MIT games—provided that he did not have to tell Congress the State Department was 'playing games.' "

Bloomfield developed an MIT form of the Rand game, while Rand, using the techniques that Marshall had pioneered, continued to concentrate on analytic games stemming from game theory. Thus two modern versions of wargaming—seminar-style interaction among human players and mathematically based analysis—emerged from Goldhamer's experiment.

Because Bloomfield believed that what he did was neither a game ("a formally structured competitive interaction with specified payoffs") nor a simulation ("explicit computer or other models with well-defined components interacting in a predictable way"), he called his games POLEX or PE for "Political Exercises." Actually, he said, since they were political-economic-psychological-cultural-intelligence-military exercises, the acronym ought to be PEPCIME. But, with great good sense, he never actually forced that acronym on anybody.

Beginning in 1958 and continuing for the next thirteen years, Bloomfield, Thomas C. Schelling of Harvard, and a few other academicians ran at least a dozen major PEs. Each game involved thirty to forty people, most of them middle-level government officials. In 1961 PE-style games, many of them run by Bloomfield and Schelling, appeared in the basement of the Pentagon under sponsorship of the Cold War Division of the Joint War Games Agency of the Joint Chiefs of Staff, which became SAGA and then today's JAD.

At Rand players staged Soviet invasions of China, Soviet aggression in the Middle East, and India-Pakistan disputes. Mostly Rand went to war against the Soviet Union, with Rand's prime customer, the Air Force, doing much of the damage. A major game, called SAFE (Strategy And Force Evaluation), began with budgeting for the purchase of weapons for a U.S.-U.S.S.R. war that would start several years later, in 1968 or 1970. There was also STRAW (*Strategic Air War*) and the acronymic descendant of Goldhamer's idea, COW (the *Cold War* Game).

Rand's gaming was played in two leagues. The "essayists" centered around Bernard Brodie, the first nuclear strategist, who looked to history for substance and endorsed seminarlike gaming. The "scientists" gathered around Albert Wohlstetter, a logician. Brodie's view of wargaming was closer to the MIT idea, although he saw gaming "even at its most elaborate" as "an austere abstraction from the real thing." Gaming, he wrote, "is a way of eliminating one kind of bias, that is, it is a means of giving the enemy his full due, and also a way of constraining weak human beings to think through systematically a number of consecutive acts or stages in a conflict."

MIT scenarios, which tended to look at small crises, often hewed close to reality. A game in 1960 examined the collapse of the Shah of Iran—"to the alarm," Bloomfield later wrote, "of State Department experts on Iran," who did, nevertheless, participate. The report on that game suggested "reexamination of the possibility of support for popular democratic forces even at the cost of annoying friendly governments." The report recommended that the United States "work more vigorously toward reform of the established order" and make "a systematic reexamination of the popular base in friendly or strategically vital countries."

The MIT games, along with those run by Rand and SAGA, filled the ranks of the government with game veterans. And Herter's admonition prevailed, for little about gaming became known beyond academic and defense circles.

Bloomfield's MIT scenarios described the emergence of a pro-Castro regime in Venezuela, insurgency in India, Chinese infiltration of Burma, and internal turmoil in Angola. The MIT games, rather than escalating to nuclear war, usually sought novel ways out of local or regional problems. In one, for example, the United States, trying to work out a solution to an Algerian-Moroccan dispute, begins by deciding on a general arms embargo and then goes on to such more imaginative ideas as asking the Soviet Union to support mediation and introducing an international observer force into a demilitarized zone between troops of the two countries.

The MIT games were more open than Rand's or the Pentagon's because from the beginning Bloomfield had insisted that all reports on the games be unclassified. By agreeing to keep the names of the players secret, Bloomfield managed to straddle the open world of academe and the secret world of national security. Still, even in the MIT games nuclear reality was inescapable. Dr. John Craven, chief scientist of the Polaris submarine missile program, once asked Bloomfield to play a game that would indicate how quickly the President might have to communicate with the missile submarine fleet.

"I demurred on the ground that our games did not produce usable numbers," Bloomfield later recalled. "Nevertheless, a specific requirement for rapid communication surprisingly surfaced in one of the MIT games, although—possibly because of my own pro-arms-control bias—the need identified was for rapid surfacing and counting for arms control purposes rather than for firing or retargeting orders." Despite that bias, Bloomfield did direct and play in the highly classified Pentagon games, and Schelling ran some at Camp David, the presidential hideaway in the Catoctin Mountains north of Washington.

In the 1960s the U.S. Arms Control and Disarmament Agency sponsored games that investigated localized conflict "and the relevance of arms control and conflict control measures aimed at minimizing the chances of small wars," especially those that might involve nuclear powers. In 1968, against the backdrop of a real example—the seemingly uncontrollable Vietnam War—Bloomfield began four CONEX (CON as in *Con*flict) games, three of which were played at Endicott House, MIT's academic refuge in Dedham, Massachusetts.

The CONEX games were not driven by crisis scenarios. "We were led to this deviation from the typical 'crisis game' model out of a conviction as citizens that better advance planning is desperately needed if preventive diplomacy is to be more than rhetoric," Bloomfield and a colleague later wrote. "We sought to ascertain if games were in fact as dependent on a high-intensity crisis environment as we, and others, had hitherto assumed."

Bloomfield's CONEX games evolved from the Blue-Red-Control game structure in vogue then—and now—in the Pentagon to a much more complex structure. In the final game, for example, there were three playing teams and several Control subteams. The playing teams were a U.S. Blue Team, a U.S. Green Team, and a Government of India Team. Control's subteams represented "interests": the United Nations, China, the Soviet Union, Pakistan, U.S. public opinion, and Congress. The subteams had influence on the playing teams, which made four moves over the two-day game period.

In the first part of the game the crisis is far enough in the future so that decisionmakers have time to sort through several choices. The cast of players was different from typical SAGA or Rand games. Instead of the traditional White House-Pentagon huddle there was a forum encompassing people from Capitol Hill—including members of Congress. Officials or staff people at agencies they represented played in the games, as did academic authorities on specific geographical or policy areas. Unlike players in the SAGA games, Blue and Green team members played roles. In one game, in fact, the President was played by "a distinguished business executive with some presidential pretension and vast firsthand experience of government." He has never been identified.

The purpose of the games was not to see how decisionmakers reacted to a crisis. The game designers had in mind certain hypotheses, such as "Preventive policy measures are more available early in a crisis than late, but the United States government typically does not take full advantage of them." The use of two U.S. teams allowed the game designers to introduce a variable in one team as an experimental control.

The most interesting variable was injected into the fourth game: One of the teams was provided with computerized information on local conflicts; the other team, more realistically, acted on experience and intuition alone. There was not much difference, except that the team with access to information seemed to focus its objectives better.

Bloomfield was running a different game from the crisis-management games that had taken hold in the basement of the Pentagon early in the Kennedy Administration. In the first CONEX game the President is more interested in avoiding a crisis than in setting one up.

A vignette about the game begins at a crucial moment: "The President signalled for silence. 'We have a hot-line message to Moscow. Yes, that's right. And it must go right away. . . .'

"He leaned back, lit a cigar, and said in the voice of a man desperately short of sleep but determined not to show it, 'That's all we can do. The U.S. is not intervening, and we can only hope the Russians don't misunderstand and start mixing in. If they do, we have no choice but to go in.'

"The Secretary of H.E.W. [Health, Education, and Welfare], whose first National Security Council meeting this was, looked up. When he spoke, his voice was perhaps a shade too loud. 'Dammit, Mr. President, we did the right thing. We probably avoided another Bay of Pigs and another Vietnam, and that's the way the American people and the Congress want it. So what if the Russians get involved? Why should we?'

"The argument that had waxed and waned for two days and part of a hectic night seemed to be starting all over again. Three members of the National Security Council Executive Committee opened their mouths to join battle once more over whether America should respond militarily when the door opened and a secretary walked into the room carrying a tray of sherry. 'Control says it's time for lunch and would you please fill out your questionnaires before you come downstairs?' "

CONEX games, like the others Bloomfield stage-managed, were more than simulations; they were parables for a time of doves and hawks. No real Secretary of HEW had ever joined the Secretaries of State and Defense in a real debate over national security. But the CONEX games were designed to see how domestic players— excluded in both real Vietnam-era debates and in Vietnam-era SAGA games—would affect debate. (In one MIT game the Secretary of HEW resigned from the team in protest over a decision.) And in the CONEX games two U.S. teams played simultaneously, each one looking at the same problem; the teams were similar enough so that their decisions varied only slightly. As a later report on the games pointed out, it was "exceedingly hard at the end of the 1960's to get any team playing the role of U.S. decision-makers to decide on unilateral U.S. military intervention anywhere. . . ."

The fourth CONEX game began with India in civil turmoil and China and Pakistan plotting military intervention. (The game runs in a fascinating counterpoint to the SAGA game that tried to start a war in Chapter 2.)

In the MIT game the Soviet Union was giving India military and economic support. The United States, while technically neutral between India and Pakistan, was concerned about Chinese military moves against India. The U.S. Blue Team decided to use quiet diplomacy "to minimize the chance of open hostilities between any of the great powers over the developing situation in South Asia." The U.S. Green Team developed a similar position. Control revealed this general U.S. policy to the subteams.

The updated scenario for the second move projected events three months forward. Civil unrest was spreading in India. As a potentially explosive domestic crisis was building in India, China announced it was going to test an intercontinental ballistic missile by launching it into the Indian Ocean. The Indian government sent an urgent message to the United States. Asserting that the Chinese test "threatens India's very existence," India asked for U.S. assurance "that you will not tolerate or permit the threat of nuclear blackmail against India." The Soviet Union told the United States that it would not intervene in the crisis, but asked for U.S. views about "counteraction with respect to the forthcoming Chinese ICBM shot."

The U.S. Blue Team warned Pakistan against trying to exploit India's troubles. The U.S. Green Team went further by threatening Pakistan with a cutoff of aid if it made any hostile moves toward India. Each U.S. team gave India the same evaluation of the Chinese ICBM test: It's just propaganda; don't worry. Meanwhile, both U.S. teams worked through the United Nations to try to defuse the crisis.

The scenario's next move pushed time forward three days. In those seventy-two hours there had been an attempt to assassinate the Indian Prime Minister, touching off more domestic violence. At the same time there were indications of a possible Soviet strike against China. While U.S. Green tried diplomacy, including the convening of an international conference, U.S. Blue told India, "If India ever has evidence of nuclear aggression against India, or the threat thereof, we would consider deployment of U.S. forces to a position supporting India should this prove necessary."

At this point the game ended—with the game designers having good reason to believe that they knew why U.S. Blue had suddenly escalated toward an indication of military intervention while U.S. Green had stayed on a course of defusing the crisis. There had been a variable: Blue was "unconstrained" and could, as in typical games, improvise through crises. But Green had to base its moves on historical information stored in MIT computers.

Before the game Bloomfield and the other designers had hypothesized that Blue and Green would have used different styles of decisionmaking. As an analysis of the game later noted, Green, required "to rationally consider previous and analogous experience," had been cautious; Blue, relying on memory and ignoring history, had been more rash. But political scientists who evaluated the game focused on the fact that neither Green nor Blue had intervened militarily because both had been "overpowered by the 'Vietnam syndrome.'"

The gaming that goes on today for sponsors in the Pentagon and the White House closely resembles the political-military gaming born in the early days of the Cold War. And, like the traditional military wargaming that sired it, political-military gaming is rarely bold. The games of the 1950s, 1960s, and 1970s reflected their eras and a future that offered only the immutability of Blue confronting Red. In game after game, decade after decade, the Soviets invade Western Europe or the superpowers hover impotently over the dynamics of the Mideast and India-Pakistan-China disputes.

The inevitability of the games sometimes inspired cynicism. Bloomfield once called them "social science fiction."

Bloomfield's MIT office looked out upon the Charles River and the backdrop of Boston's skyline. Sailboats were gliding on the Charles, just as the bigger ones do outside the windows of the Naval War College in Newport. Bloomfield, a naval officer in World War II and a civilian thereafter, seemed more interested in talking about what he was doing in the 1980s than what he had been doing in the 1950s and 1960s. He said he had retired from gaming.* "I have a very mixed view of what we were doing," he said. "Except perhaps for training purposes, for teaching purposes, for showing military people—and they tend to take it more seriously—about what the outcomes could be from blindly following certain policies."

When I asked him whether any policy discoveries from games had found their way into Washington decision machinery, he rose from his desk and walked to the small green chalkboard on the back of his office door.

---

*Not entirely. He is a member of the Board of Visitors of the National Defense University and was active in the setting up of its War Gaming and Simulation Center.

"I've spent about half of my career inside the bureaucracy and the other half of it as an academic," he said. "My picture of the bureaucracy goes like this." He picked up a piece of chalk and drew a rectangular box near the top of the board. "That's the government, and there is the top decisionmaking level." He drew two lines that dropped vertically from the box. "And here are the inside hierarchies and the outside, with advisers, consultants, RAND and MIT projects, and so on."

Another box, at the bottom of the board to the left of the vertical lines, symbolized the insiders and the outsiders. He lightly drummed the chalk on the lower box. "That's where the outside information comes in, at a sort of working level"—he wrote the words—"and it's precisely the same system that tries to move recommendations to the top"—he drew a third line from the lower box to the upper one—"and there is an absolute barrier right here"—he slashed two horizontal lines just below the upper box—"that stops them both, so that much of it is not really passed along. People are so goddamn busy they can't read the morning paper, and you hand them a printout and they say, 'Get this thing off my desk!' "

In the earliest Nixon days Bloomfield was asked by the Joint Chiefs of Staff to direct the first game of the new administration. "It was," Bloomfield recalled, "the only game I know of in which cabinet members were present. The Secretary of State was there. The Secretary of Defense was there, the deputy secretaries, the assistant secretaries, all the members of the Joint Chiefs of Staff. I was the game director. The whole thing had been planned by the bureaucracy—and I won't go into detail—which produced an outcome for the United States in which we went back into Southeast Asia, which was just insane for anyone to contemplate.

"And the Secretary of State and the Secretary of Defense and these people all looked at each other as though to say, 'What the hell am I doing wasting my time in the basement of the Pentagon listening to a gaming outcome that is unbelievably absurd? That's an example of all this Mickey Mouse that we've all been doing in the universities, and the think tanks, and the Pentagon basement.'

"It does not very often penetrate to that world"—still standing, he turned to the green chalkboard and tapped the upper box—"of the President, the national security adviser, the Secretary of State, the NSC staff, the senior staff, and the President's immediate decisionmaking family, the Meeses, the Bakers. And they generally know less about foreign affairs than the President.

"How you reach them is as much a problem for the people in the inside as it is for the people on the outside"—a chalk dot on the lower box. "If someone comes in, the director of CIA, and says, 'All my analyses say X,' then that X is the input and everything that went into X is something else again. And they don't ask. They aren't in research. That's my case."

He returned to his desk and sat down. "There'll be contracts, there'll be money, there'll be experiments, because there is the same kind of quest for an effective and persuasive wargaming capability in the same sense that there is a quest for a laser beam that will bounce off a small mirror in space. As far as I'm concerned, it has nothing to do with the crisis the United States has in real life. But that is all right, it's R and D. That's not as interesting as whether we have war or peace."

Looking back at the ephemeral quality of gaming, he said, "The learning process is scandalous. The Joint Chiefs have a box somewhere with dozens of games, some of which I ran, some of which I chaired the U.S. team, others where the services didn't trust each other and had someone in a service run a team. Henry Kissinger chaired one of the teams in that game. The results would be reported and then the game papers were classified, top secret usually. It would be nice to go through all those games and you'd find a number of things.

"First, you could make a net judgment as to what value that whole investment had been over the years. Second, you could get a much better fix on the predictive capacity of the bureaucracy because the scenarios are vetted by the desks in the bureaucracies as the best possible or the most plausible prediction and circumstances in a given future time that would initiate the kind of game you want. . . . Almost uniformly, the predictions were wrong.

"The games give you nothing, in the sense that the future is really unpredictable. If someone says this is what China is going to look like in 1991—because that's when you want this game to be: so-and-so will be dead; they'll have this, they'll have that; relations with Russia will probably look like this, and so on— empirically, that's going to be disproven, just on the record. It's not going to be like that. That's life: "Life is what takes place while you're making other plans," as someone* once put it.

"But the value of the game is not in prediction. It's an exercise in understanding what your problems are going to be."

That, however, is not how political-military gaming has been treated, especially by the people at the working level in Bloomfield's box. Because nuclear deterrence and nuclear warfare so naturally foster the image of a game—a simple Blue-Red game—gaming stays locked at the level of a never-ending duel. World War III quietly rages day after day, in game after game. But the players are not soldiers and sailors. They are analysts and their computers.

*John Lennon.

CHAPTER 8

▄▄▄▄▄▄▄

Scenario:
War with Russia

Rand began playing SAFE—Strategy And Force Evaluation game—
in 1962. The game was similar to the political-military games that
had begun in the Pentagon at that time: Blue, Red, and Control
teams playing within a scenario. The difference was that the SAFE
players started not in the midst of a crisis but in the midst of a
regular task, doing what peacetime civilians and officers actually
do: building a defense establishment over a span of time. SAFE
mirrored what was then going on in the Department of Defense.
Secretary of Defense Robert S. McNamara was introducing cost-
effectiveness and making war fiscally sound. In the McNamara
game everything began with dollars. And that was the way
McNamara's modern warriors began SAFE.

Players were given a budget, a policy statement, and a shopping
list of strategic forces to be purchased over the next decade. They
were also told the actual status of U.S. forces in 1962, along with
the estimated forces of the Soviet Union. Each move covered two
years; at least one of the games went on for two weeks. All of the
twenty-six players in the first round of six games in 1962 were
strategic analysts. Ten of them were in at least two games, playing
Blue once and Red once, though knowledge of Soviet military and
budgetary strategies did not seem to be required. Team members
did not role-play; each team acted as a high-level national defense
authority. "The purpose of the SAFE exercises," the SAFE instructions said, was "to explore the extent to which alternative sets

of strategic objectives would lead to distinguishable general war force postures. Each one of the six plays of SAFE was an instance of the implementation of a set of U.S. objectives in interaction with a particular set of SU [Soviet Union] objectives for given budget profiles."

Around the time SAFE was being developed, academicians and defense advisers in England were voicing concern about what Sir Solly Zuckerman, scientific adviser to the Minister of Defence, called U.S. "war gamesters" whose qualifications "combine those of a chess player and a soothsayer." Among the "gamesters"* named by the British were Bernard Brodie, Alain Enthoven, and Herman Kahn, all of Rand, and Henry Kissinger, then a Harvard professor serving the Kennedy Administration as a consultant. The British, who had never shown much interest in political-military games, knew that some of the U.S. gamesters had been playing with "nuclear options" in high-level, White House-sponsored games with extremely realistic scenarios.

Some of the gaming focused on how nuclear weapons could be used in a real crisis caused by Soviet threats to cut off Berlin from Western allies. In the 1948–49 Berlin crisis the Joint Chiefs of Staff had on hand a war plan, code-named Trojan, for bombing thirty Soviet cities with nuclear bombs.

In August 1961, during another Berlin crisis, McNamara met with Lord Mountbatten, head of Britain's Defence Staff. "And," McNamara later recalled, "he suggested a response to each one of the Soviet actions; and, at the end, he hadn't mentioned the use of nuclear weapons. I said, 'Lord Mountbatten, you haven't mentioned nuclear weapons.' He said, 'Are you insane?' "

The use of nuclear weapons in the Berlin crisis had been examined in a series of games at Camp David (Chapter 11). At the same time Rand analysts were showing high U.S. defense officials a paper detailing the advantages of a preemptive nuclear first-strike. The choice of action over Berlin had become a choice, by at least one British view, "between ignominious retreat and nuclear devastation."

The Berlin crisis evaporated. Then, in 1962, came the Cuban

Gamester in British usage, according to The Shorter Oxford English Dictionary, means not only a player of games but also a gambler, "a merry, frolicsome person," or someone "addicted to amorous sport; a lewd person."

missile crisis and more games. One of them, STAGE—Simulation of *Total Atomic Global Exchange*—reportedly took five months to play and proved that the United States would "prevail" in a nuclear war against the Soviet Union. The new Rand ideas—escalation, crisis management, limited war, "intrawar deterrence," bargaining during a war—were gamed. At the same time, as an Army officer later put it, the "distinction between peacetime crisis management" and "wartime destructive targeting decisions" was becoming blurred.

Gamesters were also projecting themselves into the future, looking at the world through the crystal ball of simulation. SPARC—*Space Planning Against Ranged Contingencies*—was a space-based version of SAFE. SPARC, developed for Air Force planners, doled out funds for "space arsenals" and set up probable price tags for future weapons. Putting a hydrogen bomb into orbit, for example, cost $30 million. For planners of 1960, SPARC provided a peek at the "possible worlds" of 1985.

In one imagined world the United States "surgically removed" China's ability to build nuclear weapons around 1969 and, sometime around 1981, allied with China and Europe against the Soviet Union. In another world NATO fell apart around 1969 and the United States and the Soviet Union signed a nonaggression pact. Brazil, meanwhile, was forming South America into a bloc confronting the United States while France and China were gaining support in the Middle East. By 1974 all of Africa south of the Sahara was controlled by black governments.

A "multipolar world" had France allied with West Germany and invading East Germany and the Netherlands. A "wild card world" conjured up a new, global religion that called for the massacre of all white people. The United States (where widespread race riots had put Washington under martial law), the Soviet Union (where Leningrad had been destroyed by a nuclear-tipped missile fired from India), and Europe (where nothing of note had happened) were all united against a black army of ten million men armed with spears and somehow backed by nuclear missiles.

Although such crystal-ball games declined in popularity, SAFE endured, perhaps because, in its decade-long time frame, the game had provided some insights into how money talks when the subject is war. "In every case but one," a former Defense official wrote, "the players expended the entire budget they were allocated. There seemed to be little linkage between policy statements and postures (that is, weapon systems preferred under one policy statement were likely to be preferred under any policy statement)."

A comparison of the games' make-believe decade of 1962–72 with the real decade shows little resemblance between predicted and actual history. *In the imagined decade* U.S. and Soviet budget-makers (none of them from the Navy) favored land-based inter-continental ballistic missiles over submarine-launched ballistic missiles. *In the real decade* the United States developed a strong missile submarine force, and the Soviet Navy began a massive missile-submarine buildup. *In the imagined decade* the United States and the Soviet Union spent considerable dollars and rubles on advanced, long-endurance bomber aircraft. *In the real decade* the real countries did not build such aircraft. *In the imagined decade* both countries built mobile, land-based intercontinental ballistic missiles. *In the real decade* the real countries did not build such missiles. And the Blue team spent far more on civil defense than the real United States did.

There was about these imaginary decades an air of wish fulfill-ment, a desire to translate into at least game-table reality the fears and frustrations that the players felt in their patriotic souls. The players were outside the decisionmaking system. But they were strategists-in-waiting, men always ready to switch from theorizing about policy to helping to make it. One who did was Fred Hoffman, a Rand economist who worked as the SAFE project leader. In 1964 he moved from the game world to the real world by becoming a Deputy Assistant Secretary of Defense. His work on SAFE games was later said to have led to his appointment.

Such direct promotions from simulation to reality are rarely doc-umented because so much about gaming has been kept secret down through the years. But the names of game experts that appear in professional publications often can later be found on Pentagon ros-ters. Among the publicly documented players who became gov-ernment officials are "gamesters" Kissinger and Enthoven and, among others, Rand veterans Daniel Ellsberg, Carl Kaysen, Walt Rostow, Charles Hitch, Henry Rowen, Frank Trinkl, Fred Iklé, and Andrew Marshall.

Iklé and Marshall have had long careers in government, Marshall in the Department of Defense's Office of Net Assessment, Iklé as director of the U.S. Arms Control and Disarmament Agency in the Nixon and Ford administrations and Under Secretary of Defense for Policy during the Reagan Administration. Other prominent players from think tanks and academe presided over or played Washington games but did not add government jobs to their résumés. All were members of a special club: They all had made believe that they had helped to blow up the world in a nuclear war.

SAFE evolved out of one of Rand's earliest simulations, the basing of intercontinental ballistic missiles. To Sherman Greenstein, a young analyst of today, those simulations are history, and he talks about them that way. In his office, where he worked on such projects as a naval warfare gaming system and the use of artificial intelligence in interpreting Soviet military moves of the 1980s, Greenstein talked of the value of gaming and how what he was doing had been built upon U.S.-Soviet games of the 1950s.

"Early on," Greenstein said, "the Air Force wanted to put our first ICBMs, Atlas-As, above ground. And the Atlas-A was really a pressurized balloon. You had to put gas into the skin because it was so thin it wouldn't hold itself up. Rand put together a simulation and set the ground rules: 'We're going to play the Soviets. You, the Air Force, are going to play the United States. These are your forces, and they are at these locations, and we've got 120 ICBMs and this is the population density around your cities and so on. And there's a crisis. As the crisis evolves, the game will begin here.'

"The first thing the Rand computer did was launch a [Soviet] strike against all these ICBMs. They were one quarter of a psi hard [able to withstand an increase in pressure of only one quarter of a pound per square inch]. If you got within five or ten miles of them with a nuclear blast, you'd kill them. So in one single strike the Soviet Union disarmed the U.S. retaliatory force. The Air Force cried, 'Foul! Wait! That's not fair. You didn't tell us this would happen!'

"This was when people didn't think about strategic forces. Here was a real value for this game. It gamed some insights. The Air Force said, 'We didn't think about that.' 'Well,' Rand said, 'now you're thinking about it.' "

Rand analysts had also forced military strategists to think about how many missiles were needed to destroy the Soviet Union. Rand's blueprint for reducing the Soviet nation to rubble began with the twenty largest Soviet cities, continued through the next fifty cities, and eventually reached 110 urban areas. But at the time SAFE was first being played, the declared strategy of the United States was that cities would be spared in a nuclear war. McNamara stated this in an address at the University of Michigan at Ann Arbor in June 1962. The United States, he said, "has come to the conclusion that to the extent feasible . . . principal military objectives, in the event of a nuclear war stemming from a major attack on the Alliance, should be the destruction of the enemy's military forces. . . ."

High-ranking Air Force officers were secretly advised that McNamara's no-cities declaration was not to be used for future strategic planning, apparently because McNamara's doctrine was being interpreted as a theory that made nuclear war feasible. Whatever future plans might be, McNamara's SIOP—the U.S. blueprint for total nuclear war—gave priority to the destruction of Soviet strategic retaliatory forces and air defenses not near cities. This plan superseded the nation's first SIOP, completed in 1960 and revised by the Kennedy Administration, which had targeted cities and seemed to have stemmed from the SAFE philosophy of city-busting.

So, certainly by 1963, SAFE's make-believe war appeared to be out of step with the real U.S. war plans. The game was even more detached from reality by the time of the Nixon Administration, which, in National Security Decision Memorandum 242, set forth a firm no-cities policy. And NSDM-242 itself had been war-gamed to test the feasibility of Nixon's nuclear doctrines.

Gaming, in the Pentagon of Secretary of Defense James Schlesinger, was changing, while SAFE stayed locked in the Cold War of the 1960s. As historian Gregg Herken has noted, Rand's "primitive scratch-pad war games" were being replaced "by complex, computer-generated models of the Russian economy, studied with the intent of discovering its weakest and most vulnerable spots." Soviet command and control sites were targeted for what strategists called decapitation of leadership. Nuclear war was no longer the simple matter of missile exchange played in SAFE. War planning was becoming as complex as modern war itself.

Looking at that modern war, Sherman Greenstein and other contemporary civilian analysts see unsophisticated, missile-swapping games like SAFE as antiques, interesting to look at but not for everyday use. SAFE, however, has gone on, a game of mass destruction, a game in which the world is merely Blue and Red, a game with probability tables for determining how many millions die from "prompt weapons effects" and radioactive fallout, a game with a philosophy of more-bangs-for-the-buck, cost-effective warfare.

SAFE has endured through every twist in declared and undeclared U.S. strategy. The game was still being played through 1980, at least at the Naval Postgraduate School.

The 1980 date is indisputable because that is the date that appears on an extraordinary document—a player's manual for Half-SAFE, a game derived from SAFE. The game "is designed to provide a heuristic* learning situation, a situation in which the players learn something about strategic decisionmaking by exposure to experience rather than by understanding produced by analysis."

Half-SAFE is played much the same as the traditional Pentagon game: a Blue and a Red team, each consisting of four to ten persons and each in a separate room, with a Game Director as control. The nine years of game time are divided into three three-year planning periods, with a crisis following. Each planning period corresponds to a move.

"The crisis phase may or may not conclude with a nuclear exchange," the player's manual says. "If it does not, the Game Director will call a halt to play when he decides that you and your opponents have weathered the test and resolved matters to such an extent that play has reached the point of diminishing educational returns. However, when no nuclear exchange develops naturally in the course of the crisis, the Game Director may decide that it would be useful to discover what the results of a nuclear war between the teams would have been. In that case, as in the case of

---

*Games are sometimes divided into two broad categories: *heuristic* games (from the Greek word for *discovering*) that are valuable for stimulating research and attempting to solve a problem but not for offering empirical solutions, and *stochastic* games (from the Greek word for *aiming at* or *guessing*) that proceed by guesswork and involve elements of chance. Half-SAFE uses dice or random numbers and is at least partially stochastic.

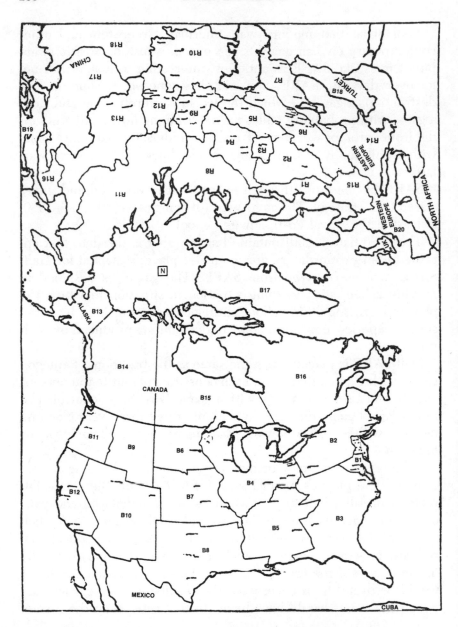

"Half-Safe," a Department of Defense war game, divides the world into B (for Blue) and R (for Red). The numbers next to *B* and *R* designate population zones. Players can keep score after a nuclear missile attack by looking up the actual population in nuked zones and then using a mortality formula to determine how many people were killed. (Credit: *Half-SAFE Player's Manual.*)

a nuclear war that develops naturally, each team selects one of its war plans for execution."*

At the beginning of each planning period each team is given a three-year budget of $15 billion, but the Game Director may vary that "to reflect political developments," such as arms control agreements. "To prevent players from enjoying an absolutely predictable budget, the Game Director subjects the base budget figure to a random adjustment up or down," based on a roll of the dice or a random-number generator. The range is usually plus or minus $400 million.

The world of Half-SAFE, which has only Blue and Red countries, is divided into twenty Blue zones, encompassing the United States, NATO nations, and Japan; eighteen Red zones include all of Eastern Europe and China. Players are given the population of each zone in terms of cities, towns, and rural areas. Red Zone No. 3, for example, contains Moscow with a population of 6.5 million and surrounding population of 6 million. Blue Zone No. 19 includes Tokyo, Osaka, Nagoya, Kyoto, Hiroshima, Nagasaki, and several other communities that make up a population of 99 million.

Equipped with a nuclear arsenal and briefed on the political situation of Year One, the teams begin the game by producing a "megatonnage study"—the number and types of weapons "your team can expect to detonate in your opponents' territory using the weapons system you inherit at the beginning of the game." (Some of the strategic information given to players in some of the Half-SAFE games is real, current data that is classified.)

The manual strongly implies that the *you* in the instructions are exclusively the military side of the U.S. and Soviet decisionmaking hierarchy. Half-SAFE unfolds in political and diplomatic isolation as stark as the Red-Blue world that is the game arena. Diplomatically, it is an unreal world. Allies are treated as passive nations that accept whatever Red or Blue wants, as long as Red or Blue pays for what is wanted. Money rules much of the game, which in its planning stages resembles Monopoly (itself a kind of war game).

The game is designed for purely nuclear war. Teams purchase or risk money only on the improvement of nuclear weapons and nuclear-weapon delivery and detection systems. There is no men-

*All quotations are from the player's manual.

tion of troops, warships (other than missile-carrying submarines), or non-nuclear arms. There is no incentive for saving money: Any leftover dollars go back to the team's national treasury, and nuclear war planners can get their hands on only 1 percent of that. So, obviously, it's better to spend all of the dollars and leave nothing for the Blue or Red national treasuries.

In Half-SAFE "people die from two effects of nuclear strikes: prompt effects (blast, shock and prompt radiation), and fall-out." Prompt effects "kill only the urban population"—numbers that are determined through tables that show "the percentage of the still-living populations of cities of various sizes which would *survive* the prompt effects of weapons of various yields." Tables help to determine the number of people who die from the effects of fallout; the number depends on how many shelters the budget planners had decided to build.

The hierarchy of the armed services includes many men—for war planning is a man's game—who played SAFE and similar missile-exchange games during the formative years of their military careers. As these men of the SAFE generation retire, they are replaced by the Half-SAFE generation, players who may have tried their hand at Half-SAFE but also could play B-1 Bomber on a home computer or may have blasted thousands—or millions—of space invaders in an arcade bunker. To them a bombardment satellite sounds quaintly realistic. And the career paths of some of them will take them to jobs involved with a visionary project, a defense system whose existence will remain more plan than fact for years and whose existence is so imaginary that at birth it earned itself a fictional name, Star Wars.

I have been told authoritatively that the Strategic Defense Initiative (SDI) was gamed prior to March 23, 1983, when President Reagan announced a "comprehensive and intensive effort" to create a space defense system. But I could find no documentation. Reliable, though anonymous, accounts of the events leading up to the Star Wars speech indicate that the far-reaching decision was impromptu, impulsive—a projection of reality similar to the kind of projection that arises from game tables.

Since the speech, as planners have worked to make SDI a reality,

Star War games have been played, but details of the games have not been revealed because the data on which the games are based comes from highly secret U.S. intelligence sources. But an anonymous description did leak from the SDI gaming. A player, described as a nongovernmental Soviet affairs expert, told a *New York Times* reporter, "We found we were playing against defense contractor personnel who know nothing about Soviet doctrine. It took our whole team, the Red Team, less than twenty minutes to agree that our first counter to 'Star Wars' would be to increase offensive missile numbers. Their team, the Blue Team, said, 'No, that is not how the Soviets think.' Every step we took surprised them."

Wondering how effective SDI gaming would be, compared, say, to Half-SAFE gaming, I asked Sherman Greenstein about it. He is very much of the Star Wars generation. He took my question one step further—past the point where Bloomfield had said that gamed ideas are usually blocked. In Greenstein's vision, thoroughly gamed SDI concepts have gotten past the policy blockade and are now in the Oval Office. The President, the Secretary of Defense, and other high officials have been invited to an SDI presentation.

"You get them in a room and you say, 'We're going to postulate an SDI system that has this kind of capability: It's 99 percent effective in detecting launches. It's 98 percent effective in the boost phase—killing all the boosters that are coming up.' And you describe it. It has all these satellites, and so forth.

"Then you give these people ambiguous conditions. You tell them, 'There are no detections of Soviet launches, but we have three nuclear detonations in the United States—Chicago, Los Angeles, and New York. And, after that, one detonation in Kiev. And then the Soviets start knocking out our SDI capability. Neither system, theirs or ours, has detected anything.' "

Greenstein paused and smiled at the scenario he had just made up. "You could examine that," he resumed. "It has nothing to do with modeling the capabilities of the SDI, but it has a lot to do with seeing what those decisionmakers might do under those conditions. The point I'm making is that, for the most part, really important policy studies, on which you might base games or simulations, don't have to be supported by great complex computer systems. Will an SDI system really work? How much time will the

President have to make a decision? You might want computer support for those questions, but the decision the President will make is almost irrelevant to a game."

Ambiguity is a word often heard when defense analysts discuss the modern problems of nuclear war. One analyst offered this one to me: "Suppose our satellites see a lot of Red Army activity suddenly starting. Long lines of trucks are showing up on highways, heading for what appears to be mobilization points. Is it mobilization? Or are the trucks getting new carburetors and some army bureaucrat decided to do it this way? That 'warning' intelligence from the satellite may be no warning at all. But how do you know this? How do you decide to act on ambiguous information?"

Ambiguity was not an element in SAFE, which Rand developed in simpler times. Nor has ambiguity entered SAFE's modern version, Half-SAFE, which continues the traditional view of a simple, nuclear-spasm war in a political vacuum. Ambiguity spoils games, which are played by rules and governed by stylistic moves. The real world produces squishy problems. In the real world national leaders do not look at a game board; they look at whatever they can pull together in the form of advice and intelligence and then grope toward a solution.

The typical young analysts of today—the presumptive advice-givers of tomorrow—do not play games like Half-SAFE. They believe, as an article of faith, that if nuclear war begins, it will begin in ambiguity and will become far more complex than something that can be settled with a U.S.-U.S.S.R. missile exchange.

In the 1980s one such analyst, William Martel, along with retired Army officer Paul Savage, a professor of political science at St. Anselm's College in New Hampshire, developed a brink-of-war scenario for a nuclear war game that students could play. Soon after introducing the scenario at St. Anselm's and Harvard, Martel became an analyst for Rand. He is part of a Rand team developing an automated U.S.-U.S.S.R. war game (Chapter 16).

The Martel-Savage game was designed to teach students what to do in a crisis. "Does the other side have a set of instructions that says this certain action on my side will lead to some certain action on their side?" Martel asked during a discussion about the game. "In Vietnam we bombed selected areas in the North to indicate our resolve, but we were saying, 'We don't want you to think we're escalating. We don't want you to think we're pushing you into a corner.' Well, the North Vietnamese either understood the theory but didn't believe it or didn't understand the theory and still didn't believe it.

"Things are ambiguous. There's an explosion somewhere. Meanwhile, some military asset is lost. One side is up the escalation ladder on DefCon.* Do you attack? Does the first side know what's going on? Or is it taking an opportunity to go first? Or you scramble your bombers to protect them and the other side sees the scrambling and says, 'Ah-ha! That guy's trying to get his bombers out, trying to disperse his forces so that. . .' Well, maybe some of this is unresolvable."

In the game played at St. Anselm's, Martel and Savage, as Martel put it, "threw the students a series of uncontrolled events." One of the events, Martel recalled, "was an eruption in the Middle East, where the Israelis, in response to a Syrian nuclear attack, hit Odessa. Then the Libyans blew away part of Boston with a smuggled low-yield weapon. This pushed the students very quickly in the direction of a nuclear exchange and a Soviet need to respond. One of the students in the room had been sort of enjoying the whole process. Then, with a very serious look on his face, he said, '*Boston?* My family lives in Boston. Now I have no family.' "

Martel and Savage decided to make the war game at Harvard not quite as complicated or spasmodic as the one at St. Anselm's. At a briefing the night before, Martel had told the teams, "You'll be in a relatively serious situation. But nothing will be beyond your control."

But the game masters, who played Control in the games, were trying to teach lessons, not worry about delicate psyches. Martel knew the students needed "a series of catastrophic decisions." Unless they walked into a severe crisis in which the nuclear threshold already had been crossed, "the kids could play around for days and

*U.S. forces are kept at levels of alert designated as DefCons (Defense Conditions). Most forces most of the time are at the lowest alert level, DefCon V. Strategic Air Command bombers are usually maintained at the next level, DefCon IV. DefCon III increases readiness but does not indicate that war is likely. DefCon II is ordered when attack is believed to be imminent. DefCon II and DefCon III are signals, in games and reality, of severe crisis.

days and never have to come to decisions." At the Friday night briefing "they had no conception of how serious the scenario would be."

The Harvard game began at ten o'clock on the morning of Saturday, September 29, 1984, four days after the crisis had begun. "They had the Apple computers with the software right in front of them," Martel remembered. "People say, 'Well, sure. Let's drop a nuclear weapon on Chicago'—or Moscow or Leningrad. When they say that with the software, they get casualties, loss of forces. They get collateral damage. It adds tremendous realism to it.

"There was a high level of excitement in the room about fighting a nuclear war. After I gave them the copies of the scenarios, their emotions flip-flopped. Depression. Some of them were just staring blankly at the wall."

In the nuclear war game played at Harvard the U.S. and Soviet teams were given essentially the same scenario. The game assumes that both the United States and the Soviet Union are getting just about the same information on the crisis from their respective intelligence sources. The scenario used here is the U.S. team's version. Excerpts from the Soviet version are interleafed as italicized statements in parentheses whenever there are any substantial differences between the two scenarios.

On September 25, 1984, a Soviet Whiskey-class submarine (*on routine maneuvers*) is detected in Swedish territorial waters near a sensitive naval installation. The reported intrusion is one of several similar incidents over the past several years. [Western intelligence services assign code-letter designations to Soviet submarines; the letters are generally expressed as phonetic words, such as Alpha* for A-class and Whiskey for W-class. The Soviet Whiskey-class submarine, armed only with torpedoes, has diesel-electric propulsion, carries a crew of fifty to fifty-five men, and has a range of about 6,000 nautical miles.]

After detection of the submarine, Swedish naval units are placed on high readiness. Concurrently, Swedish authorities advise their ambassadors in Baltic and European nations to warn that Sweden will react strongly to this "provocation."

---

*Usually misspelled as Alfa.

U.S. satellites confirm that Soviet warships have left the ports of Murmansk and Vladivostok. (*Soviet satellites confirm that a U.S. fleet has left the ports of Norfolk and Portsmouth, Virginia*). Abnormally high levels of communication are detected between the Soviet Defense Ministry and the naval task force. (*Abnormally high levels of communication are detected between the U.S. Defense Department and the naval task force.*) Soviet Foreign Minister Andrei Gromyko denies a Swedish "allegation" that the submarine in Swedish waters is Soviet and urges "all parties to act with caution and restraint." The Soviet government says that Soviet fleet operations are "simply in response to Swedish actions."

On September 26 Swedish naval forces attack the submarine with conventional depth charges. But reports are that the submarine has not been destroyed. Meanwhile, U.S. intelligence detects large-scale Soviet naval operations in the North Atlantic and Baltic Sea. To counter this presence U.S. naval forces are alerted and a number of allied surface ships and submarines are sent to the areas where the Soviet naval activities are taking place.

Later on September 26 the Swedish Navy detects the submarine again and again attacks with conventional depth charges. The submarine fires a nuclear torpedo that detonates, sinking eight Swedish ships. (*Soviets suspect that the submarine commander is not mentally competent.*) Swedish casualties total 2,000 dead on the ships and ashore.

U.S. and NATO leaders express "collective outrage" at the Soviet action. (*East Germany mobilizes its reserve units. The Soviet Union prepares for full-scale mobilization.*) On September 27 and 28 there are incidents between U.S. and NATO forces and Soviet and Warsaw Pact forces in the Persian Gulf and Iran. The Iranians blockade the Straits of Hormuz. The United States responds by occupying the southwest coast of Iran to keep the straits open. Other U.S. forces enter key passes of the Zagros Mountains. Forward elements of the U.S. 101st Airborne and 82nd Airborne encounter elements of the Soviet Army that, after the U.S. action, have entered Iran at Iranian request.

There is fighting in four passes, and U.S. forces, cut off from supplies and reinforcements, are suffering heavy casualties. (*U.S. resistance is fading quickly.*)

On September 28 U.S. intelligence confirms that the Royal Navy has destroyed a Soviet reconnaissance aircraft as it approached a British carrier task force. Soviet naval forces retaliated by destroying several British aircraft and damaging a destroyer. U.S. intelligence also reports a large Soviet naval task force in the North Atlantic. U.S. attack submarines have established "positive trails" on Soviet missile submarines in the Atlantic. (*Soviet hunter-killer submarines have established "positive trails" on several U.S. missile submarines.*)

The U.S. and Soviet governments issue separate statements urging "all parties to act with care" to "avoid actions that will precipitate further reprisals."

At 5 A.M. (EST) on September 29 a Soviet attack submarine destroys a U.S. *Benjamin Franklin*-class Poseidon missile submarine. [This class of nuclear-powered U.S. ballistic-missile submarine (SSBN) carries sixteen nuclear-warhead missiles and a crew of some 165 men.] U.S. nuclear forces are placed at readiness level DefCon II, one step below a wartime state. Indications are that Warsaw Pact forces have also stepped up their readiness levels.

At 7 A.M. Warsaw Pact forces cross the West German border at Hof, Fulda, and Lauenburg south of Hamburg. (*"Indigenous political groups in West Germany requested Soviet intervention to assist in removing U.S. nuclear forces from their soil."*) The U.S. Seventh Army is driven back more than fifty kilometers, losing 20 percent of its forces and one third of its vehicles. The President gives authority for the release of tactical nuclear weapons. Fifteen low-yield weapons of twenty kilotons or less are launched against Warsaw Pact forward and rear elements. [A twenty-kiloton weapon's destructive power is equivalent to 20,000 tons of TNT, the approximate power of the nuclear bomb exploded over Hiroshima.]

U.S. intelligence determines that all Soviet nuclear forces are now on a high readiness level. Although a preemptive Soviet strategic attack is not considered likely, U.S. satellites and radar sensors are put in a state of high readiness. (*A preemptive U.S. attack is not considered likely.*)

At 7:20 A.M. a U.S. naval task force, including the aircraft carrier *Nimitz*, is destroyed by several nuclear weapons. U.S. satellites confirm the detection of nuclear explosions. U.S. casualties are estimated to be 30,000.

Shortly after the destruction of the task force, a U.S. attack

submarine, using nuclear-tipped rockets, destroys the Soviet aircraft carrier *Kiev* and several support ships. Soviet casualties: 25,000
dead.

At 7:50 A.M. Soviet nuclear weapons destroy air bases and ports
in West Germany and explode in forward NATO positions and rear
supply areas. U.S. nuclear forces go on a full wartime readiness
level, DefCon I. Soviet nuclear forces are placed on a similar level.

At 8:40 A.M. three Soviet submarine-launched ballistic missiles
hit three Virginia sites—Norfolk, Hampton Roads, and Portsmouth—where U.S. naval facilities are located. Nuclear weapons
explode at San Diego and Bremerton, Washington—West Coast
ports with Navy facilities. The United States immediately retaliates
against the Soviet ports of Murmansk and Vladivostok (*and Odessa
and Sverdlovsk*). There is heavy loss of life on both sides.

U.S. intelligence indicates that Soviet nuclear bombers have
been scrambled and large numbers of Soviet officials are leaving
the Kremlin. (*Soviet leadership is on a ten-minute alert to leave
the Kremlin.*) U.S. Strategic Air Command bombers are launched
as a precautionary measure against preemptive attack. (*Soviet
bombers are scrambled.*) The plane that serves as the National
Emergency Airborne Command Post is on ten-minute alert at Andrews Air Force Base near Washington, D.C.

U.S. satellites show that Soviet intercontinental ballistic-missile
fields are at full readiness and more Soviet missiles are being rapidly
moved to launch points. A large-scale Soviet nuclear attack against
the United States is anticipated. (*A U.S. nuclear attack against the
Soviet Union is anticipated.*)

All U.S. nuclear forces are on full wartime alert. Fifty percent
of U.S. ballistic submarines are at sea. (*All Soviet nuclear forces
are on full alert. Fifty percent of Soviet ballistic submarines are
at sea.*)

That was the crisis facing the Blue and Red teams, made up of
Harvard undergraduates, on Saturday morning. The teams separately contemplated the crisis until the lunch break, when one of
the players told Martel, "We're going to delay Armageddon now
while we have lunch." Soon after lunch it looked as if Armageddon
was being postponed indefinitely. "The thing began to defuse,"
Martel remembered, "and the kids were beginning to deal with

more or less mechanical things: 'How do we negotiate X? How do we negotiate Y? Let's put together a treaty of some sort, and what will be the talking points?'

"About three o'clock we told them we would terminate at four, so whatever happened at four o'clock was up to them. At four o'clock we gave them a short respite, and then put them all in the Soviet room. The two teams picked from their councils three members who would conduct peace negotiations for about thirty minutes. The teams were pretty recalcitrant. I had expected that there would be a coming together of Comrade and Friend. But it wasn't that. They had very strict talking points and very strict limits on the negotiability of the positions they had been given.

"This went on for about thirty minutes, and then we went into what I thought was the most interesting event of the game: the debriefing. What I basically did was moderate, saying, 'Let me tell you what we wanted to do and then you can tell us what we wanted to do, and then we'll hassle it out from there. My perception right from the beginning was that I was marked as the bad guy because the crisis was much worse than you people expected.' " Martel laughed as he remembered. "And in fact they *were* doing a lot of mumbling about being misled.

"I said, 'I wanted to put you people in the most serious crisis I could conceivably do. I wanted to put you on the brink, where either a nuclear response on some level or negotiations for termination was a reasonable consequence. The reality—even though you people may not have believed it during the game—is that Control could have forced you to fight a nuclear war whenever we really wanted to.'

"And they said, 'Oh, no. You couldn't possibly have done that.' And I said, 'Well, consider. Suppose I came to the Red Team and said, "We aren't supposed to tell you this, but we think it is in your interest to know that the United States has launched a full counterforce [military targets] attack on the Soviet Union. You people have five minutes to respond. And in ten minutes warheads will be landing." What would you do?' And someone responded, 'Well, I probably would have retaliated.' And I said, 'See. There you go. There's your nuclear war.' "

Martel said that in his experience with games, both those for students and those run for professional analysts, "People skid through

games without hitting the threshold. Even in the worst circumstances, in situations where Control is pushing as hard as it can,
teams avoid using nuclear weapons." And he added something I
had heard dozens of times, from civilian analysts, from retired
military officers, from officers on active duty: It is difficult for game
masters to get players to start a nuclear war. And, in Martel's words,
"military people tend to be more conservative than civilian people."

Perhaps that is why military gamers, unlike civilian gamesters,
tend to look at the small and bloody picture of the battlefield rather
than the big and exciting picture of nuclear war.

CHAPTER 9

Gaming Guerrilla War

The first formula for fighting against guerrillas in Vietnam came not from the Office of Systems Analysis in the Pentagon or from an analyst at Rand but from a French Army officer, Colonel Gabriel Bonnett, writing from the perspective of France's anti-insurgent warfare in Indochina in the 1950s. The formula was RW = G + P, "where revolutionary warfare [RW] results from the application of guerrilla-warfare methods [G] and psychological-political operations [P] for the purpose of establishing a competing ideological system or political structure."

In April 1964, six years after Bonnett published his formula—and at a time when Vietnam was beginning to obsess American policymakers—Alfred Blumstein of the Institute of Defense Analyses presented another formula to the Thirteenth Military Research Symposium in Washington. The source of Blumstein's formula was Ngo Dinh Nhu, brother of Vietnam President Ngo Dign Diem and head of the secret police.

Nhu's formula was TT + TG − TN, with TT representing ideological, logistical, and technological self-sufficiency; TG representing health, conduct, and creative initiative; and TN representing attributes of an obscure philosophy known as "personalism," which Nhu had discovered when he lived in France. These attributes centered on a respect for a person and were analogous to what became known as winning the hearts and minds of the people. Ngo Dinh Nhu and Ngo Dign Diem were assassinated in 1963 during a military coup.

Blumstein proposed that Western analysts try to develop guer-
rilla-fighting models derived from such formulas. The proposal was
inspired by a groping for a new set of Lanchester Equations for a
new kind of war, a war whose murky ways challenged analysts.
One of the earliest attempts was made by a group at the Douglas
Aircraft Corporation. A member of that group was Janice B. Fain,
a physicist who worked with her husband, William Fain. In 1965
the group was asked to look into what war in Thailand would be
like in 1972.

Janice Fain was still looking at ways to put war on a computer
two decades later when I talked to her in her office in Defense
Systems, Inc., one of the numerous Pentagon-supported think tanks
scattered around the Washington Beltway. "They wanted us to
investigate high-speed transportation of troops. We came up with
a hypothetical war," she recalled. "I was entirely innocent. I went
to the *Encyclopaedia Britannica* and discovered that Thailand had
historic invasion routes. I allowed my imagination to flow freely.
It was positively written and not intended as a game. No one had
inside military knowledge. We had a basic idea of how to put history
into a computer."

The history was what war gamers called synthetic history, and
the hypothetical war in Thailand in 1972 was both a look back at
the fighting of World War II and a look forward to prospective
fighting in Vietnam. Both the look back and the look forward saw
essentially the same kind of fighting. The Thai conflict model has
the Chinese invading Thailand, with casualties determined by
Lanchester-like formulas applied to such factors as the type of
terrain where a battle is fought and the type of units doing the
fighting. When casualties cut a unit down to a certain size, the
computer put the unit out of action; if defensive units suffered a
certain, disabling number of casualties, the computer sounded retreat.

The heart of the model was a timetable that showed what happened to Thailand if U.S. troops arrived at various times after the
Chinese invasion. The scenario with the happiest ending got all
the American forces to Thailand in fifteen days—borne in swift
transport planes presumedly built by Douglas—and resulted in a
Chinese withdrawal. In other Day D-plus scenarios, the Americans

got to Thailand later and later—presumably flying in fewer transports or slower ones not built by Douglas—and more and more
Chinese Communists got closer and closer to Bangkok and killed
more and more Thais. (Without any American rescue, Bangkok fell
on D-plus 26.)

The Douglas model lived on in many forms, Janice Fain said. A
NATO group adopted it and added air defense and air strikes to
it. The Canadian defense establishment used the model for reproducing tabletop war games on a computer. The Center for Naval
Analyses adapted it for amphibious warfare. "It was later picked
up by the Air Force, which modified it, called it Talon, and added
so much to it that it became unwieldy," she said. "So they threw
it away, and I lost track. A model is like a skeleton. It puts on
different clothes."

As irregular, unconventional warfare loomed on the American
military horizon, traditional wargaming suffered another setback.
Models, which are to war games what solitaire is to bridge, were
on the rise. Analysts at computers replaced military officers gathered around tabletops. For the study of the stealthy, unpredictable
world of guerrilla war, a model was especially attractive to what
one expert on such warfare called "the American intellectual theo-
reticians of order."

A model's ingredients could be manipulated and replicated by
the men of order, and its discoveries, unlike the verbal ramblings
of gamesters, could be neatly preserved. A veteran of the Pentagon's Joint War Gaming Agency, retired Army Colonel James Y.
Adams, went to the Stanford Research Institute, where he worked
on a guerrilla game that he hoped would combine men-around-a-
table gaming with computerized simulations. The games he envisioned would compare hypothetical combat with real data, primarily gathered from Vietnam warfare.

The Stanford project eventually was dropped, but the quest
continued for ways to comprehend antiguerrilla warfare. The rules
seemed to be changing in the real game of war, and so the military
bureaucracy, especially the Army, began changing the rules for
the simulating of war. Suddenly antiguerrilla warfare—more formally called counterinsurgency—was an idea that rated a band-
wagon and a salute.

"Some appreciation of the effect," wrote infantry Colonel Harry

Summers, "can be gained from such publications as the March 1962 *Army*, the influential publication of the Association of the US Army, which was devoted to (in its own words) 'spreading the gospel' of counterinsurgency." (*Army* reported that the Army Chief of Staff had questioned the White House-directed emphasis on counterinsurgency—and six months later had been replaced by General Earle G. Wheeler. It was a lesson in the unexpected career effects of antiguerrilla warfare.)

Good old Theaterspiel, which had been developed by the Operations Research Office at Johns Hopkins back in the 1950s, spawned a Cold War Model that modified military doctrines for use in guerrilla war and added new, nonmilitary subjects—economics, psychology, politics, and sociology—to a repertoire that previously had been strictly military. The most extraordinary aspect of the revised Theaterspiel was the way, even as early as 1964, it anticipated the attempts that would be made to quantify the chaos of the Vietnam War.

The political part of the model, according to the semiofficial history of operational research in that era, "recognizes the five power factors that contribute to the success or failure of insurgency warfare: geographic, military, economic, psychological-sociological, and political. The quantification of these factors is a major research problem, but the Cold War Model moved boldly into these areas and applied arbitrary units of measure based on informed opinions and limited data . . . in devising a gaming model for cold war."

Instead of a game built around the traditional military ideas of front lines and force-on-force contact, this updated Theaterspiel measured how Blue and Red won political control of population centers, particularly "strategic hamlets." The model looked at South Vietnam in terms of Blue and Red control, shown in a political-military spectrum: from *BB* (an area in which Blue has both supporters and an armed village) through *BS* (only Blue supporters), *BA* (Blue armed village), and *BC* (Blue controlled) to *NN* neutral, and then on through the enemy's *RC*, *RA*, and *RB*.

The idea in Theaterspiel's Cold War Model gaming was to change the *R*s to *B*s, maintain the Blue supply lines, and disrupt the Red supply lines. Winning or losing was based on the accumulation of points for such matters as how many nonmilitary personnel Blue and Red assigned to work at tasks that won hearts and minds. The work of these helpers—such as dentists (Blue 25, Red 0), school builders (Blue 900, Red 67), and teachers (Blue 30,600, Red 1,310)—was assessed not only in terms of how many there were but also the amount of time they spent in an area.

The transformation of Tacspiel, an old war game with rigid rules, into a much more fluid guerrilla war game, led to the addition of sixty-six pages of new rules to the previous Tacspiel manual of about four hundred pages. Players in the standard Tacspiel game, for example, moved an armored force at the company level across grids one-kilometer square, with each move made every thirty minutes (in game time). Guerrilla war players moved their units at the lower platoon or squad level and, because of mountainous or jungle terrain, movements were made at fifteen-minute intervals across grids a quarter of a kilometer square.

While researchers tinkered with the guerrilla-war version of Tacspiel, the Army was developing the idea of an aerial cavalry. To look at possible tactics for the AirCav, a game was designed using the 1965 guerrilla model of Tacspiel. The scenario had an air cavalry squadron supporting a South Vietnamese Army division in combat against a gradually increasing Viet Cong force in the central highlands. Reality soon produced an analogous scenario, and the director of a Tacspiel game in a 1966 report noted "the similarity of combat operations occurring in the central highlands of Vietnam and those occurring on the game board."

That same year war games were played in Vietnam to work out new tactics against the Viet Cong and the North Vietnamese Army. "Actual tactical changes, made as a result of these games," an Army report said, "later paid off in an impressive fashion." The report, which looked into the use of operations research in Vietnam, is studded with tables that sound like they came from Theaterspiel's Cold War Model: progress indicated through the measurement of such nonmilitary activities as the amount of U.S. labor on civic-action projects, the number of orphanages assisted, the number of patients treated by U.S. medical personnel, the number of bridges and the kilometers of roads built or repaired, and the number of hours that propaganda was broadcast over loudspeakers.

Operations research reports often mention the MOE, for the measure of effectiveness—a way to prove, or at least to indicate, that a model, or some aspect of a model, has actually worked. Since the model itself is an abstraction, the MOE is too. A typical MOE is *Pk*, shorthand for probability of kill. The Pk of a perfect weapon is 1.0: Every time the weapon is fired it works perfectly and kills what is aimed at. In a tactical war model the MOE may be expressed in, among other factors, the Pk of weapons and combat units (with the Pks sometimes based on *other* models designed to determine the weapon or unit Pk). To the despair of military operations research people updating Tacspiel, the MOEs of guerrilla war were as elusive as the guerrillas themselves.

"An obvious solution," said an account of Tacspiel's evolution, "is a system for upgrading or downgrading the effectiveness of each side in relation to the factors of guerrilla warfare. By this means terrorist acts such as the elimination (assassination) of a village chief, a teacher, or medic; the burning of an undefended hamlet; the destruction of crops or livestock; and the abduction of village males or females are actions that could be built into expanded models." The account does not make clear whether these intriguing MOEs ever were put into Tacspiel.

During the war the talky, old-style war games went on in the Pentagon, but the Joint War Games Agency, which was running the games, saw the need for something more up to date. The agency funded an enormous computerized model called TEMPER (*T*echnological, *E*conomic, *M*ilitary, *P*olitical *E*valuation *R*outine). Scant public information about Temper emerged, but much about it became known in the gaming community.

Temper was an incredibly ambitious undertaking—"an effort literally to model the world," one report said. The model was packed with information on 117 nations—from a country's population and gross national product to the size of the military forces and the level of citizens' patriotism. Temper researchers assessed each nation's politics, culture, ideology, psychology, science, economics, and military establishment and assigned each category a numerical value. These categories, reduced to numbers, became the basis of the nations' foreign policy.

The Temper nations, as "actors" driven by psychological, ideological, or cultural motivations, played their roles on a computerized planet divided into twenty "conflict regions." Actor-nations joined power blocs, formed coalitions, and based peace or war decisions on such cues as ideal sensing, reality sensing, and ideal-to-real discrepancy measuring. In Temper's world there could be wars all along the scale, from guerrilla to nuclear.

Every national decision was literally a measured response, based on calibrations—the cost effectiveness of a weapon needed for an attack on another nation, the trade-off between the purchase of military hardware and the spending of tax money on social programs, the choice between spending defense funds on research and development or on more weapons already in production. The originator of Temper was Raytheon, one of the nation's major defense contractors.

Temper could simulate a decade of international relations in about half an hour. But Clark Apt, principal designer of Temper, wanted people to be able to play a high-speed version after a few minutes of training. "What we would like someday," Apt was quoted as telling an interagency group on strategic studies, "is a system with fifteen buttons, five for each of the military political functions. One button might control the variable the user wanted to operate on, another might control the geographic regions, and so on. With a few dials and a map display it might then be possible to represent most of the complexities of the model."

Apt's automated wargaming, seemingly improbable in the 1960s, anticipated the computerized nuclear warfare gaming system that Andy Marshall sought—and got—in the 1980s (Chapter 17).

Temper, picked up by the Industrial College of the Armed Forces in 1966, was eventually abandoned by the Joint Chiefs' gamers after two outside contractors gave sharply differing evaluations of Temper's potential as a useful gaming device. One problem was that its computer programs stepped over the line from objective to subjective judgments. The "range, accuracy, and reliability of missile systems can be ascertained with reasonable levels of confidence," a critique said. "The will to use missiles and the circumstances under which they will be used cannot be determined with nearly the same confidence level."

But Pentagon faith in computerized gaming models persisted. In the 1960s, as now, the Department of Defense's Advanced Research Projects Agency, by calling for bids on certain studies, produced waves in think tanks and signaled new needs to the defense industry. ARPA* specifically wanted a modern, computerized approach to fighting guerrilla wars. In 1964, when ARPA put out a request for a computer model that simulated "major aspects of internal revolutionary conflict," Temper's designer, Clark Apt, responded. Apt had formed his own think tank, Apt Associates, in the academic talent pool of Cambridge, and, shifting his talents from global affairs to guerrilla war, went to work.

Apt's researchers determined that the most important elements in defeating counterinsurgency were "information," "loyalty," and "effective military force." Apt used the struggle for these elements between government officials and insurgents as the basis for a game at first called Agile-coin (counterinsurgency) and then shortened to Agile. In the first run of the game Apt researchers playing government representatives visited six "villagers," one by one, and offered each one food or security in the form of playing cards symbolizing soldiers, food, and offers of harvesters. The villagers were also visited by people playing insurgents, who, to win, had to get and keep the loyalty of four villagers in three consecutive moves.

Apt abandoned this version because it lacked the pressure of terror. In the next version the search for loyalty through food was changed to a demand for loyalty through threats of terror by the insurgents and promises of protection by the government. The game evolved through fifteen iterations, each adding more complications, such as a need to train villagers to fight and the possibility that the guerrilla players could assassinate government administrators.

"By the fifteenth play," wrote British military correspondent Andrew Wilson, "the game had moved outdoors and messages were delivered by a courier system that was supposed to introduce the kind of delays that might be found in the field. There were also provisions for conflict within villages; military engagements with different degrees of surprise; the calculation of casualties; ambushes; counterambushes; the making of propaganda." A switch in loyalty was "monitored by means of 'village report forms,' filled out after each move and containing questions such as 'Who do you

*Now called DARPA, for Defense Advanced Research Projects Agency.

want to win?' 'Who do you *think* is winning?' and 'How do you estimate the loyalties of the rest of the village?'" This manual version of Agile was used for training members of Special Forces, many of whom would play a real form of the game in South Vietnam.

Apt next developed a computerized version of Agile that tracked guerrillas by using computer programs:

Belligerent move start >

—Belligerent makes list of villages to be visited and apportions troops according to intended actions >

Approach village that is next on the list >

Village decision whether to warn belligerent of ambush >

Ambush? >

Yes >

Engage forces, calculate losses [*or*]

    No >
    Village decision to resist/argue/accept entry of belligerent

Ultimately the belligerent had to decide what steps could be taken to "change village loyalty": *Terrorize. Bribe. Communicate.*

Agile and Temper signified a shift in classic approaches to fighting war—and to gauging the effectiveness of new ways of fighting. Like biological changes that evolve and die out but point to what will come, Agile and Temper anticipated a movement toward the rationalizing of warfare. The Vietnam War would become a laboratory for systems analysis, a laboratory that acquired data about the way soldiers and civilians acted under the stress of a war without front lines. Ideas engendered by Agile-style research eventually found their way into the villages of South Vietnam.

Rand researchers, deployed across South Vietnam to analyze the war, reported on the effects of bombing, on the morale of the Viet Cong, on the village pacification programs, and on the need to step up psychological warfare, or what a Rand researcher called a "psywar weapons system." A Rand report on Viet Cong morale, based on interviews of 850 captured Viet Cong, defectors, and refugees, contains concepts quite similar to those that inspired the Agile games.

"A central question in this analysis," the report says, "was the influence of friendly operations on the Viet Cong's will to fight or willingness to obey orders." Just like the Agile game visitors with their food and soldier cards, Rand interviewers wanted to know whether the "friendly operations" of South Vietnamese soldiers had any effect on Viet Cong food supplies: "Did the interviewee ever go hungry while in the Viet Cong?" And the analysts wanted to know whether harassment by military forces and by bombing inspired a shift in loyalty: "Did the interviewee indicate that 'hard life' in the Viet Cong as a reason for defection?" A Military Pressure Index measured how morale held up when food supply was disrupted (a 0.28 correlation) and when threatened by ambush (0.00 correlation).

The Simulmatics Corporation, which had evaluated Temper for the Pentagon and had compared it in importance to the Wright Brothers' first flight, was hired by ARPA to study the villagers of South Vietnam. Simulmatics found them the "social theater of operations" for guerrilla warfare.

Interviewers showed villagers a drawing of a ladder and said, "Let's suppose the top of the ladder is the best possible life for you . . . and the bottom represents the worst possible life for you." The villagers were then asked what rung they thought they had been on five years previously, what rung they saw themselves on at the time of the interview, and what rung they expected to be on five years hence. The investigators also found, from questioning the villagers, that they dreamed little about sex but did often dream about what the reports called aggressive social interaction.

From about the late 1960s to the early 1970s the Pentagon's systems analysis office published "The Southeast Asia Analysis Report," in which MOEs were used to show progress in such matters as the anti-infrastructure campaign and the hamlet evaluation system. Variations on Lanchester's equations were used to analyze Viet Cong ambushes, but, since those equations depended upon "constants" not found in the Vietnam War, the data on ambushes did not prove to be very useful.

In McNamara's Pentagon the war seemed to be a war of numbers and charts. There were charts on defectors, on the enemy-friendly kill ratio, on South Vietnamese weapons losses, on incidents of sabotage, terror, harassment. And there were, week after week, the body counts.

"People have criticized me for stressing this very brutal concept of body count," McNamara said in 1984. "It was, in a sense, a terrible thing. But if you're Secretary of Defense and you're concerned about whether you are progressing militarily, and it is said to be a, quote, War of Attrition, unquote, in which ultimately the friendly forces would prevail by attriting the enemy, and you attrit them . . . then it is important to try to understand whether you are accomplishing the attrition or not. . . . And that is why I put as much emphasis as I did on understanding what was happening on the infiltration routes and understanding on the battlefield, i. e., the body count. And I became very skeptical of both."

McNamara said that, for the first time, as a witness for General William C. Westmoreland in the general's unsuccessful libel trial against CBS, which had charged that there had been a conspiracy about the count of enemy forces in Vietnam. McNamara conceded that infiltration estimates could be off by at least as much as 50 percent. He also said that the "figures relating to the Vietnam War, whether it was rice exports from the Delta to Saigon, or killed in action, or level of infiltration, or enemy strength, or whatever it might be, were in constant controversy."

But if the numbers were in controversy and often wrong, what of the analyses based on them? Can a war be modeled—or conducted—by the numbers? "There are two views of war," Janice Fain said, looking back over more than twenty-five years of modeling combat. "You can regard it as a human activity carried out through the use of certain tools or as an interplay of mechanical, electronic systems operated by fallible human beings. You are in firmer control when you are on the side of analyzing mechanical and electronic interactions. I was nearly thrown out of a conference when I said that models do not increase knowledge. Human beings gain knowledge by experience and their own senses."

A postwar analysis that was itself an analytical spiral—an analysis of the systems analysis office's monthly report on how analysis was being used in the war—showed that there usually was not much to report about the use of analysis in combat. Most of the topics covered in the monthly reports "lay outside the scope of traditional military operations," according to the analysis of the analyses by Clayton Thomas, a former president of the Military Operations Research Society and a veteran specialist in Air Force operations research. "For this kind of struggle," he wrote, "there existed fewer combat models."

There is an eerie air about these reports on reports, these analyses of analyses, these orderly models of disorderly worlds. It is as if the dispensers of these abstractions know something, but they cannot measure it and they cannot describe it. They may be seeing, as Janice Fain did, that models do not increase our knowledge of the real world we all live in.

Thomas, pondering the paucity of models and operations research during a war, found this peculiar truth: A real war blots out attempts to model war. "With a 'war on,' " he wrote, "there is less incentive for, and less interest in, the investment of great effort in building a massive computer model to 'substitute' for the war. On the one hand, the war itself serves as an 'overall model,' a collection of 'test ranges,' etc. On the other hand, the complexity of a war; the variety of operations; the quality of the data (real but often 'dirty,' partly as a result of enemy concealment and 'disinformation'); and the obvious importance of leadership, morale, courage, and luck all serve to remind us vividly of how hard it is to model combat on the grand scale."

If models could not help to solve the mysteries of the Vietnam War, there was always human experience to fall back on. In the basement of the Pentagon the Joint War Gaming Agency was still in business, with men gathered around a table talking about how Blue and Red should fight a war.

CHAPTER 10

███████████

Vietnam's Prophetic Scenarios

Sigma II-64 is more than a top-secret war game. It is a piece of the Vietnam War puzzle that fits precisely between the Tonkin Gulf incident and the sending of U.S. combat troops to Vietnam. The officials who played in Sigma II-64 knew that many of the moves they made in the Pentagon's Room BC942A would soon be repeated in Vietnam. And at least one policy issue pondered in the game would hover, unseen, over the war: Blue used nuclear weapons against Red.

The scenario for Sigma II-64 was a vision of the long war to come. On tranquil September days in the Washington of 1964, players grappled with make-believe problems that would flare into reality in the future. And in the Saigon of 1964 there were Sigma II-64 game problems that were already grimly real—the tottering South Vietnamese government, the growing numbers of North Vietnamese troops in South Vietnam. Documents from the game show that many future problems were accurately perceived: the inevitability of escalation, the futility of bombing, the lack of broad domestic support, the need for the President to order at least partial mobilization. Sigma II-64 showed that even in a game the architects of the Vietnam War could see that they were attempting to create the improbable. And yet they played on.

Sigma I-64, played in April 1964, was a prelude for Sigma II-

64. The first game revolved around a covert "tit-for-tat" concept: The United States, in a strategy not publicly revealed, would respond with ever-increasing military pressure to North Vietnam's military activity against South Vietnam. One of the U.S. ratchets of response was a secret bombing campaign. (U.S. Air Force "training" missions, with South Vietnamese crewmen aboard, were already being flown when the game was played.)

The summary report of Sigma I-64 sound more like a warning of what to avoid in future reality than an account of a war game. If "the cover" for the U.S. bombing is "so thin as to resemble cynical aggression," the report says, "we may encounter severe problems in the UN and before the bar of world public opinion." And a "significant vocal proportion of US opinion may join in the hue and cry." Looking beyond Southeast Asia, the anonymous report points out that the "same principle" of U.S. intervention in South Vietnam "would lay a basis" for "direct" U.S. action "against Cuba for its subversive activities in Bolivia, Venezuela, Columbia [sic], etc. . . . The Soviets, CPR [China], and Bloc might be expected to oppose such an attempted change in the 'ground rules' of Cold War with all the means at their disposal and with great tenacity."

In the game, which had the artificial starting date of June 15, 1964, a U.S. flier is shot down and "the cover" is exposed. The incident, the report says, "heavily influenced" world opinion. (In amazingly coincidental reality a Navy pilot *was* shot down in June 1964, was taken prisoner by the Pathet Lao, and the growing, direct American involvement in Southeast Asia could no longer be denied.)

Discussing the game's covert bombing campaign, a member of the Yellow Team (China), says, "It would be fun to see or try to get some analysis of what the country would look like to a North Vietnamese if every single target that could be taken out by whatever means we have to take them out were, in fact, taken out. . . ."

Blue wonders about getting "these people to go on a long march." Amid laughter, someone adds, "Right back to Yellowland."

In a critique of the game, Seymour Weiss, Deputy Under Secretary of State, wrote, "The eventual capture of a US airman is a high probability and would give 'hard' evidence of US involvement (if US press coverage from SEA [Southeast Asia] doesn't disclose the thinly covered US participation even before such a capture). The notion that US pilots are only in a 'training' capacity when B-57 aircraft are on a bombing mission, over NVM [North Vietnam] and against the industrial complex, could not be made persuasive."

Weiss concluded his Sigma I-64 critique by noting, "The world has been accustomed to the 'cold war.'. . . Both sides are assumed to conduct such operations within the rules of the cold war game. Exactly the opposite is the case with regard to 'overt' aggression. The use of regular military forces . . . fulfills the established and accepted criteria of war." An overt war against the Viet Cong, Weiss wrote, would be waged "without convincing justification" because what the Viet Cong were doing "falls within the accepted bounds of cold war activity."

On August 2, 1964, convincing justification was found. On that day North Vietnamese torpedo boats made what the U.S. Navy called an "unprovoked" attack on the U.S. destroyer *Maddox* in the Gulf of Tonkin, a bay whose waters lapped the North Vietnamese ports of Haiphong and Vinh and the Chinese island of Hainan. The *Maddox* and another destroyer, the *Turner Joy*, were affirming the Navy's right to sail in the international waters of the gulf. And the *Maddox* was electronically eavesdropping on North Vietnamese military radio traffic. Planes from the U.S. aircraft carrier *Ticonderoga* attacked the patrol boats.

A second PT-boat attack on the two destroyers was reported on August 4. Although there was a question whether this attack ever took place, President Johnson, in the midst of an election campaign against Senator Barry Goldwater, went on television late on the night of August 4, reported the two attacks, and announced that he had ordered a retaliatory air strike against the patrol boats' ports.

On August 7, with a unanimous vote in the House and an 80-to-2 vote in the Senate, Congress passed the Southeast Asia Resolution (usually called the Tonkin Bay [or Gulf] Resolution), empowering the President, whom Goldwater had been calling soft on defense, to "take all necessary measures to repel an armed attack against the forces of the United States and to prevent further aggression."

The resolution committed the nation to war in Vietnam. Sigma II-64, staged a month after the passage of the resolution, made recommendations on how to begin that war.

Sigma II-64 differed in design from typical games produced by the Joint War Games Agency. Besides Control, there were Blue and Red Action Teams and Blue and Red Senior Policy Teams. The action teams began the game on Tuesday, September 8. Each team was assigned to one of the two game rooms formed from the large Room BC942A in the Red Seal Area of the Pentagon, so named because it was sealed for security reasons at the end of each day. The action teams considered the scenario and policy questions presented to them at the first meeting and later briefed their respective senior policy teams. The game was to end on Thursday, September 17.

The policy teams, which consisted of high-level civilian officials and military officers, were asked to consider "the major political and military questions that should be answered prior to making the decision to commit substantial contingents of US Armed forces to combat in Southeast Asia." The players were to determine whether the Commander in Chief of Pacific Forces should be authorized "to employ tactical nuclear weapons in the event of overwhelming attack"; whether a partial or total U.S. war mobilization was going to be necessary; and whether Chiang Kai-shek's army on Taiwan should be sent into combat on the Asian mainland.

Policy team players were drawn from the State Department, the Central Intelligence Agency, the Department of Defense, and the U.S. Information Agency. The military officials included representatives of the U.S. Military Assistance Command in Vietnam, the Joint Chiefs of Staff, the Pacific Command, the Defense Intelligence Agency, and the Office of the Secretary of Defense. Action team members came from all of those sources plus the White House. Team members were "familiar with the Southeast Asia objective, policies, and directives of their respective agencies," along with "the number and type of personnel" and "financial outlays" being made in the area.

In the still-censored records of Sigma II-64 the identities of Blue and Red and Director are blacked out, as are all the names of the team members. (One page that emerged unsanitized is full of anonymous doodles: a football player, an Asian man with a star on his chest and his right index finger and thumb turned upward, a person bound with rope and grazed by an arrow labeled ZULU, a cartoon of a startled-looking, high-domed man labeled Political-Military Gamer.)

Some game records have been declassified with little sanitizing. These records contain the names of the some of the officials invited to Sigma II-64, and by strong inference indicate that they did play. Members of the Blue Senior Policy Team were McGeorge Bundy, White House adviser on national security; Assistant Secretary of State William Bundy, Air Force General Curtis E. LeMay, Commander of the Strategic Air Command; John A. McCone, Director of Central Intelligence; Assistant Secretary of Defense John McNaughton; Admiral Horacio Rivero, Jr., Deputy Chief of Naval Operations; Deputy Secretary of Defense Cyrus R. Vance, and General Earle G. Wheeler, who had become the chairman of the Joint Chiefs two months before. The only Red Senior Policy Team member named in the records was Ray S. Cline, CIA's deputy director for intelligence. McGeorge Bundy, McCone, Wheeler, LeMay, and McNaughton had all been in Sigma I-64.

Senior team members met at five o'clock on three afternoons (September 10, 15, and 17) to be briefed by their respective action team each day. Each policy team, working from the "scenario projection" that Control produced following each move by the action teams, had to "provide policy guidance" to its action team. Control played the unusual role of national-command level authority for both Blue and Red. Manipulating both sides from on high, Control later admitted that, sometimes playing through the mind of Chinese leader Mao Tse-tung, it had pushed Red farther by forcing the Red team "to draft a plan for overrunning SEA with ground forces."

In the game dialogue, nameless *Blue* is clearly the United States acting alone, although South Vietnam is supposedly part of the Blue Senior Policy Team and Blue Action Team. Red, officially North Vietnam, the Viet Cong, the Pathet Lao, and the People's Republic of China, appears to be dominated by Chinese, rather than North Vietnamese, attitudes. This reflected the U.S. expectation that China would be a major player reacting to the real-world American actions in South Vietnam. Both Blue and Red are representing governments "except for highest executive authority."

"Blue" and "Red" are obviously more than one person, as can be seen when, in a postgame discussion, one Blue chided another

Blue about plans for an amphibious landing in Vietnam: "If you're going to land some troops, there's no use fooling around with Vinh, you rascal. Go into Haiphong and Hanoi." Vinh, a port that often tempted amphibious warfare tacticians during the real war, frequently was invaded in war games.

Background information provided the action teams included a chronology of Vietnam history going back to 111 B.C. and a recent *New York Times* article that said North Vietnam was "sending more and more regular army members into combat beside the Viet Cong" and that the United States had announced that it "intends to enlarge its advisory force in South Vietnam by an additional 5,000 men, and more military equipment." The player packet also contained long, detailed instructions about Pentagon parking regulations.

In game time Sigma II-64 started on April 12, 1965. *During the early fall of 1964*, the scenario began, *the number of RVN* [South Vietnam] *air, naval, and ground raids on targets in NVM* [North Vietnam] *mounted with mixed results. . . .*

North Vietnamese reinforcements had raised the number of Viet Cong in South Vietnam to 40,000. China had established a military mission in Cambodia and increased its support to North Vietnam and to the Pathet Lao in Laos. Chinese weapons and technicians strengthened North Vietnam's air force and air defenses. Road building was stepped up between the Chinese border and northern Laos.

The political situation in South Vietnam was shaky. Premier Nguyen Khanh had accomplished little to stiffen the war effort. In Paris, President Charles de Gaulle was asking for a Southeast Asia peace conference. China was attempting to influence Japan to curtail ties with the United States, implying that the Japanese Communist Party could interdict the U.S. line of communication from Japan to Southeast Asia. In Burma the Chinese were attempting to coerce the Ne Win government into a policy of passive cooperation.

More and more U.S. troops were being killed or wounded in South Vietnam; more advisers and Special Forces were being sent in. Chiang Kai-shek's offer of a force of up to three divisions was temporarily rejected. U.S. and South Vietnamese air and ground strikes hit the Viet Cong in Laos.

In November announcements of joint Chinese–North Korean maneuvers were followed by a series of division-size exercises south of the Yalu River. During the winter of 1964–65 the Chinese introduced Migs, flown by "volunteer" fliers, into Laos and North Vietnam. Three North Vietnamese Army brigades infiltrated into the six northern provinces of South Vietnam.

On Christmas Day the Viet Cong shelled Saigon. In January the National Liberation Front established a provisional people's government, which was recognized by all Socialist countries.

The U.S. Seventh Fleet mined Haiphong and other harbors; a Polish freighter sank in Haiphong, apparently after hitting a mine. In late February a U.S. destroyer entering Saigon harbor was sunk by a North Vietnamese mine.

On February 26 the President announced the debarkation of a Marine Expeditionary Force and the establishment of a permanent U.S. base at Danang, together with the planned airlift of an Army brigade to Thailand. He also announced that in the future the United States would strike selected industrial and military targets in North Vietnam in retaliation for major Viet Cong terrorism and sabotage in South Vietnam.

The Soviet Ambassador to the United States stated that the Soviet Union had no alternative but to provide moral and some matériel support to the North Vietnamese.

On March 15 the Soviet Union and East Germany announced that they would sign a peace treaty in May; East Germany announced that it would assume full control of the access routes to Berlin on May Day.

On April 1 Buddhist rioting in South Vietnam precipitated a sharp division among the military and there were demands for Khanh to resign.

Confronted by this scenario at the start of the game, Blue was asked to make the first move after considering the following proposals:

- The President announces that the "defense of freedom" in South Vietnam was "a shared responsibility of all free peoples, that Saigon and Berlin be declared co-equal symbols of Western determination and that the United States was undertaking partial mobilization."

- The "immediate and rapid deployment" of logistical support units to Thailand and South Vietnam "to prepare for further combat deployments."

- The "immediate alert of the 11th Air Assault Division" for deployment to Southeast Asia.

- "The authorization of CINCPAC [Commander in Chief, Pacific] to use limited tactical nuclear weapons to preclude the destruction of major US or friendly forces in an emergency."

- Development of this objective: "to compel the enemy to cease support of insurgencies, to assist local forces as necessary in the elimination of the insurgents who thereafter persist, to reunify Vietnam and to achieve the independence and security of friendly nations in the area."

Blue decided to openly state its principal objective as the maintenance of "the independence and security of allied and neutral nations in Southeast Asia," but Blue secretly decided to "accept a divided Laos." Blue dispatched a U.S. Army task force to Saigon to assist the faltering government there, began moving three divisions plus three Air Force squadrons to Thailand, and set in motion the deployment, within sixty days, of an Army airborne brigade and an infantry division and a full Marine Expeditionary Force toward South Vietnam. Two carrier groups were dispatched to the western Pacific. The Pacific Commander in Chief's request for "delegation of authority for use of tactical nuclear weapons" was denied. (The game commentary remarked that Blue's actions "closely parallel those in existing contingency plans.")

Red planned to avoid direct confrontation with the United States while "collapsing the South Vietnamese political base" and pressuring the Thai government to lessen its support of the United States. Red also moved a Chinese infantry division and support forces into North Vietnam and alerted three more divisions for possible deployment in North Vietnam. The Viet Cong shelled airfields at Danang and Bienhoa, destroying a dozen warplanes.

As the game progressed, it began to conjure up a prophetic sense of what the real war would be like. A few Blues, because of "frustration with the chaos" in South Vietnam, stepped up the bombing of North Vietnam, discovered it had little effect—and then argued with Control's appraisal. "Control accepted the comments philosophically," said the game commentary.

"Neither the stoic attitude of a regimented oriental population with its philosophic view toward life and death, nor the racial undertones of war in which US pilots kill women and children was overlooked in estimating the effect of a major bombing program," the anonymous commentary added. "It was also recognized as most unlikely that US game participants had a real insight into probable reactions of a country where self-immolation is practiced to 'sell' a point." (This was a reference to the form of protest used by anti-government Buddhists who staged protests by setting themselves on fire before television cameras.)

Blue moved quickly to punish the North Vietnamese, wiping out every industrial target that Control's data base had supplied and, in a demonstration of resolve, calling up more than six divisions (about 90,000 men). Under the real U.S. general war plans then in effect, this could be done only through a presidential declaration of a national emergency and at least a partial mobilization. "A question was raised regarding how long the American public, with the current flavor of press coverage, would support the commitment of ground forces in such strength," the game commentary says. But Blue plunged on, commandeering civilian airliners and commercial merchant ships to get the ever-increasing cargoes of men and supplies to Southeast Asia.

The frenzied escalation was expected to frighten Red enough to stop the buildup in South Vietnam. But Red, unimpressed, did not even mobilize. Red said the North Vietnamese population "has stoically assumed that the worst is still to come and continues to improve and expand passive air defense facilities and organization while propaganda agencies make the most out of photographs, eyewitness reports and editorials on barbaric imperialist air strikes."

Red's anti-American language is not commented on, but there is no doubt that it annoyed Blue and sometimes enflamed Blue's patriotism. John McNaughton, a player in Sigma I and II, was known as a Blue who carried his real-life attitudes out of the game room. Once, when a State Department official admitted pessimism about the war, McNaughton disdainfully said, "Spoken like a true member of the Red Team."

The question of just how Red was Red did come up in the postgame wrap-up. "I feel," a Blue said, "that the Red Team understood most of the things that we did and when we tried to communicate something to them, they read our communication and interpreted it exactly the way we wanted them to interpret it. I would present for your consideration, however, they all speak English. They went to the same schools we went to. They have the same background that we have, and the same pressures on them." The same question would haunt the real war. "Signaling," a concept that had emerged from political-military gaming, was American-to-American in games and cross-cultural in reality. American signals did not seem to translate into Vietnamese.

In the game, as in the war itself, Blue had no fear that U.S. actions in Southeast Asia would involve the Soviet Union and result in a confrontation or a war between the superpowers. Control did help to keep the countries at a safe distance from each other; the two merchant ships sunk in a mined Haiphong Harbor, for example, were Polish ships, not Soviet ones, and Soviet threats about cutting off Berlin never reached the crisis stage.

Control also seemed to design events so as to keep out of the game both great hordes of Chinese troops and the nuclear weapons Blue had threatened to use against them. The commentary cautiously said that decisions in the game "indicated that present policies in contingency planning are correct; i.e., plans should be militarily and logistically feasible based on no use of nuclear weapons, but with alternate plans which assume authority for their use."

After the initial Blue and Red moves the scenario was projected forward to April 15. By now it was September 14 in the real world, and Sigma II seemed to have a reality of its own, for though McGeorge Bundy was enmeshed in the real problem of another aborted military coup in Saigon, he still showed an interest in Sigma II, and game officials saw to it that a report on the projection was sent to him in the White House. The White House frustration over the balky, squabbling leaders in Saigon was being reflected in the game, where Blue was insisting that the South Vietnamese government "must be amenable to U.S. control.")

The projection showed that U.S. planes had bombed airfields at Haiphong and Phuc Yen; U.S. warnings of the possible use of nuclear weapons in Korea had leaked to the world press and caused an uproar in Japan; the Viet Cong had shelled the Marine airstrip at Danang; and Haiphong Harbor had been mined.

Blue increased the intensity of the bombing, knocking out North Vietnamese railways and bridges, leveling Vietnam's cement plants and the country's only machine-tool factory. The bombing brought no decrease in Red infiltration of South Vietnam.

Then, on May 24, on a highway about ten miles west of the Laotian town of Tchepone, an American battalion task force was ambushed. "It was a disaster," the commentary said. "Approximately 350 men were killed before the survivors were able to regroup and break out, leaving the bulk of their wounded and many weapons. A relief force found out that the enemy had murdered approximately 25 of the US wounded and had desecrated their bodies."

The game formally ended with Blue engaged in full-scale war planning: an invasion of North Vietnam, with amphibious landings at Vinh and near the Haiphong-Hanoi area; a confrontation with Chinese troops in northern Laos; and "concurrently with confrontation, to bomb selected industrial and military targets in China." The Blue Team "did not reach consensus" on whether to bomb China's nuclear facilities* with nuclear bombs and "execute a general nuclear attack against principal targets in Communist China."

The Director asked whether Red had anticipated that Blue would contemplate the use of nuclear weapons not only against Chinese troops in Laos but also against China itself. "We had a general estimate of U.S. policy that they would be reluctant to cross the nuclear threshold," Red replied. "We had quite a bit of room for maneuver. . . . Our whole effort has been to try and stay out of this thing. Present no targets, present no excuse, present no provocation or justification. Without adequate justification, Blue actions have raised the international sound level and pressures for international conferences. . . .

"We were surprised that the Americans were capable of com-

*U.S. intelligence knew that the explosion of China's first nuclear bomb was imminent. The bomb was exploded a month later, on October 16.

mitting such heinous crimes. Their persistence did cause us a little
bit of worry, particularly when the Americans on the second day
were willing to go after our cities the way they did without any
justification. That did raise some concern that they might do this
against China. This was the one thing that gave us cause to worry.
We were not concerned about nuclear war."

Blue, referring to Red as "Father Ho," asked about Soviet intervention, pressing Father Ho to say whether he "really believes
that the Soviet Union would risk the destruction of the Soviet Union
in order to preserve Red China, North Vietnam, or any other
conceivable area that you could mention." Red said that "Ho doesn't
control this thing any more. We've started moving Chinese divisions in there and I think he's pretty well our puppet. . . . And
. . . we're Mao—we're Chinese."

"The situation in South Vietnam seems to be a real tease," the
Director admitted, adding that without "some success in the . . .
immediate future—we have no reason to be supporting South Vietnamese." The North Vietnamese "can continue, under the cover
that they now have and which is more or less accepted by the
world, all of the actions that they are now doing and it is probably
not going to get anybody but us upset. . . ."

But the real tease in South Vietnam was indeed real. A national
commitment was rapidly building in the world outside the game
room. Events played as moves in the game were already in motion
as real moves by a real nation called the United States, not Blue.
The Director, in fact, made the game sound like an advance rati-
fication of what was about to happen. The game, he said, was staged
to "find out if our plans were reasonable. Can we get the forces
moving that we have written down and studied? The group of forces
that Blue deployed here is not actually in plans, but in magnitude
is very close to the magnitude of the deployment that is planned."

The game had shown "that it will take a declaration of national
emergency or its equivalent to generate all the forces that would
let this thing be done. The question is, what is going to have to
happen—how bad is the situation going to have to be, to justify
such a thing?"

Later he raised the issue again. "The question that I would like
for someone to discuss is will we ever be in a position where a
decision would be made to get this many U.S. forces going over

there without a direct act on the part of the other side, where we could say, 'All right, you crossed the border with X number of divisions—here we come.' In other words, was our initial scenario completely beyond belief?"

After some rambling discussion, Blue said, "We had 350 troops killed yesterday in one engagement. That's a sizable fracas. Multiplied by four, then the whole damn battalion would have been wiped out."

"Now what I suggest," Red interjected, "is that precisely this 350 troop killing is exactly the kind of thing that would trigger the next step."

"Right," Blue said.

"You're there. You're committed," Red continued. "Your honor is at stake. Now you've got to do something."

Again, Blue joined reality to the game. "I think," he said, "that the kinds of things that we have been planning to do, and in real life plan to do, will in fact seriously disrupt the capability of the North Vietnamese to do all the things that they are doing now or that they plan to do. And I think that they would stop. If we do not dry them up completely we will just create so many problems for them that they will have to really reduce their activities in South Vietnam instead of intensifying them. . . ."

"Since the game [Sigma I-64] was played," the Director remarked, "reconnaissance has given greater knowledge of what's to be hit, and there are some appraisals of what such hits would do. And it would hurt them. There are no two ways about it. . . ."

The Director said that Blue's decision to send combat forces to Vietnam and Thailand "does bring up the question" whether "the United States must have specific acts of provocation to put such a movement in force. . . ."

"Or something like the Gulf of Tonkin again?" Blue asked.

"Yes," the Director replied, "a Gulf of Tonkin incident can be parlayed into something that would require a national reaction."

"Would there be a move to impeach the President . . . ?" Red asked.

Reality and game talk began to mingle when Blue referred to the Southeast Asia Resolution. With the resolution, "and with sufficient provocation," Blue said, "he [President Johnson] could start to move everything that the Blue Team has moved here."

"And more," a player added.

The scenario was incredibly prophetic, as shown in a comparison of game events (in italics) and real events:

The political situation in South Vietnam was shaky:–Riots swept the country in November 1964; Premier Nguyen Khanh fled South Vietnam in February 1965; a military regime under Air Vice Marshal Nguyen Cao Ky assumed power in June.

On Christmas Day the Viet Cong shelled Saigon:–On Christmas Eve 1964 the Viet Cong did bomb a Saigon hotel where American officers were billeted; two were killed and fifty-eight wounded.

The Viet Cong shelled airfields at Danang and Bienhoa, destroying a dozen warplanes:–On November 1, 1964, in a Viet Cong attack on the Bienhoa air base, five U.S. servicemen were killed and six U.S. B-57s destroyed.

The mining of Haiphong Harbor:–Ordered by President Nixon in May 1972.

President Charles de Gaulle asks for a peace conference:–In September 1966 he called for U.S. withdrawal from Vietnam.

Ambush in the Laotian town of Tchepone:–South Vietnamese troops sent in to attack Tchepone in 1971 suffered more than 3,000 casualties. They were driven out of the town and lost more men in a disorderly retreat.

On February 26 the President announced the debarkation of a Marine expeditionary force and the establishment of permanent U.S. base facilities at Danang:–On March 8, 1965, two Marine battalions arrived to defend the airfield at Danang. They were the first U.S. combat troops sent to Vietnam.

The Danang landing was itself a sequel to the most realistic kind of war game—field exercises with troops and equipment. "Silver Lance," the 1964 exercises of the Fleet Marine Force, Pacific, were, according to Lieutenant General Victor H. Krulak, the commanding general, "patterned as closely as possible upon the emerging situation in Vietnam. . . . We added realism to the exercise by having Marines, carefully rehearsed for their roles, take the parts of friendly and hostile native forces as well as of our own political and diplomatic personnel." About a third of the Marines who had been in the Silver Lance exercises were at sea off California "when the decision was made to land at Danang." The brigade was sent to Okinawa and later to Vietnam, as were all the units that had taken part in Silver Lance. One battalion on Okinawa "wargamed a landing at Danang only two weeks before the landing took place."

Two years after the prophetic Sigma II-64, the Pentagon players assembled for the most unusual exercise in the Vietnam series. After six major games involving escalation in Southeast Asia, the Joint War Games Agency set up a game looking at what would happen if *peace* suddenly broke out.

Sigma I-66 opened with Ho Chi Minh privately informing the United States that he would withdraw all North Vietnamese forces from South Vietnam and would request the Viet Cong to cease hostilities. In return the United States was to announce a cease-fire, stop the bombing in the North, begin withdrawal of its own forces, and guarantee free elections in South Vietnam.

The premise was not that far-fetched, said the anonymous briefer in the video-tape summary of Sigma I-66. "In our pregame discussions in Washington, Honolulu, and Vietnam," he said, "most of the people interviewed believed the conflict would most likely end with the enemy fading away after he had been thoroughly defeated, much as he did in Malaya and Greece."

The scenario had Ho withdrawing because the steadily escalating bombing had devastated North Vietnam; because North Vietnamese forces had been "mauled" and were losing the ground war in the South; and because "an act of God—a typhoon—wiped out the rice crop in North Vietnam."

Whatever may have been the scenario's optimistic belief about Ho fading away, the team members, in their cautious language, express disbelief. Control has to force Blue to stop the air attacks against Brown (North Vietnam). Yellow (China) has to be restrained from marching across the border on Brown. The only team that sounds happy is Black (the Viet Cong).

"I have certainly enjoyed participating in this exercise," the Viet Cong representative says. He then questions the meaning of a cease-fire. "We bandy this term of cease-fire around Washington without the foggiest idea of what it means in Vietnam. . . . If we really mean what the English word says, everybody stops shooting.

"That's like the cops giving up shooting in agreement with the robbers. You're turning the countryside back over to the Viet Cong. . . . You have not explained to me how you're going to get rid of a hundred thousand, roughly, armed Viet Cong left in Vietnam as this thing terminates."

The senior discussion of Sigma I-66 climaxes with the words of Earle G. Wheeler, Chairman of the Joint Chiefs of Staff (identified by his title but not by his name in the Pentagon-censored document). The general says that after listening to the discussion about the unsolvable problems of peace, "I'm afraid some cynics might be tempted to rediscover that there are things worse than war."

CHAPTER 11

Red and Blue in the White House

One of America's best-kept secrets for a quarter of a century has been presidential wargaming. There have been occasional glimpses that blurred reality. Kennedy in the White House Situation Room, his eyes locked on a plotting board showing U.S. and Soviet ships nearing confrontation during the Cuban missile crisis. Lyndon Johnson in that same room, hunched over a sand table, a colossus looking down upon a Vietnam battlefield. Ronald Reagan making a conference call to command centers, thanking the players for their work in a game war that had killed a President and leveled both Washington and Moscow.

Johnson was photographed at the sand table, surrounded by aides, tensely contemplating a miniature Khesanh, where little banners in the sand symbolized a few thousand U.S. Marines and the 40,000 North Vietnamese surrounding them. At the same time an Army historian named Colonel R. W. Argo shuttled between Saigon and Khesanh, between past and present, learning about Dienbienphu, France's catastrophic final battle in Indochina—and studying Khesanh, which had become a presidential obsession. That sand table was in the White House because Johnson, fearing an American Dienbienphu, wanted his own model of battle to better understand reality elsewhere.

Through historic analysis, Argo tracked Khesanh against Dienbienphu, U.S. tactics against French tactics, Communist strategy in Indochina in 1954 against North Vietnamese strategy in 1968. "I projected their timetable and predicted their attack to the day— a week or ten days before it happened," Argo told me. "I used history. I did not game. But in Washington they said they were role-playing and they had an Oriental in State playing Ho Chi Minh." (Such role-playing at the State Department often has been reported in anecdotes but never officially admitted.)

Johnson at the sand table and Kennedy at the plotting board were playing a real presidential game. In that special kind of game what happens in the real world is transmitted to the world of the Situation Room, where decisionmaking becomes what gamers call a tabletop game. There may not often be a sand table or a simulation of the Atlantic to aid the President, but there will be briefing papers, scenarios of possible events—the real-life versions of the props used when lower-level officials play in the war colleges and the Pentagon's basement gaming room.

The only difference is that crisis management in a game is usually more orderly than crisis management in the Situation Room. Dr. Robert H. Kupperman, who was involved in crises in the White House and became an impresario of gaming for a prestigious think tank, once said that "at a time of deep trouble," a President "grabs his closest advisers who may know absolutely nothing about the crisis, or its resolution, simply for emotional support." The White House, Kupperman said, never has "a planning horizon beyond ten minutes. It is simply always in trouble."

In games players usually do not make believe they are actual national leaders. Social scientists who have studied presidential decisionmaking frown on such role-playing. They say that only an American President can be an American President, only a Soviet General Secretary can be a Soviet General Secretary. If surrogates do role-play the chiefs of state in a game, one study suggested, the environment should be as realistic as possible—with an overloaded, unreliable communication system "providing ambiguous, threatening, and probabilistic situations." At best, the study said, the setting for such a game is a laboratory, and what happens there does not necessarily happen in real life.

The study pointed out that Freud, after observing apes in the London Zoo, decided that a sexual drive dominated their behavior.

Freud then extended this to all primates, including human ones. Decades later, behaviorists studying apes in their natural environment saw that the apes' sexual activity was nowhere near the level Freud had observed in the zoo. Freud's inferences, the study of game behavior concludes, "were incorrect because the nature of his laboratory distorted the behavior of his subjects. The same problem arises when human subjects are placed in a laboratory setting that severely restricts their range of behavior."

In most games the abstract Blue plays the abstract Red. Blue or Red at its highest level is often referred to in games as the National Command Authority (NCA).

The NCA is not an abstraction. It is a real concept embodied in military doctrine, especially the doctrine that governs control over nuclear weapons. If the President is unable to employ his authority to order the use of nuclear weapons, the nuclear "release authority," as it is called, does not follow the Constitution's line of succession. The release authority passes from a disabled or missing President to the Secretary of Defense, and then, if necessary, to the Deputy Secretary of Defense.

Details about the NCA are vague and contradictory. One source suggests that "calculated ambiguity" cloaks revelations about presidentially delegated retaliatory authority. Another says that definitive information about the NCA is one of the nation's closely guarded secrets. No two authorities I consulted agreed on the passing of authority beyond the Deputy Secretary of Defense. But it can be imagined as a movement down and down, through catastrophe upon catastrophe, to the Secretary of the Army . . . the Navy . . . the Air Force—and ultimately, if all of them are gone, to an Air Force brigadier general in a command plane code-named Looking Glass flying somewhere over a leaderless nation devastated by nuclear war.

The design of U.S. nuclear decisionmaking, Lincoln Bloomfield says, "has walked the line since the 1940s between a different pair of competing pressures: on one hand greater centralization so the President can retain tight control over any use of nuclear weapons, and *de*centralization so the system is not paralyzed by a communications snafu or a terrified (or incinerated) President."

The NCA is the decentralizing answer to the problem of nuclear release authority. In a real nuclear crisis the NCA is a real human being—the one person who makes the nuclear release decision. In a game, however, the NCA is nobody at all. A game's NCA, although a flesh-and-blood player, is actually a device designed to prevent role-playing. The game NCA is supposed to purify the decision of Blue or Red, making them more useful for analysis because they do not represent a real or contrived personality.

Either way—in game or in reality—the concept of an NCA also reflects the probability that a nuclear war will begin with the killing of the President. The theory stems from modern strategic concepts about starting a war by what nuclear strategists call the "decapitation" doctrine: A nuclear attack upon the NCA to eliminate civilian leadership and the nation's ability to continue fighting the war.

This probability was at the core of Ivy League, a White House war game based on a scenario that wiped out Washington—but not before a thousand or more officials from the Pentagon and civilian agencies were flown from Washington to underground emergency government centers operated by the Department of Defense and the Federal Emergency Management Agency.

The White House underground Situation Room was both a nerve center and a target in Ivy League, which was staged in 1982. In this role-playing game former Secretary of State William P. Rogers stood in for President Reagan and former Director of Central Intelligence Richard Helms was Vice President. Other players formed a simulated National Security Council. They included Fred C. Iklé, Under Secretary of Defense for Policy; Thomas Reed, a former Secretary of the Air Force; Air Force General James E. Dalton, staff director of the office of the Joint Chiefs of Staff, and Deputy Secretary of State Walter Stoessel.

Under the NCA doctrine the President or his replacement runs the war through the Secretary of Defense and the Joint Chiefs of Staff. An ordinary war game would be played in the Joint Analysis Directorate game suite in the basement of the Pentagon. But Ivy League was no ordinary game, for part of the scenario called for realistic tests of the military command, control, communications, and intelligence system—C^3I (pronounced "see-cubed-eye") in military jargon.

The crisis began with the role players in their places: the President (Rogers) and the simulated National Security Council in the

Situation Room; the actual Joint Chiefs and their staff at the National Military Command Center, the complex of Pentagon offices that make up the nation's primary war room. The players had been told that weeks of tension had brought the Soviet Union and the United States to the brink of war. The minutes to nuclear midnight were already ticking away.

To test the Joint Emergency Evacuation Plan, helicopters and transport planes took "selected personnel," including, by one report, Commerce Secretary Malcolm Baldridge and Interior Secretary James Watt,* to alternative command posts far from Washington. The destinations of the evacuees were reported to be the Alternate Military Command Center in southern Pennsylvania and "special facilities" at Mount Weather in northern Virginia, in Massachusetts, and in Texas. More than one thousand other persons were moved hundreds of miles from Washington to underground bunkers that are stocked—"desks, yellow pads, that sort of thing," an evacuee in a similar exercise told me—for the creation of a postwar government. Government units "continued operating from hundreds of locations scattered throughout the U.S.," according to the *The Wall Street Journal*'s extremely knowledgeable account of the top-secret game.

As the game got under way the stand-in President took the precaution of ordering "Vice President" Helms to be ready to take a helicopter to Andrews Air Force Base, about ten miles from the White House, where he was to board an Air Force Boeing 747 jammed with communications equipment. Aloft, the plane—one of four, all identically equipped—becomes the National Emergency Airborne Command Post. The military name for the plane is Kneecap (an attempt to pronounce the acronym NEACP); another name is the Doomsday Plane.

The game began, as many Pentagon-designed games have begun, with a North Korean attack on South Korea. The Joint Chiefs ordered the Defense Readiness Condition for Pacific forces notched up from its usual DefCon V to III and then to I—a state of readiness

---

*Presidential succession, after the Vice President, Speaker of the House, and President Pro Tempore of the Senate, is Secretary of State, Secretary of the Treasury, Secretary of Defense, Attorney General, secretaries of Interior, Agriculture, Commerce, Labor, Health and Human Services, Housing and Urban Development, Transportation, Energy, and Education.

at the level of a wartime alert, which is aptly labeled "Cocked Pistol." The U.S. air defense warning was moved up from Yellow for "probable attack" to Red for attack "is imminent or is in progress."

On the evening of the third day of the game, the U.S. destroyer *Spruance* in the North Atlantic sent a flash message—*NUDET* (*nu*clear *det*onation)—and fell silent, presumably struck by a nuclear-tipped Soviet missile and sunk with all hands, some three hundred officers and men. Then came news of a poison-gas attack on a U.S. Army tank regiment near the West German town of Fulda, site of the strategists' Fulda Gap, which long had been singled out as a Warsaw Pact invasion corridor from East Germany. This attack killed 745 men.

The President authorized the Army in Europe to respond to the chemical attack with tactical nuclear weapons. Similar orders went out when the President, acting on the advice of military advisers, believed a nuclear response was warranted elsewhere after Soviet-U.S. clashes.

At the same time the President used a simulated version of the hot line in a vain attempt to explain each escalating move to the Soviets. (The hot line is a telex machine communications system opened in 1963, after the Cuban missile crisis. It was originally a cable link between Washington and Moscow via London, Copenhagen, Stockholm, and Helsinki, with a backup radio circuit through Tangier. Under a 1971 agreement, signed by Rogers as Secretary of State, the hot line primarily operates through Soviet and American satellites. It is a telex, not a telephone, as it often is depicted in fiction.)

When signals suddenly stopped coming from some U.S. satellites, the simulated information coming from them was denied to the players until the Air Force launched simulated replacements and put in motion plans that had been worked out to patch over the gaps. Enough communications survived for the North American Aerospace Command to report a "decapitating" Soviet missile strike aimed at Washington as part of a 5,000-megaton attack on the United States. (A megaton has the explosive force equivalent to one million tons of TNT. The bomb dropped on Hiroshima had an explosive force equal to 20,000 tons of TNT.)

The White House was destroyed and the President killed. The

NCA instantly became the former Vice President—now the new President—in the Kneecap aircraft high over Ohio, trying to manage the nation through the Minimum Essential Emergency Communications Network (MEECN), a lash-up connecting other airborne command planes and whatever military and commercial transmitters and satellites still survived. (Other command posts were also set up at emergency command centers and at a U.S. Embassy in Europe to test the game players' ability to improvise ways of preserving communications for the NCA.)

The ex-Vice President ended the five-day game by ordering what is known as the SIOP's Major Attack Option—a massive retaliatory nuclear barrage that would leave the Soviet Union decapitated and laid waste. Major Attack, designed to destroy all centers of military and political command, along with most cities, is one of the choices the NCA has under SIOP. (The others are Selected, Limited, and Regional Attacks. The NCA can also order Launch on Warning, meaning an attack based on warnings given the NCA by the U.S. warning network but not confirmed by the actual impact of nuclear warheads, and Launch Under Attack, ordered after an attack has been confirmed.)

After the game, according to the most authoritative report on Ivy League, President Reagan "placed a conference call to all the command centers used during the exercise, telling the players:

" 'While we pray to God that we will never have to use the procedures you have tested the past week, the nation is better off for what has been done.'

"President Reagan added that 'the exercise will not only improve our ability to respond to such a critical emergency—but, more importantly, the lessons learned will ultimately help us prove that our adversaries have nothing to gain by such an attack.' "

Other reports said that there was applause after Helms made the decision to launch the massive nuclear attack on the Soviet Union. President Reagan was said to have asked questions and watched avidly, but as a kibitzer, not a player. "No president should ever disclose his hand, even in a war game," an official said.

Although not a single official word of confirmation came forth about Ivy League, details about it were not kept secret. Information emerged in what appeared to be a campaign of leaks. The President's "lessons learned" speech was reported without open authorization—but was not denied. By tolerating an extraordinary flood of details, the White House allowed the assumption to spread that Ivy League was an instrument of military diplomacy—a demonstration of Washington's ability to react to a decapitating attack.

The story of this not-so-secret game begins in the December 1981 issue of *Armed Forces Journal International*. Editor Benjamin F. Schemmer accompanied an article on wargaming with a side story that began, "For decades, the American planners and policy officials who are custodians of US nuclear policy—and who might have to use its weapons—have tried to be meticulous in their precision and purposely vague in their ambiguity about the conditions under which this country might resort to the use of nuclear weapons." While trying to "create uncertainty," he noted, they had always tried to dispel confusion. He then reported on what had happened in the previous few weeks.

On October 16, 1981, President Reagan told a group of editors, "I could see where you could have the exchange of tactical weapons against troops in the field without it bringing either one of the major powers pushing the button." Next, on November 4, then-Secretary of State Alexander Haig, testifying before the Senate Foreign Relations Committee, said, "There are contingency plans in the NATO doctrine to fire a nuclear weapon for demonstrative purposes" to warn the Soviet Union that the United States is prepared to cross the nuclear threshold.

The next day Secretary of Defense Caspar W. Weinberger told the Senate Armed Services Committee, "There is nothing in any plan that I know of that contains anything remotely resembling that, nor should it." Then, later in that day, the White House and the Pentagon issued statements saying that, some years before, NATO planners had discussed a nuclear demonstration as a possibility, but with "significant doubts" about its value.

On November 11, at a White House press conference, President Reagan said his previous statement about tactical nuclear weapons had appeared in print "in an entirely different context." Then he added, "But I could see where both sides could still be deterred from . . . exchange of strategic weapons if there had been battlefield weapons, troop-to-troop exchange . . . there's high risk, there's no question of that."

Schemmer concluded the litany of events by commenting, "The

past few weeks have vividly shown that Ronald Reagan and his top advisors might benefit from 'gaming' nuclear war more and talking less." (I asked Schemmer, a highly respected military journalist with superb Pentagon connections, whether he had triggered Ivy League with his suggestion or whether what happened the following March was a coincidence. He said he thought it was a coincidence. "In this town," he added, "it's hard to find cause and effect for anything.")

Less than a month after Schemmer's remarks were published, columnist Jack Anderson quoted from what he said was "a secret Pentagon briefing paper" about a nuclear war game code-named Ivy League. Anderson quoted the paper as saying that the worldwide exercise "could show strength of purpose" but that it could also "be perceived by some as making preparations for the actual use of nuclear weapons and thus acting inconsistently with our position that such exist for the purpose of deterrence."

The anonymous briefing paper author also pointed out, according to Anderson, that the exercise "could be exploited by certain elements in Europe to increase fears that we are planning for the conduct of a nuclear conflict limited to the continent." (Reagan's remarks about a limited nuclear war had rekindled European worries about U.S. nuclear policies and had given new energy to the antinuclear movement both in Europe and the United States.)

It seemed unusual that, as the briefing paper implied, a debate was going on in the administration about conducting a major, worldwide war game. Such regularly scheduled games, which went back to at least 1963, were top secret. If Europeans or anyone else had heard about one, it would be because of a grave breach of security.

There should not have been any worry about such a breach; a minor leak could be ignored or denied, and a major one would constitute misconduct that warranted serious punishment in an administration obsessed about the leaking of sensitive information. As later events showed, the debate inside the administration was not about having the game but about whether to reveal what happened in the game.

The Department of Defense ignored the Anderson leak until February 26, 1982, when a four-paragraph Pentagon news release announced that "a routine worldwide command post exercise," code-named Ivy League, would be held for five days, beginning

March 1. The release said that the scenario for Ivy League "is a fictitious series of worldwide events leading to increased tension and conflicts among nations." The exercise's "overall objective" would be the testing of "command and control procedures" and "the interaction of national, departmental and agency plans and policies." The announcement also said that some unnamed Pentagon officials, as part of the exercise, would be moved "to other command post locations."

The announcement said that the exercise "does not involve movement of any other personnel." This would turn out to be patently *untrue*. Nor was it true that this was a routine command post exercise—Pentagonese for a strictly military drill. It seemed highly likely that the news release was some kind of compromise between those who wanted to keep Ivy League a secret—which would mean *no* announcement at all—and those who wanted to get out some word about U.S. nuclear war-fighting exercises so as to tell the Soviet Union and the rest of the world that decapitation would not stop the United States from massive retaliation.

It also seems likely that, if there had been plans in February to later expand upon the terse news release in the form of an authorized leak, the gratuitous misstatement about "the movement of any other personnel" would not have been made. Why say something was *not* going to happen when, in fact, it *was* going to happen? The insiders' game about the game appeared to have gone awry.

Within days after the March 5 end of the game, a high-level decision was made to reveal a great deal about Ivy League. The most thorough story was leaked to *Wall Street Journal* reporter John J. Fialka, who had been a well-known defense correspondent on the defunct *Washington Star* before joining the *Journal*. His knowledge of the Pentagon was legendary. Additional details about Ivy League were doled out later to other newspapers and to *Newsweek*, which credited the *Journal* for the basic story.

A few months later, more nuggets of information appeared in *S.I.O.P.: The Secret U.S. Plan for Nuclear War* by Peter Pringle, a Washington correspondent for the London *Observer* and a former London *Sunday Times* foreign correspondent, and William Arkin, a fellow at the Institute of Policy Studies, a liberal think tank, and a director of the Arms Race and Nuclear Weapons Research Project. Although certainly not a member of the conservative military journalism establishment, Arkin, a former Army intelligence analyst, is highly respected by people who would not agree with his political and philosophical views.

The leaking of Ivy League by an administration frequently critical of leakers was a phenomenon worth contemplating in terms of the reality of White House gaming. Ironically, "Vice President" Helms, when he was Director of Central Intelligence, had once resisted an open study of national security decisionmaking because "hostile elements would find [it] very useful in guiding penetration attempts, planting disinformation or timing information related to crisis situations." Yet here he was, one of the players of a game that went public.

One result of the game was practical, and would have happened whether or not Ivy League had been made public. During the game the Situation Room and the National Military Command Center were linked by the Worldwide Military Command and Control System (WWMCCS; pronounced "Wim-ex" by Pentagon acronym speakers), which also is connected to the headquarters of the Strategic Air Command in Omaha, Nebraska; to the commanders in chief of regional commands (CINCs, pronounced "sinks") and other key military unit commands throughout the world; to such essential C^{3}I centers as the National Warning Center at Colorado Springs, Colorado; the Alternate National Warning Center at Olney, Maryland; the Alternate National Military Command Center in southern Pennsylvania; to the State Department's Operations Center and the Central Intelligence Agency's Operations Center.

Coincidentally, as Ivy League was going on, the House Committee on Government Operations was about to issue a report calling WWMCCS unresponsive, unreliable, "incapable of transferring data and information efficiently," mismanaged for over fifteen years, and a "billion-dollar failure." The report was similar to others that had been issued in the past. This time, though, there would be action. High civilian officials had actually tried to use WWMCCS during Ivy League—and had found out about its flaws. Within six months after Ivy League the President issued a directive making the improvement of national security telecommunications a matter of high national priority.

The disclosures about Ivy League were a safe, effective way to tell Americans, foreign friends, and prospective foes that the nation's nuclear warfare machinery worked. The administration apparently hoped that the Soviet Union would receive the principal educational benefits of the Ivy League revelations.

The most important kibitzer, Ronald Reagan, may have gained some insights, but he will keep them to himself. Presidents do not talk about gaming. The reason for this, I have been told, is that knowledge of actual presidential thoughts about nuclear warfare would be the greatest secret an adversary could learn. "I do not honestly know if any President has gamed or not gamed," a longtime Pentagon official told me. "And if I did know, I certainly would not tell you or anyone else." The dogma on this seems quite simple: Do not ever discuss presidential gaming. And a corollary seems to be: Never say that a President has ever gamed.

High-level games remain enwrapped in secrecy for decades and, unlike many sensitive government documents, they are not always automatically released after the passage of a certain period of time. Presumably the records on Ivy League will not be opened until at least the next century.

Two of the earliest high-level games for which there are available records were played in 1961, around the time the Joint War Gaming Agency began its games in the Pentagon. But these games—designed to examine ways to handle a crisis over Berlin, then going on—were so secret they were not played at the Pentagon.

At least one was played at Camp David, the presidential hideaway in Maryland that is guarded by Marines and impenetrable to the media or other unwanted guests. The other appears to have been played at a tightly secured military building in Virginia. The nuclear "warning shot" issue, which would be raised over Haig's testimony in the fall of 1981, apparently had its origin in these games. They were directed by Thomas C. Schelling, a Rand strategist and a professor at Harvard. (Schelling, a pioneer writer on game-based negotiating, in an interview once gave this example of how a bold move may not actually be one because it is not tested: "I have a gun, intending to shoot you if you show disrespect to me; but you don't show disrespect *this* time. Then you don't see how bold I am.")

Records show who was invited to the Camp David game—veterans of Rand's "scratch-pad" wars, representatives from Mc-

Namara's office, including a specialist on nuclear strategy; McGeorge Bundy, Henry Kissinger, Carl Kaysen, and Robert Komer from the White House; a contingent from the State Department, three officers from the office of the Joint Chiefs, a man from the U.S. Information Agency—but not who played on what team. (Kennedy himself was conspicuously not there. He was at the family compound at Hyannis Port, Massachusetts, on both weekends.)

The language of the Berlin scenarios was shrill and melodramatic, unlike the cool, dispassionate style of the scenarios of later years and today. But in these early examples of the scenarist's art can be heard the escalating tone that still would be clanging in scripts written two decades later. *Signaling, escalating, challenging, counter-challenging*—the vocabulary of the Vietnam War—was already being practiced in that war game. Kennedy's policymakers were pondering U.S. responses to the erection of the Berlin Wall in August and the threat by Soviet leader Nikita Khrushchev to sign a separate peace treaty with East Germany, sealing off Berlin from West Germany. Some of Kennedy's advisers were taking this threat seriously enough to examine the use of nuclear weapons.

In a typical Berlin game scenario, rioting breaks out in a dozen East German cities. East German troops "sent in to support the police defected in several instances, joined the uprisings, took up defensive positions, and used local radio stations to call for general revolt." Soon most cities are in rebel hands and Soviet Army units are "using fire, tanks, and aircraft to massacre the people of East Germany."

The scenario deepens the fiction, moving from a description of the uprising to a sermon on the need for U.S. intervention: "Only help from the West, within hours, could prevent the bloody suppression of East German self-determination. The West, and especially Germans west of the iron curtain, could not this time shirk their civilized obligations to intervene and demand Soviet withdrawal. The alternative is disaster at Soviet hands. . . ."

In the real world the crisis did not escalate, Khrushchev abandoned his plan for a separate peace, and there was no East German uprising. But in a year the United States and the Soviet Union would be in another, deadlier new crisis, and this time the wargaming would be for real stakes. The use of wargaming props— the little ships and the wooden ocean for the President in the

Situation Room—intensified the chessboard thinking behind both gaming and the real U.S. strategy for nuclear war.

No thirteen days in American history have been analyzed more than those days in October 1962 when the United States and the Soviet Union went to the brink of nuclear war over the crisis created by the presence of Soviet missiles in Cuba. Political scientists have focused much of their analysis on the handling of the crisis by President Kennedy's advisory Ex Comm (Executive Committee of the National Security Council). Working in isolation from their staffs, forming informal coalitions to propose or oppose options, weighing risks with an Olympian sangfroid, Ex Comm had an air of Blue playing Red.

"We normally met in [Under Secretary of State] George Ball's conference room, and we had no stenographers or any other outside people present," Deputy Under Secretary of State U. Alexis Johnson, a member of Ex Comm, recalled. "In fact, we did not even inform our own secretaries, doing most of our work in long-hand. . . . One of our methods for working was to take a proposition which seemed to commend itself, have the proponents of that proposition, war-game it, you might call it, that is, work out a plan of action, put up the arguments for it, and then let the others shoot at it, and examine it for its strengths and weaknesses. . . .

"As far as my own role was concerned, having become a proponent of the quarantine approach, instead of drafting long papers of pros and cons, as some of our group were seeking to do, I had drafted what I called a 'scenario,' set forth in very simple terms the exact action, both military and political, to be taken in various stages. Of course I drew on the discussions of the group in doing this up. This scenario was discussed, and with some modifications was the paper on this that was finally shown to the President. This scenario, which was only a little over three pages, contained all of the major elements of all the actions that were taken to implement the quarantine action, both on the political and military side."

Johnson said the wargaming, scenario-driven approach to crisis management "is a model" of how to "blend a political and a military course of action together. . . . As I have told many War College classes since that time, the end of military power of course is not with shooting; the end is using it to accomplish national purposes."

Johnson's endorsement of the wargaming "model" showed how

well gaming was already entrenched among the veterans of high-level government. When Johnson recorded his recollections of the Cuban missile crisis, the political-military "model" was operating in South Vietnam, where he was the Deputy U.S. Ambassador.

During the crisis itself the Navy was wargaming at sea. A long-scheduled Navy exercise, Philbriglex-62, was being conducted in the Caribbean. The purpose of the exercise was the liberation of the mythical Republic of Vicques from the tyranny of a dictator named Ortsac, *Castro* spelled backward. The concept of exercising was so deeply etched in the minds of Navy officers that even ships engaged in the real crisis took the opportunity to play at war.

Navy ships had suddenly become real pieces on the board, but, as President Kennedy and Secretary of Defense McNamara learned, the pieces did not always move the way the players wanted them to move. McNamara argued with Admiral George W. Anderson, the Chief of Naval Operations, about details of the Navy's actions in setting up rules for a quarantine of Cuban waters. The Navy pieces, Anderson informed McNamara, were under the control of the Navy. The Navy gratuitously took the opportunity to force six Soviet submarines to surface before the President learned of the action and ordered restrictions on antisubmarine tactics.

The Navy had seen the crisis as a chance for an exercise in antisubmarine warfare. Soon after the crisis Anderson boasted of this, saying in a speech to a Navy League banquet, "The presence of many Russian submarines in Caribbean and Atlantic waters provided perhaps the finest opportunity since World War II for U.S. Naval Anti-Submarine Warfare forces to exercise their trade, to perfect their skills and to manifest their capability to detect and follow submarines of another nation."

In 1965 a top-secret study was made of the command and control resources available to the President, especially during a crisis. The study analyzed the handling of actual crises at the presidential level—the Soviet invasion of Hungary, the British and French seizure of the Suez Canal, crises over Berlin, the Cuban missile crisis—and "war games that have been conducted by the Joint War Games Agency dealing with Berlin, East Germany and Southeast Asia. . . ."

In this side-by-side examination of fictional scenarios and real life wargaming was cited for its value in teaching policymakers how to take steps on the path leading to war. Noting that gaming attracted senior officials from the Pentagon, Department of State, the CIA, and the White House, the report recommended that lower-level officials use gaming as a technique for making interagency reviews of contingency plans. "Key decisions," the report said, "tend to be made by small groups and not by large gatherings of people in a conference room."

While generally praising wargaming, the study found that in "very intense crises" veterans of games may move too swiftly toward the brink. War-game scenarios usually kick off with an intense crisis and then rush the players toward war. And "since the object is to examine the general war situation, the penultimate crisis gets quick and perfunctory treatment and the real scrutiny is reserved for the general war phase." But "national decision makers in real life may be far more concerned with the preliminaries." In the still-censored copy of this report I read in the Lyndon B. Johnson Library, someone had underlined the next sentence: *Indeed, the hope of staving off general war may rest heavily on skillful handling of intense crises.*

To the outsider who gets an occasional glimpse at what goes on inside, the skillful handling of crises seems more casual than, say, the way a professional football team's coaching staff might plan a major game. One glimpse comes from reading the observations of Richard Helms, whose life in the secret world encompasses spying in the wartime OSS and the CIA—and acting as Vice President in Ivy League. The President of the United States, Helms said in a recollection recorded in 1969, "does not make his important decisions in an orderly way or the way the political scientists say they should be done or the way the organization experts would like to see them done or, in fact, the way 99 percent of the American people understand that they are done."

Helms illustrated this remark with a glance into the Situation Room in June 1967, when Egyptian President Gamal Abdel Nasser invaded the Sinai and Israel responded with an attack against Egypt, Jordan, and Syria. "[Soviet leader Aleksei] Kosygin came on the hot line. I don't have the precise recollection of the entire message, but the message that came along was to the general effect that unless the United States made its influence with Israel felt and weighed in to stop this war, to bring about a cease-fire, that the Soviet Union was going to have to take whatever actions it had within its capacities, including military actions.

"Well, it was a rather somber group that was sitting down in the Situation Room of the White House. The President was eating his breakfast, and McNamara was there. . . . A message was sent back and the morning wore along and everybody was speaking in a very low tone of voice, and the Russian text was examined very carefully to be sure that the Soviets did say that they intended to take any actions within their capacity, including military—and sure enough, the word was military when translated by [Llewellyn] Thompson [U.S. Ambassador to the Soviet Union] and everybody else.

"After this had been going on for an hour or so, the President finished his breakfast and excused himself for some reason; went out of the room. While he was out of the room . . . McNamara turned to Thompson and said—and everybody was speaking in a low voice—'Don't you think it would be useful if the Sixth Fleet which is simply orbiting around Sicily [you know, they go so many miles in this direction, so many miles in that; I mean that's the way they cruised in those days], that in light of this Russian threat and so forth, that we sort of make it clear to them that we don't intend to take this and take it lying down. Wouldn't it be a good idea to simply turn the Sixth Fleet and head those two aircraft carriers and their accompanying ships to the Eastern Mediterranean?'

"Thompson said, 'Well, the Soviets will get the message right away because they've got some fleet units in the Mediterranean and they're sure watching that Sixth Fleet like a hawk with their various electronic devices and others. Once they line up and start to go in that direction, the message is going to get back to Moscow in a hurry.'

"Both Thompson and McNamara certainly agreed with that. So when the President came back in the room, McNamara said, 'We've been talking about this and we'd like to recommend that we head the fleet toward the Eastern Mediterranean.'

"The President smiled and said, 'That's a good idea.' So McNamara went to the telephone and the fleet got headed for the Eastern Mediterranean.

". . . I mean this was a decision that was made literally from one minute to the next. There were no papers; there was no direct organization; there was no estimate; there was no contingency plan; there was nothing."

Civilians who war-game in real crises—pushing real fleets around and talking of nations as "we" and "they"—are members of a special breed that political scientists like Graham Allison have dubbed Rational Actors. As actors they strut a world stage and play the roles assigned them. They may even look upon their nation as an embodiment of political drama: the United States or the Soviet Union as National Actor. The semantics of theater and game are similar—roles, players, scenarios, cues—and when the Rational Actors walk into the Situation Room, they change from theatrical players to game players. Allison, analyzing the Cuban missile crisis, cast Kennedy and Khrushchev as "partners in a game against nuclear disaster."

Allison's views on the relationship of game and crisis became particularly significant in 1985 when he was made consultant to Secretary of Defense Weinberger. Allison was also dean of the John F. Kennedy School of Government at Harvard, which had a contract to teach management skills to Weinberger's senior officials. Allison had previously been a consultant to Rand, the Department of State, the Office of Management and Budget, and the CIA.

"Games," Allison wrote in 1971, "proceed neither at random nor at leisure. Regular channels structure the game; deadlines force issues." A decision to use nuclear weapons is less likely to come from a game in which the President "has most of the chips than from a game in which the military have most of the chips."

In a real confrontation between the United States and the Soviet Union, Allison wrote, neither leader "will be overly impressed by differences between the death of one million and one hundred million of his own citizens when choosing to take, or to refrain from taking, a risk. Each will be more sensitive to the other's problem than is any other member of the central game. Both may well appreciate the extent to which the 'kings' are partners in the game against nuclear disaster. Both will be interested in private communication with each other. If channels can be arranged, such communication offers the most promising prospect of resolution of a crisis."

The idea that crisis management is essentially a game and a piece of theater has attracted not only television producers but some of the real-life political actors as well. Bloomfield, former U.S. Ambassador to NATO Robert Ellsworth, and Paul Warnke, former general counsel of the Defense Department and chief U.S. negotiator at SALT, all played themselves in a Home Box Office "speculative fiction" drama about a nuclear confrontation, "Countdown to Looking Glass."

But the most elaborate televised drama drew in real decisionmakers, who were introduced to a national television audience as "players" who "have acted similar parts on the world stage. They have sat with real Presidents in the White House Situation Room during real crises."

Welcome to "The Crisis Game," telecast as an entertaining news show—and, because it was so realistic, reincarnated as a training film for real future decisionmakers who will walk not only corridors of power but also the halls of mirrors where gaming and crisis management are reflections of each other.

CHAPTER 12

Playing the Crisis Game: An Historical Scenario

"I would say," the Director of Central Intelligence told the President, "this is probably the most military activity we have ever seen on the Soviet side, certainly since the invasion of Afghanistan in '79." The President turned to his national security adviser, who said gravely, "It seems to me the ultimate question here is what is the best way to protect our interests without getting into a world war."

The scene was in a television version of the Situation Room and the crisis was fictional—an ABC News docudrama called "The Crisis Game." But the actors in the drama were not professional actors. They were people who had been in government and had dealt with crises. James Schlesinger, as Secretary of Defense, played a role he had played in real life. Others played the bosses of agencies they had served in. And the President, Edmund S. Muskie, was playing a role that he had auditioned for in real life when, in 1968, he unsuccessfully ran as a candidate for the Democratic presidential nomination.

The Director of Central Intelligence was William C. Hyland, a former career CIA official who had been deputy national security adviser to President Ford and, as an Assistant Secretary of State, had been the director of the Bureau of Intelligence and Research. Muskie's national security adviser was Winston Lord, former director of the Department of State's policy planning staff and former assistant to Henry Kissinger when he was the assistant to the President for national security affairs.

The other players were: as Secretary of State, Clark M. Clifford, who had been Secretary of Defense in the Johnson Administration and had been an adviser to Presidents Truman and Kennedy; as Deputy Secretary of State, Richard C. Holbrooke, who had been an Assistant Secretary of State in the Carter Administration; as Deputy Secretary of Defense, Antonia H. Chayes, who had been Under Secretary of the Air Force in the Carter Administration (and the first woman to hold that post); as Chairman of the Joint Chiefs of Staff, General Edward C. Meyer, who had been Army Chief of Staff. (Meyer, who had retired, wanted to appear on the show in civilian clothes. "We finally got Clifford and Schlesinger to talk to him," Bill Moore, the producer of the show, told me. "He wouldn't wear the uniform *to* the show. He'd change in here.")

President Muskie had as senior advisers Hodding Carter III, an Assistant Secretary of State and State Department spokesman during the Carter Administration, and Richard E. Pipes, a professor of Russian history at Harvard and former director of East European and Soviet affairs on the National Security Council.

Behind the scenes of "The Crisis Game" was a Control Team consisting of a retired Air Force general and other players experienced in real crisis management. As in standard war games, Control acted as everyone not in the game, from the Soviet Union to Congress. A week before the game each player had been given thick briefing books that included elaborate background papers, ranging from an assessment of the world energy outlook to "The Domestic Scene in the U.S.S.R." On camera, they sometimes rambled, and their musings did not have the polish of briefing papers. But they sounded human and real.

As anchorman Ted Koppel put it, the issues aired in the game "are very real indeed," though "the situation is not." The game was serious enough and sophisticated enough for Graham Allison, who would later become an adviser to Secretary of Defense Caspar Weinberger, to acquire rights to the tapes for use by Harvard's Kennedy School of Government. "The Crisis Game" is a short course in high-level decisionmaking.

One breach of reality was the inclusion of a woman in that decisionmaking. Former U.S. Ambassador to the United Nations Jeane J. Kirkpatrick recalled a meeting in the Situation Room when "I looked across the room and saw a mouse, making its way slowly with composure across the room." And later, she said, she thought to herself, "It might be that the mouse was no more surprising a creature to see in the Situation Room than I. . . . I may be the only woman who has ever regularly sat at the table in the Situation Room."

The glossy Situation Room created for television was not the kind of place one would expect to see a mouse. The Situation Room was at the ABC News building in Washington and was more "televisionish" than the real one, which Moore said "was a little dull." But much about the off-screen game has the ring of authenticity— the memos to the President, the cool analyses of Soviet intentions, the authoritative assessments of military forces. The on-screen dialogue sounds real because it was real. There was no script.

The game, played in 1983, projects to 1985. The briefing materials reflect a recurrent premise of war-game scenarios: A general crisis involves more than one trouble spot. In real war games (to differentiate them from televised war games) and in real crises, decisionmakers search not only for solutions but also for an understanding of exactly what is at stake. What is threatened? Like many scenarios, this one links the NATO alliance to geopolitical moves by the Soviet Union.

When the Soviet Union crosses the border into Iran, President Muskie's adviser, Pipes, a real-life hardliner, puts the case, saying that "if the Soviet Union were to gain control over Iran and therefore over the Persian Gulf, because there'd be no force able to stand up to it, then in fact it would have an ability to cut off oil from Europe, and this would have the profoundest effect on the survivability of NATO."

Schlesinger, from his real-world experience as Secretary of Defense and Director of Central Intelligence, sees more than a border crossing. He tells Muskie that the Soviets "have established a scene in which they are holding at risk the assets of the United States and the West. It is indispensable, I think, that we create in their minds the belief that their own assets are seriously endangered if they pursue this course recklessly."

"The Crisis Game" dramatized the basics of long-term American strategy, such as Europe's dependence on U.S. preservation of free-flowing Middle East Gulf oil. This reality had been one of the driving forces in the Naval War College's Persian Gulf game, and it shows up again and again in scenarios. The televised game also gave insights into how that strategy is maintained and perpetuated through the camaraderie of gaming.

Most of the actors on screen were members of the U.S. foreign-policy establishment. They were working from highly detailed briefing papers that had been composed by off-screen members of that establishment. The players' strategic thinking thus was ratified on- and off-screen by the establishment. And, through Allison's propagation of these televised articles of faith, U.S. strategic dogma is passed to new and future members of the establishment.

In the game rooms players use the world not so much as a stage as a laboratory, where experiments can be carried out, in anticipation of real action that may not come for years. The fascination of the televised "Crisis Game" often was in the way it revealed the banality of power: Casual conversations became policy, offhand remarks became orders to armies. The players who are saying "we" and "they" in the game will speak just as authoritatively in actual crisis. We and They are the prime words of power.

Games act out or anticipate crises on the agenda of the foreign-policy establishment, but not necessarily on the public or congressional agenda. A list of proposed games in fiscal year 1966, for example, reveals the real concerns of the inner government of that era: Southeast Asia again and again, "NATO problems," the spread of "Castroism" into Bolivia, "Sino nuke capability" by the year 1970, Arab-Israeli water rights, a Turkish-Greek confrontation over Cyprus, a "long look at cybernetic society and technology" in the 1970–80 decade. The secret list, obtained through the Freedom of Information Act, was sent to McGeorge Bundy. Games involved high-level players then. They still do. The televised "Crisis Game" drew players from the ranks of the people who get those secret invitations to the games in the Pentagon basement. To listen to them on television is to eavesdrop on the psychology of power at the highest level.

The crucible for "The Crisis Game" is Iran, following the death of the Ayatollah Khomeini. The war with Iraq had ended the year before. As the briefing book sets the scene, by March 15 "the vestiges of central authority crumbled in Iran, with the establishment of a leftist revolutionary government in Azerbaijan and Kuristan and the coincident rallying of anti-Soviet forces around Colonel Ibrahim Shirazi in the oil-rich province of Khuzestan." At the same time, in East Germany, "popular discontent with rising food prices and other manifestations of the failed East European economies was clearly on the rise. . . . Heavy-handed government efforts to suppress the movement seemed only to rally additional citizens to its cause" and "the demonstrations increasingly assumed political overtones, with a nascent German nationalism not far beneath the surface of the workers' economic complaints. The difficulties in East Germany also raised echoes in Poland, where the vestiges of Solidarity maintained an active underground network." The war in Afghanistan ground on, with the Pakistan border essentially closed off under Soviet pressure.

U.S. intelligence monitors Soviet military activity along the Soviet-Iranian border, and a countdown to crisis begins. On April 10 the U.S. forces stage an "exercise" centered in Egypt. On April 18 Moscow announces that extended Warsaw Pact exercises will take place in May. On April 20, as elements of the U.S. Rapid Deployment Force are put on alert, NATO officials hold a special meeting to discuss the mounting crisis, which is still not yet fully visible.

The crisis surfaces on April 21 when the Secretary of State says he will not rule out U.S. military intervention in Iran if the situation there worsens. On April 22, as the Director of Central Intelligence flies to Riyadh to brief the Saudis, twenty-three Soviet military advisers are killed during a raid by Shirazi's guerrillas. In Washington administration officials leak the news of the Soviet mobilization: All twenty Soviet divisions in the Caucasus are now at full strength. The aircraft carrier *Ranger* joins the carrier *Kitty Hawk* on patrol in the Arabian Sea.

On May 1 the assistant to the President for national security affairs notifies the National Security Council of a meeting at four o'clock to discuss the worsening situation in Iran and growing unrest in Eastern Europe. The President distributes to the council members a memorandum from the Director of Central Intelligence stating that at least five Soviet divisions have crossed the border into Iran and a second anticarrier task force has been deployed to the Arabian Sea. In Europe, the CIA memorandum says, the Warsaw Pact exercises have begun, with no signs of mobilization or "other preparations necessary for a major offensive operation." The CIA has no hard evidence as to whether there is a global Soviet strategy behind the two crises.

In the Situation Room the meeting begins. After Muskie gets the feel of the discussion, he leaves. This is a realistic touch based on historical precedent. Most of the Ex Comm deliberations were outside of Kennedy's presence, as were discussions by Schlesinger, Henry Kissinger, and others during Nixon's crises. The meeting continues with Clifford, as Secretary of State, taking over and looking for more intelligence information. "We're flying blind," he says.

"Well, Mr. Secretary," the Director of Central Intelligence says, "you're asking for intelligence about Soviet intentions, and unfortunately we can't take a picture of it or put it in any real concrete form."

Schlesinger, as Secretary of Defense, speaks up. "All that we know, Mr. Chairman, about the Soviet behavior suggests that they will push, they will see whether there is any resistance. If there is little resistance, they will roll to the Gulf. So at this stage it is necessary for us to interpose a barrier so that they do not achieve what we all agree is vital to the American and Western interests."

"Well," Clifford responds, "I suppose we get down to the analysis that the reason Iran is so important to us is because of the oil. Now, we as a nation are really no longer dependent upon Middle Eastern oil. We could get along probably pretty well without Middle East oil. It's Great Britain and Belgium and Italy and Germany and Japan and others who are dependent on the oil. Now, if they should tell us, notwithstanding that fact, they are not prepared to support us or to join with any troops in going over and protecting Iran, does it seem feasible for us to do it on our own when as a matter of fact we were really doing it, in the main, for their interests?"

"That, Mr. Chairman," Schlesinger says, "is the responsibility of a great power. In circumstances of this sort the United States, as the only superpower in the West, must take a world view. And, regrettably, our allies are able to look at things more from a parochial view than we are."

Muskie returns and seems to surprise his advisers when he says he has decided that "this is not a situation in which either the Soviet Union or we have a nuclear option." Quickly (through the machinations of Control) word of the President's decision is leaked to the media, and this sparks another discussion between Muskie and Schlesinger, who says, although he supports the decision about holding back on nuclear weapons in the Middle East, "there is enough trouble in Eastern Europe at the present time, enough concern in Western Europe, that we will have to make a public statement that the use of nuclear weapons will not be precluded in the event of a Soviet move westward into Western Europe. Otherwise we will stir up more concern among our allies than we will resolve."

"I have overlooked something in this respect," Muskie replies, "and I need to be reminded. But if we don't have an option for dealing with that, you know, whatever happens to our nuclear policy if we should reconsider it here today—whatever happens, I think one of the great risks with respect to the nuclear issue is that when we undertake to exert our power in an area where we cannot with conventional forces achieve our objectives, we escalate the possibility of using the nuclear option."

"There is that risk," Schlesinger agrees. "I think it's a minimal risk. Since World War II in each and every circumstance where a line has been drawn between the United States and the Soviet Union, they have respected it. They have moved into this area with, I think, tentative objectives. I do not think that they will wish to precipitate a clash with the United States with the implicit risk that you suggest."

With these remarks, which encapsulate the hard-line side of the American foreign-policy establishment, the game suddenly becomes a stage for a clash between Schlesinger, the real-life hard-liner, and Clifford, the real-life liberal. "If we flounder around," Schlesinger says, as he pushes for sending troops into Iran, ". . . that in itself is a signal of a weakness of will. . . ." Clifford, in response, says swift action could be disastrous.

Clifford, who in real life as Secretary of Defense worked to end the Vietnam War, now raises that threat: "We get into the conflict, we lose a thousand men, then we lose 10,000 men, and then we get up to what we lost in Vietnam, we lose 50,000 men."

Muskie decides on a "limited insertion of troops."

Control moves the calendar ahead two weeks. The Soviet Ambassador to the United States opens a back channel for possible negotiations by inviting to lunch a former U.S. Ambassador to the Soviet Union. At lunch he says that it is possible to work out a political arrangement that will allow for the withdrawal of U.S. troops to avoid a direct clash between the superpowers.

While the President's advisers ponder the ambassadorial meeting, General Meyer says he has just received information that an unarmed U.S. airborne warning and control system (AWACS) aircraft has been downed, with the loss of nineteen U.S. servicemen. (The idea of shooting down an unarmed plane is attractive to scenarists. An AWACS plane is also shot down in Home Box Office's "Countdown to Looking Glass." During a real minor crisis, a U.S.-Libyan confrontation in 1986, an unarmed Navy surveillance plane was buzzed by Libyan warplanes.)

The AWACS plane apparently was shot down by Soviet fighters in Turkish air space, but when Hyland, the Director of Central Intelligence, wonders if the United States can safely conclude that a Soviet plane shot down the U.S. plane, Schlesinger says, "I do not think we can accept the loss of one aircraft in this manner without responding," and suggests that the U.S. Navy shoot a Soviet reconnaissance plane.

"Are you suggesting this without any protests, without any communication?" Schlesinger's deputy secretary, Antonia Chayes, asks. He replies by saying that shooting down a plane "is a form of communication." (On September 1, two and a half months before the game was televised, the Soviet Union had shot down a South Korean airliner that had entered Soviet air space. That form of communication had killed all 269 people aboard the plane.)

Soon the advisers are discussing the possibility of escalating and sending in more U.S. troops. Pipes points out that if the Soviet Union moves against the troops already there, "it's unleashing a war with the United States. This is how far this will go. Unless we are willing to withdraw these troops unilaterally, this is what we have committed ourselves to."

As Clifford and Pipes argue about the future results of the commitment, Clifford, tripping over his syntax, makes one of those slips that shows a hidden moment of the mind: "Are what you saying is that whatever it takes in South Viet—in South Iran—that we're prepared to do?"

Muskie's national security adviser, Winston Lord, says he thinks he has detected "a hidden agenda"—are nuclear weapons necessary to prevent the destruction of U.S. forces? Deputy Secretary of State Richard Holbrooke puts the question directly to Muskie: "Are we prepared now—since we're on the edge of the point where for the first time since World War II, U.S. and Soviet troops could be shooting at each other—to back our troops up with nuclear weapons if that's necessary?"

Muskie tries to brush the question aside by saying, "We're not at that point." Schlesinger, surprisingly, agrees. And Clifford makes a little speech. "I cannot picture," he says, "an American President ever being the first to use either a tactical or a strategic nuclear weapon."

Schlesinger speaks up with a speech of his own because what Clifford said "is fraught with the deepest significance for our alliance. . . . The policy of no-first-use is detrimental to our position in Europe. . . ." Clifford responds, and an argument begins, but Muskie stops it.

Clifford raises another issue: a declaration of war by Congress. The War Powers Act, which theoretically restrains the President's use of troops in a crisis, is not discussed. Such disregard of Congress is typical of crisis scenarios.

A discussion about a declaration of war is interrupted when the Director of Central Intelligence reveals information he has just received: "Some Soviet ballistic-missile submarines have moved out of port, not on a schedule that we would have anticipated." There are also indications that Soviet troops, who outnumber U.S. forces about seven to one, are moving southward toward the oil fields. All that stands between the superpowers is the ragtag, pro-U.S. army of Colonel Ibrahim Shirazi.

At this point "The Crisis Game" suddenly becomes serious. There is a quick change in the tone of the talk. No longer are the players discussing pure strategy. Now Soviet forces may be heading toward U.S. troops—and, outnumbered as they are, they face annihilation or surrender. It is finally time to hear from the soldier.

General Meyer talks around the problem in his rambling way, and then he says, "If you wait, if you wait, then they will be able

to amass a larger force down to the south and it makes the problems
of defending our forces in the south far more difficult." He is a
soldier who has looked at the maps, and he has examined the
terrain. He has found the place where "they're able to move out
into the plain."

Meyer offers two possibilities. "One, which is less effective and
that is—effective, but less effective—and that is by increasing the
number of Americans and Iranians that we have in the area, and
being able to disrupt through clandestine means the various routes
through that area, or to even give them permission to attack convoys as they're coming south in that move. The second—"

"Excuse me." Lord, the national security adviser, interrupts. "Is
that American troops attacking the Russian convoys?"

"American unconventional warfare forces with the indigenous
forces," Meyer replies. "The combination of—"

Lord interrupts again: "That is a step beyond—"

"It *is* a step beyond," Meyer says.

"That is direct Soviet-U.S.—I just want to get the—"

"I am making this as clear as I can," Meyer says, interrupting
Lord. "And the second is to do that with air power, with precision-
guided munitions to blow up the area and to do that."

Lord, the national security adviser, is clearly upset. "Let me get
one straight thing—one thing straight—on the air power. Does
that have to be American air power, or can it be Shirazi air power?"

"In my judgment at the present time it has to be American air
power, because we're—because it's only our capability to do that.
Even if it were Iranian air power, it would have to have a cap from
the U.S. and would have to have, you know, additional assistance
provided by the U.S. for them to be successful."

Hodding Carter, as a senior adviser, does not, like most of the
other players, represent an institution or a constituency. He interrupts the Lord-Meyer dialogue. "Well, let me just put in my
dissent right now," he says. "It seems to me we've made an awful
quick leap here as to how fast you have to go to that, and I want
to go back—"

"And I was very careful that I didn't give you a time frame,"
Meyer points out.

"Yes," says Carter. He was a Marine Corps lieutenant, and he
adds, "sir."

"We can't ignore the commander of the Joint Chiefs of Staff when he says my troops have got to have support," Holbrooke says.

"I agree," Carter says.

"I'm not going to support an option," Holbrooke says, "in which they get—they run the risk of being wiped out by the Russians. Either they are going to get—"

Carter interrupts: "Well, now, if we all agree that we do not wish the American troops—"

"Hodding!" Holbrooke exclaims. "Let me finish."

But Carter continues, as if he hadn't heard Holbrooke: "—to be wiped out, those are not the only alternatives."

"Hodding! Let me finish," Holbrooke repeats. "If I understand General Meyer correctly, he's saying either pull my troops out or put them in a tenable position. Now, if he's saying—"

"Wait a second," Lord breaks in. "He said it would be prudent. I don't think he was quite that crisp."

Meyer, who is used to having to be patient with his civilian masters, tries again. "I was saying, at this point in time you are going through a very critical time warp, where, if you wait to make the decision and the Soviets have consciously made a decision to continue to the south, if you wait to make a decision and part of their plan is to continue their attack to the south, then we will be at a disadvantage."

"Well," Lord says, "the dilemma is we don't know they're going to go to the south."

"As I recall," Carter says, "as I recall—"

The CIA's Hyland suddenly enters the discussion: "Is anybody in this room recommending we use American forces against Soviet?"

"Is anybody for that?" Lord asks.

"I'm not," Hyland answers.

"In which area?" Meyer asks. "In the north?"

"In Iran," Hyland says irritably, just as Holbrooke tries to be helpful with, "In the north."

"I'm saying," Meyer says, "that if we had a clear indication that their intention is to continue to the south, you better do that."

The State Department's Holbrooke decides it's time to look at a map to find out just how far U.S. and Soviet troops are from each other.

Meyer comes up with "about two hundred kilometers—one hundred and fifty to two hundred kilometers."

"And reducing," Holbrooke adds.

"And reducing," Meyer repeats. "Correct."

Pipes looks over the situation and suggests "horizontal escalation"—an academic strategic term that means raising the ante, usually by making a move that does not necessarily raise the level of seriousness in the principal crisis. Pipes's choice is Afghanistan—supplying the anti-Soviet guerrillas there with surface-to-air missiles. He also mentions Aden, South Yemen, "one could retake with very small forces. It would not be more than a Grenadian operation, probably." And then he mentions Cuba.

Holbrooke makes a crack about "horizontal escalation" being a Kissinger idea, but he accepts a move against the Soviets in Afghanistan. Hyland sniffs that horizontal escalation now is "second-rate stuff," but he likes putting "military pressure" on Cuba.

"What do you mean by pressure?" Holbrooke asks. "Do you mean overt, covert?"

"I don't have a scenario for putting pressure on Cuba," Hyland replies, "but it shouldn't take much imagination to know what they can do."

"You want to invade Cuba with American troops?"

"Not yet. But I want to be prepared to put military pressure on Cuba."

"What Bill now explains," Holbrooke says to the rest of the room, "he means, and now that I understand it, I would like to associate myself with him, is a rapid increase of our strength in a highly visible manner—in Florida, B-52s moving down there; putting more Marines in the Guantanamo [Naval Base in Cuba]. I think it's an integral part of our actions."

The general tries to be patient again. Instead of maps, he talks about logistics. He does not think all this horizontal escalation is going to be possible "with the forces we have in being at the present time." If *horizontal escalation* is an academic term, *in being* is a military one that goes back to Clausewitz and probably to Caesar.

"This is Alice in Wonderland, gentlemen," Carter says, forgetting the presence of Antonia Chayes. "You are all talking about our being able to endure on four fronts."

"Well," Lord says, "I think it's for General Meyer and the experts to decide."

"He just said it," Carter retorts.

"He said maybe," Lord says.

"No, no, no," Hyland chides Lord, giving Lord, the president of the Council on Foreign Relations, a quick lesson in strategic etiquette: "This a major strategic political decision, and it is not for General Meyer to make that decision." (Hyland would become editor of *Foreign Affairs*, the prestigious journal of the Council.)

In the hubbub of argument, Lord insists, "I didn't say he should judge whether he should do it." He looks down his nose at Carter: "And not senior adviser Mr. Carter. He doesn't know enough about military matters. Now, look, we don't have an unlimited amount of time—excuse me—if you let me move you along. . . ."

At about this point Muskie returns to the Situation Room and vetoes what Hyland calls a "demonstration" in Cuba. (A demonstration—originally, the demonstration of a nuclear weapon by detonating it at some "harmless" site—is another academic contribution to nuclear crises. The idea goes back to Harvard's Thomas Schelling in the Berlin games at Camp David.)

Clifford, as Secretary of State, meanwhile, has met with the Soviet Ambassador. "The old affability that had developed over a relationship of close to thirty years," Clifford sadly reports, "was entirely gone. Formal, cold, exceedingly distant." All the ambassador "wanted to talk about was plans for our withdrawal." But the ambassador has also passed another message through Horelick, who is playing the former U.S. Ambassador to Moscow.

Pipes interprets the message on what is called the "Horelick channel" as "giving us, really, a choice between being humiliated or fighting." But Hyland has a different interpretation. He thinks that the Soviets do not want to fight; they want to negotiate.

Schlesinger and Clifford weigh the possible interpretations, against their own experiences, against even Lenin's writings.

Then, suddenly, Carter bursts out, "Mr. President, you were not elected, you were not elected by the people that elected you to substitute intellectualization for what your own experience in politics and in the world at large has taught you. And one of the things is that to hypothesize when direct action can find out for yourself what really is, is madness. And it is doubly madness when hypothesizing leads you to the brink of a confrontation which you do not have to encounter."

"What are you talking about?" a startled Holbrooke asks Carter.

"What I'm talking about precisely is, I happen to endorse the notion at this moment for the direct contact at your level, Mr. President, with a message which speaks directly for a mutually satisfactory resolution. Now to stand back and to theorize about what it is they may or may not have on their minds, when you have an opportunity, on an opening provided by this unofficial contact, to say we understand that we have a mutual interest here, would be wrong."

The other advisers back off from use of the hot line or other direct contacts, although Pipes reveals that "during the Lebanese crisis in 1982 we in fact had weekly exchanges between President Reagan and Mr. Brezhnev. Messages went back and forth. And that channel is not the last resort; it's a perfectly normal channel, and it seems to me we have used it in situations which are far less grave than the present one."

Schlesinger unexpectedly puts another issue on the table: He wants to raise the alert status of U.S. strategic forces and call up air defense reservists. The game's extensive briefing books have the Soviet Union keeping about fifteen missile-carrying submarines at sea. As the crisis has worsened the number of submarines has increased to thirty. Schlesinger says this is "largely intended for theatrical effect," but he feels the need to send a "countersignal" by stepping up the defense alert level. (A naval analyst who was intrigued by the reality of the game told me, "That number— fifteen Soviet submarines—is true only in the game. But increasing the number of submarines like that in a crisis is what the Soviets would probably do.")

By the twenty-fifth day of the crisis U.S. forces in Iran are openly aiding the anti-Soviet Iranians—and the Soviets have not pushed farther southward. Some NATO forces have been called up, Japan has joined the European allies in leveling economic sanctions against the Soviet Union, and the debate goes on about "the hot-line option."

Muskie decides to use the hot line—and raise the alert status of U.S. nuclear forces. In the message that he orders transmitted to the Soviet leaders, he says, "The only stable settlement of the crisis [is] withdrawal of both our nations' military forces and recognition of the continued integrity of Iran." He suggests that Clifford and the Soviet foreign minister "meet immediately to seek a resolution of this crisis."

But someone has leaked to the media the fact that the U.S. commander in Iran has requested authority to introduce tactical nuclear weapons in response to the Soviet introduction of SS-21 battlefield missiles. The United States does not know whether the missiles have nuclear warheads. The National Security Council unanimously turns down the request. Schlesinger recalls that in October 1973, during the Arab-Israeli Yom Kippur War, "the Soviets introduced nuclear weapons into the Middle East [and] we did not overreact."

(Schlesinger, who really was Secretary of Defense at the time, ordered a general worldwide alert at the level of DefCon III. The 82nd Airborne Division at Fort Bragg, North Carolina, went on alert; B-52s, the Strategic Air Command's nuclear bombers, were recalled from Guam; the Military Airlift Command was put on alert for continuous airlift, and the Sixth Fleet in the Mediterranean was put on a readiness alert in what the Chief of Naval Operations later called the most tense situation since World War II.)

The game ended abruptly with the arrival of a hot-line response from Moscow that offered an immediate beginning of negotiations. Perhaps because this was, after all, television, the end came quickly. And the European crisis that seemed to be boiling on the first night of "The Crisis Game" seemed forgotten by the fourth and final night.

In the *Nightline* version of the hot washup, Schlesinger said that the game was "quite similar to the real-world decision process." Clifford agreed, adding that the game "carried me back to thoughts of other meetings. There was really the most startling similarity. . . . I sat in the meetings thirty-five years ago in the Truman Administration, when the Soviets informed us that they were blockading Berlin [and] the two major military advisers of President Truman advised that they arm a train and send the armed train through to Berlin. If it got through we would have broken it [the blockade]. If the Soviets stopped it, we would then consider it an act of war and declare war on the Soviet Union.

"We went through day after day, very much of the same kind of searing experience that this group has gone through. President Truman decided that he was not going to follow that advice; he started the airlift. In a number of weeks it broke the back of the blockade; he succeeded, and we had no war."

Recalling the Cuban missile crisis, Clifford said that the rhythm of the meetings had been similar to those of "The Crisis Game." For the first three or four days "the group discussed what parts of Cuba would be bombed and how Cuba would be destroyed. . . . Finally, by the time the decision came out, it was a much better and more thought-out decision."

When the U.S. electronic spy ship *Pueblo* was seized by North Koreans in 1968, Clifford said, the decision at first was "to attack the capital of North Korea" and even "drop one nuclear device" on the capital. But those "early emotional factors were repelled and suppressed, and ultimately, through a long series of negotiations, we got our ship back and we got our men back."

The game Clifford had called "searing" and Muskie had called "this sometime agonizing experience" had lasted eighteen hours over Saturday and Sunday, November 12 and 13. The videotape was edited down to four hours and telecast on November 22 through 25. At the end of the fourth night Clifford and Schlesinger were brought on live to talk to Koppel. Muskie was not there. Two days after videotaping was completed, he had a heart attack.

CHAPTER 13

The Perils of Make-believe

Edmund Muskie's heart attack may not have had anything to do with stress produced by "The Crisis Game," but his sudden illness (from which he soon recovered) dramatized one of the dilemmas of realistic gaming. The make-believe crisis can become so real that decisionmakers feel real pressure—and yet it is this very pressure that critics of gaming claim is missing from decisionmaking games.

Muskie and his National Security Council put on a highly realistic demonstration of high-level decisionmaking, and the scenario they confronted was true to geopolitical life in the 1980s. The technical moves—"signaling" with heightened military alerts, opening back-channel communication, controlling escalation—all had the ring of authenticity.

In fact, a major defense think tank, Science Applications, Inc. (SAI), had arranged the docudrama, using techniques acquired in real-life games. Christopher Makins, who performed as an on-screen "expert," commented on the game as if it were a real crisis that a real television station would cover by calling in a real expert like him. In one of those odd crossovers that blur the make-believe of high-level gaming, Makins *was* an expert on crisis management and had been involved in a Rand-SAI competition that was ushering in a new era of wargaming (Chapter 16).

"The Crisis Game" was so well tuned to reality that players even acted as if they believed what they were doing. There were occasional flares of temper, but, except for Hodding Carter's impatience with the mandarin ways of nuclear brinksmanship, there was no indication of human concern over the potential deaths of millions of people.

McGeorge Bundy, veteran of Ex Comm, perceived this when he wrote of the "enormous gulf between what political leaders really think about nuclear weapons and what is assumed in . . . simulated strategic warfare. In the real world of real political leaders—whether here or in the Soviet Union—a decision that would bring one hydrogen bomb on one city of one's country would be recognized in advance as a catastrophic blunder; ten bombs on ten cities would be a disaster beyond history. . . ."

Yet, when President Reagan praised the workers in Ivy League, he said that "the nation is better off" as a result of lessons learned from the game. Perhaps what he meant was that *he* was better off because he had learned, through the game, how harsh and fast reality can be in a genuine crisis.

At the highest level, game playing can be a dress rehearsal for this reality of crisis. In a game, as in a crisis, participants feel time and events tightening on them, until they reach what a study of nuclear crisis management calls "the point where they will have only a few and extreme options left." Game playing gives potential crisis managers that desperate feeling, at no cost to their nation or the world, and at little cost to their own psyches.

In a game, as in a crisis, participants depend upon themselves rather than upon their staffs and the normal flow of information. In a game, as in a crisis, snap judgments—even hunches—drive decisions. Governmental bureaucracies are not set up that way. What appears to be the *unreality* of gaming—the vise of time, the lack of solid knowledge, the dominating influence of Control—is the *reality* of crisis. Substitute Fate for Control, substitute the shooting down of a real Korean Air Lines plane for a make-believe AWACS plane, substitute garbled messages for the terse typewritten slips from Control, and the result is the reality of crisis.

The reason that critics perceive high-level games as unreal is because they do not reflect what appears to be the normal ways of government: the day-to-day calm of bureaucracies, the neatness of organization charts, the precision with which meetings are scheduled, the orderliness of agendas. Crisis is as devastatingly simple as a game. It's *We* versus *They* and what are *We* going to do?

Three veterans of Pentagon and National Defense University war games have proposed that gaming be built directly into the National Security Council staff. Their idea is to give high officials, including the President, games befitting their rank. These games would be based on real, over-the-horizon potential crises. By playing them out as rehearsals, decisionmakers could learn the strengths and weaknesses in the crisis staff that is supporting them, and the strengths and weaknesses of their own souls.

The reality of life-threatening tension during a crisis is demonstrated in another drama of synthetic history, Home Box Office's war game, "Countdown to Looking Glass," in which the Secretary of Defense has a fatal heart attack during a meeting in the Situation Room. This gives a cue to Lincoln Bloomfield, who is playing himself as an expert on crisis management—and an advocate of improvement. Although supposedly speaking fiction, he echoes what he said to me in an interview and what he has frequently written about: "It's not possible to count on completely normal behavior. I remember in my own service, one very high-ranking government official coming out of a crisis meeting and throwing up in a wastebasket from tension. Another one was rushed to a hospital in the middle of another Middle East war. We spend billions on machinery for crisis management, but what is inside people's heads is the same old equipment, and I think this is a terrible problem."

Psychiatrists, academicians, and professional soldiers, for differing reasons, have in recent years voiced a growing concern about the effects of gaming and other forms of simulation on concepts of reality, especially nuclear reality. One modest proposal for imposing a haunting reality came from a law professor who suggested that the President's nuclear-release coding device, which always accompanies him, be placed in a capsule and surgically implanted next to the heart of a volunteer war starter, who would also carry a knife. When the President decided to launch missiles, the war starter would hand him the knife and the President would first have to do more than push the proverbial button. He would have

to kill. "He has to look at someone and realize what death is—
what an innocent death is," explained the originator of the idea,
Roger Fisher, professor of law at Harvard.

In nuclear war simulations, death is merely rows of numbers in
the millions or tens of millions—numbers which themselves are
the products of other nuclear war games. These games especially
appeal "to the talented, imaginative gamesmen" who become leaders and advisers to leaders and, while making their rational decisions, do not face the "real horrors of a nuclear war," Dr. Morton
Deutsch said in his 1982 presidential address to the International
Society for Political Psychology.

"Leaders of the nuclear nations climbed to power before nuclear
weapons burst upon us, so they try to deal with these weapons as
if they were conventional ones, despite an intellectual awareness
that they are not," Dr. Jerome D. Frank, professor emeritus of
psychiatry at Johns Hopkins University School of Medicine, has
observed. "Intellectually they may be in the nuclear age but emo-
tionally they are still back in the days of spears and clubs. They
are experts at the old international game of deterrence and war—
lethal games in a nuclear world, but the only games they know
how to play."

A leader under the stress of a nuclear decision, according to
Frank, "perceives fewer alternatives, simplifies issues and focuses
exclusively on combatting the immediate threat without considering remote or long-term consequences. . . . Nothing is harder
when under emotional stress than doing nothing. . . . The power
of a threat to psychological integrity—that is, to the self-image of
the individual or group—is generally underestimated. . . . In the
1929 crash some financiers chose suicide, even though they still
had plenty of money left, apparently because they could not bear
to face the implication that their judgment was faulty."

The real feelings of real leaders are difficult to replicate in simulations and are often difficult to find and examine in the aftermath
of real crises. Researchers who have analyzed how people act under
stress generally agree that a certain amount of pressure can sharpen
performance. People can develop what psychologists call a coping
pattern, a rationalization of fears and anxiety that prepares the ego
for the ratcheting stress that accompanies a building crisis. That is
the standard psychological view of how healthy people handle stress.

But when this view is applied to political leaders in a crisis, the picture changes. The psychological studies that produced the standard view of stress were, for the most part, based on the reactions of college students, not mature politicians.

Alexander L. George, of the Center for International Security and Arms Control at Stanford University, an authority on stress in crisis, has cautiously written about the psychological collapse of two unnamed high-ranking officials in the Kennedy Administration during the Cuban missile crisis. Theodore Sorensen, special counsel to the President, said during the crisis, "I saw first-hand . . . how brutally physical and mental fatigue can numb the good sense as well as the senses of normally articulate men." No one, even after two decades, publicly mentions details of that real stress in a real crisis. Such reality is too embarrassing and too revealing. We want iron men at the button.

Although little is known publicly about the behavior of American Presidents under stress, a remarkable amount of information exists about the way some other world leaders have acted during crises. George lists Stalin, who "evidently suffered a temporary depression after the Nazi invasion of the Soviet Union"; British Prime Minister Anthony Eden, who suffered "a near physical and emotional collapse" after the debacle of the Suez invasion in 1956; Israeli Army Chief of Staff Yitzhak Rabin, who was temporarily relieved of his duties because he was "visibly over-stressed and/or suffering from nicotine poisoning" during the crisis that preceded the Six Day War in 1967. Both Jawaharlal Nehru of India in 1962 and Gamal Abdel Nasser of Egypt in 1967 became "mentally and physically incapacitated" after their nations had military setbacks. George also cites the mental illness of James Forrestal, the first Secretary of Defense, who committed suicide shortly after resigning.

Steve R. Pieczenik is a psychiatrist and a former Deputy Assistant Secretary of State who served in the Nixon, Ford, and Carter administrations. "The conflicts were very real in 'The Crisis Game'— the territorial imperatives, the bureaucratic imperatives, the personalities," Pieczenik told me in a critique of the docudrama. He saw something else in the game that he had seen before in real crises. When an expert enters the game—or the crisis—"we're not working off the crisis. We're working off his mind. And that makes it hard for me because I have to find out what his distortions are, and I can't, even with my own psychiatric skills. But, more importantly, I then have to convince the system that the expert doesn't have credible eyes on the thing. So you're spending half of your time trying to convince the system and the other half figuring out what he's saying, what his line is."

Pieczenik, who became a crisis management specialist in the Department of State, endorsed gaming as a technique for learning crisis decisionmaking. But he said he had lost faith in most gaming because its value was not recognized by high-level leaders who send low-level, untrained aides off to the games.

Speaking of them, he said, "Gaming or simulation—it's like doing ballet practice when you don't know how to walk. Gaming is of a high order of magnitude. But primarily it gives jobs to a lot of academicians and a lot of colonels who are never going to see action. The products of the game go into the bureaucratic wasteland." Gaming, he said, should be "the necessary means by which people who are really involved in crisis continuously game out their skills."

In the Pentagon games (which, like most players I talked to, he still called "SAGA games"), "the military is not the strongest advocate of force. It's usually the wishy-washy liberal or the political appointee who comes in and feels very desperate and resorts to force. The ones who do frighten me are the scientists. God almighty! Those nuclear boys, you know, the ones from the labs. They're *deadly*. They love to think in megatons this and kilowatts that.

"In the SAGA games, it's sometimes unknown why some people are chosen. In a real-life situation, that's not the case. Some hot shot comes in from the White House who thinks he has handled something. You know, like a little Indian riot."

As an example of the kind of incident that occurs in real life but is never gamed, Pieczenik, implying that it came from his own experiences, gave this: You are handling a hostage situation—the seizure of an airliner in a foreign country. And then you are told, "The President has a friend in that country and he wants to get *his* friend off the plane." Pieczenik paused. "And," he asked, "are you going to do something about it? That never comes into those so-called bureaucratic models that Allison's writing for graduate students."

Pieczenik, as a ranking Department of State official, was fighting the image of bureaucratic model in his own gaming and in his management of crises, which he thrived on. "It's not only exciting," he said. "It's addictive. You become addicted to crises. *I* became addicted. I was a psychiatrist and I *knew* I was becoming addicted. I needed a higher and higher amount of that action in order to satiate a sense of accomplishment and challenge I wasn't going to get from a tax crisis. So each crisis became increasingly more complex and more interesting—and conversely, each day in the bureaucracy became increasingly more boring and dull."

Pieczenik represents one of the constituencies of political-military games: the government civilian who has been in enough crises to know that even good gaming can only mildly mimic the grinding reality of an actual crisis. Unconnected to government or think tank, he speaks candidly, with the independence of a psychiatrist in private practice and a writer (whose first novel, *The Mind Palace*, set in the Soviet Union, earned fan letters from Richard Nixon and Secretary of State George Shultz). Pieczenik is a rare free speaker. Former government officials are usually reluctant to talk openly about gaming because there is always a chance they will say something that will hurt their chances of being brought back into a game or into a real crisis.

Military players who talk are rarer still. One of the articulate exceptions is retired Major General Edward Atkeson, the colleague of Trevor Dupuy who wrote anonymously about what was wrong with games and, after retirement, revealed his authorship (Chapter 3). During his career, besides being director of the Army's indigenous think tank, the Concepts Analysis Agency, Atkeson was the CIA's National Intelligence Officer for general purposes forces. He was also a player in numerous games across the simulation spectrum, which he saw from the viewpoint of a soldier.

"At one end of the spectrum," he said, "is the reality of war. Then next to that you have exercises, and you begin to say, 'You really don't need all the soldiers out there—they're just cluttering up the landscape—so let's just have a command-post exercise. Let's just have the staff there.' And then you say, 'You only need the operational bit of the staff for that.' Then people say, 'Well, you really don't have to do that because the communications break down and you waste a lot of time. So let's all of us just sit around here in the basketball court and play it on maps.'

"And so right through this whole spectrum you're moving away from realism to higher and higher levels of abstraction. And finally you begin to work your way over to the deterministic area. You get away from the stochastic play to the deterministic area, where the real analysts come into play. They want something that is scientifically replicable. They'll say that if you have to roll the dice, or look for expected values or something, you can't depend on that.

"And finally you get to very abstract simulations, and the beauty of them is that on a computer you can do them very, very quickly. You can run World War Three in four hours, and run it over and over again. That end of the spectrum is most attractive to the guy who's concerned with resources. He wants to know, if I have a hundred billion to spend, how many tanks do I buy? How many airplanes do I buy? He wants to try it with different mixes. He wants to do sensitivity analyses: Suppose I added four tanks instead of three tanks to a platoon? The force developer wants to see whether we should have a twelve-man or nine-man squad. How did it work when so many guys had rifles? And when a different number had them? And so forth: insights, likelihoods, history in parallel.

"The Army has a big model improvement program now. TRA-SANA—the Training, Doctrine, and Systems Analysis Agency. They do very high-resolution models, tank on tank, platoon on platoon. And what they are basically trying to find out is, How can we fight outnumbered and win?

"I'll tell you how: You ambush the other guy. And then you'll get very favorable exchange ratios. Well, the Army wants to have a consistency. We're teaching guys down at our platoon and com-pany level—and what happens at battalion?* You get a whole bunch of ambushers. Well, what happens at brigade—and corps? And theater? You end up with a World War Three that is a *thousand ambushes*. It means that somehow we have to find a way to fight World War Three where *we* pick the terrain. *We* pick the phase of the moon, the time of the day, the season of the year.

"Well, that's not the way the enemy usually likes to play. He doesn't ask, 'Well, where do you want me to be sucked in for an ambush?' *He's* going to pick the phase of the moon, the time of the day, the season of the year—and the stacking of the forces. The initiative is on him. But we can't deal with that primarily

*For sizes of military units, see note on page 356.

because we in the Army don't know how to think about anything above corps level. We have locked ourselves into a very narrow view of battle."

Atkeson told a story about an analyst and a general. "They were arguing about how World War Three was going to come out, and finally the general, in exasperation, says to the kid, who has all kinds of PhDs, 'Well, goddamn it, we'll just have to start the war and see how it goes.' And the analyst said, 'That would be totally invalid. You'd only get one run of the experiment.'"

For the typical officer in any service, the political-military game is on a familiar part of the simulation spectrum. The game is not a field exercise, with real troops or real ships, but it is at least a chance to sit around the basketball court or a room in the Pentagon with real people and work through a problem. The most troublesome spots on the spectrum are found just past the political-military gaming and the command-post exercises. At that point, shades of reality begin to fade into models. The spectrum begins with war as the supreme reality, and it ends with analytic models as the supreme abstraction.

At least one operations research pioneer, Jacinto Steinhardt, has questioned the growing use of models and simulations, saying that a simulation was proof that the analyst who built it had failed to find a real solution to the problem being simulated. "To many trained in the model-experiment theory interplay of the natural sciences," he said, "feelings range from discomfort to disbelief that there are educated and seemingly intelligent people who have the temerity to create mathematical models of a phenomenon as grandiose and complex as 'warfare.'"

The U.S. General Accounting Office, which investigates issues for Congress, was asked in the late 1970s to find out whether the Department of Defense was getting its money's worth for the multitude of expensive models filling up the abstract end of the simulation spectrum.* A catalogue published in 1982 listed 363 war games, simulations, exercises, and models—some so complex that they had taken seven years to develop.

"Different analysts, with apparently identical knowledge of a real

---

*The report's critique of NATO models is covered in Chapter 4.

world problem," the GAO report said, "may develop plausible formulations that lead to very different conclusions—none of which are verifiable or refutable. . . ." To expect such models "to produce 'objective,' scientifically valid results is no more reasonable than to expect that a particular brush will produce fine paintings, or a particular knife fine carvings."

Much defense planning is based on theater-level combat models, which, like weapons, are usually produced by outside contractors. The GAO report looked at four theater-level models, each of which "uses different mathematical criteria to fight the same real world battle" and renders "profoundly different interpretations of the same real world phenomena." The phenomena included numbers of troops, types of weapons and equipment, weather, and terrain. The models assigned variables to such factors as the firing rate of weapons, the probability of hits, the distance a target can be seen and fired at effectively, weapons' line-of-sight in different kinds of terrain, and the amount of time it takes for troops to get into and out of armored personnel carriers.

Models varied greatly in granularity, the model-maker's term for a model's fineness of detail. One theater-level model was based on 40,000 bits of data, another on 212,250. Deployment and firing of antitank missiles in one model was based on data that showed no obstacles between a gun and its target. On the actual terrain where a battle would be fought, the antitank missiles would have more likely hit trees and hills than tanks.

Models are incredibly complicated, and going from a model of a rifle company to a NATO-size model is not just a matter of multiplying. "We have not yet got to the point that we can relate our battalion models to our brigade models, to our division models, to our theater models," Trevor Dupuy said in response to what analysts call the aggregation problem—and to what Atkeson saw as the ambush problem. "We can't take the results of the lower-level models and make them coincide with the results of the higher-level models. You run into very difficult technical and conceptual problems: At what level do you model the Vietnam War? If you model it at the highest level, you can't relate that to the results of a company- or a battalion-level engagement."

Many military people and many civilians associated with military work do not like the GAO. But most gamers conceded that much of the GAO's criticism was justified. "Combat simulation in the United States and the western world today is in a state of chaos," Donald S. Marshall, one of Dupuy's seekers of a theory of combat, said. "Few combat models will take the same input and provide comparable output results. There is generally no agreed methodology for determining attrition and advance rates. . . . Yet, with the increasing complexities of warfare, [U.S. defense] officials may be, in an ever-larger part, dependent upon use of models to help them make decisions."

When those officials reach into the model bag and pull out something to help them in the real world, they rarely know what they have in their hands. Models are universally described as "black boxes," contents unknown.

The results of some of the games and the data displays of some of the models are patently unrealistic, but not all of these wind up scattered about the bureaucratic wasteland. They find their way into the contingency files of the Joint Chiefs of Staff and the National Security Council. In a real emergency these decisions and plans would be whipped out of the files and brought to life as actions to be carried out in the nation's name. Suppose, though, that in a crisis the Joint Chiefs called up a plan and only then discovered that the plan was a fantasy. The nation's blueprint for going to war may well be one of those fantasies.

On the Atkeson wargaming spectrum, a mobilization exercise is a version of a CPX, a Command Post Exercise, and is located somewhere around his basketball-court gaming. Mobilization exercises, which are like maneuvers without troops, are for learning how to move troops and what military logistics experts call the bullets and beans.

The exercises usually do not get much attention, even inside the Pentagon, because, as a veteran of those exercises, William K. Brehm, put it, "If you drive around the parks of Washington, you'll never find, as far as I know, a statue that is dedicated to a committee or to the champion of spare parts. There is no champion for spare parts, for ammunition, bandages, syringes." Brehm is chairman of the board of Systems Research & Applications Corporation, a major defense think tank whose founding in 1978 was tied to a study of a massive mobilization exercise called Nifty Nugget.

Secret mobilization exercises began during the Kennedy Administration and have been conducted about every two years since. But rarely do high-level officials get involved. The problem had been recognized as far back as the Johnson Administration, when a secret study of presidential decisionmaking said, "In order to get participation by top level leaders, exercises must be realistic and interesting. . . . Furthermore, the exercises should be designed in a manner which considers the motivations and concerns of high level people. For example, a high level official might not want to become involved in an exercise in which he is performing before many people. . . . He might not want to seem unprepared to feel committed as a result of impromptu action. . . . One solution that has been suggested is to use stand-ins for these key leaders and let the key leaders watch and observe how these stand-ins behave." (This apparently was done to some extent during the Ivy League game.)

By 1978 high-level officials were still not sending themselves or even stand-ins to the exercises, which had been going on regularly, with lower-ranking officers getting their chance to play at being higher-ranking officers. A large bureaucracy had grown up around the bullets-and-beans drills, and many people in that bureaucracy knew that the drills were exercises not in make-believe but in deceptive make-believe.

In the paper exercises the nation was smoothly mobilized, the troops were sent off to war with all their equipment, and they arrived on time at the places the scripts wanted them to be. All of this was, of course, imaginary, for no troops or supplies really materialize in a mobilization exercise. But the reality that the exercises were supposed to demonstrate was a lie. The happy ending in the mobilization scripts was absurd, the assumptions of the mobilization foolish, the expected results phony.

Planners, playing the exercise game far below the managerial or decisionmaking levels of the Pentagon, knew about the phony make-believe. But they went on treating each exercise as a real blueprint for a real war. Some of the more dedicated workers tried to let people know what was going on. Slowly the knowledge of the charade made its way upward through the bureaucracy, and it finally reached the Office of the Secretary of Defense.

In the summer of 1978 Brehm got a call from that office, a place that Brehm, the consummate industrial-military complex executive, was very familiar with. From 1964 to 1977 he had shuttled between stints in defense-oriented civilian jobs and Pentagon jobs, serving at various times as Deputy Assistant Secretary for Land Forces Programs in the Office of the Secretary of Defense; as Assistant Secretary of the Army for Manpower and Reserve Affairs, and as Assistant Secretary of Defense for Manpower and Reserve Affairs.

Brehm said the man in the OSD, as the Pentagon powerhouse is usually called, "was inquiring whether I'd put together a senior team to evaluate the OSD staff in a major exercise that was coming up in November. So I inquired a bit and got seven or eight people— a four-star rep from each of the services, plus a couple of other specialists. Very senior people."

The top-secret exercise was code-named Nifty Nugget, a global, three-week imitation of war that would involve every military command and every major Pentagon agency, along with thirty-one civilian departments and agencies, ranging from the Department of Agriculture to the U.S. Postal Service. The then Under Secretary of Defense for policy, Robert W. Komer, veteran of the CIA, the Kennedy and Johnson White House, and Vietnam War councils, called Nifty Nugget and two related exercises "the most ambitious tests of mobilization ever undertaken in this country and perhaps the world." The exercise was to simulate the transportation of 400,000 men and 350,000 tons of supplies across the Atlantic in the wake of a surprise invasion of Western Europe by Warsaw Pact forces.

"The scenario was cast in concrete before we got into it," Brehm recalled. "The scenario was kind of typical, a fairly quickly developing crisis preceded by a long period of tension buildup, but not so much so that [nuclear] thresholds are crossed."

The focus was on mobilization, but since there was a war going on, some war-fighting was built into the scenario. "We found in Nifty Nugget that if you start playing war," Brehm said, "the generals and the admirals would a lot rather play moving troops around— and ships—than mobilizing, which is kind of boring. You know, 'Can we call up this many? Can we draft this many?' Well, the focus started drifting toward moving troops and moving ships, and [Lieutenant] General Wickham would keep bringing them back: 'Hey, guys, this is a mobilization exercise. We'll let the commander in chief in Europe fight the war. Let's worry about the mobilization issues.'"

Army Colonel Jerry J. Burcham, the senior Department of Defense planner for Nifty Nugget, was one of the officers who knew the truth about mobilization plans of the past. "We didn't find out anything *we* didn't know. The difference was that the big guys were there when we found this out again. We highlighted a lot of things that the senior people hadn't known," he recalled. "And, really, when you lump them all together, it was a horror story." There were, for instance, "a lot of dual-assigned forces," Burcham said. "The 82nd Airborne is probably in every war plan there is. Everybody wants that airborne division."

By the time Nifty Nugget was played, the NATO-Warsaw Pact war had been planned longer than the Thirty Years War had been fought. Yet, as exercises had revealed year after year, many of the fundamental realities of the anticipated war had been ignored. On paper in the 1950s and in computer programs in the 1970s, all went well: The U.S. reinforcements arrived on time, along with their equipment. And, magically, there were no problems about the thousands of dependents of U.S. servicemen in Europe, no panicky American tourists running about the war's landscape, no refugees clogging the roads leading to and from the front lines.

In Nifty Nugget realists insisted that the civilians be honestly put in the exercise. The result was chaos. Nearly a million American civilians jammed the European commercial airports that were to receive the first contingent of airlifted reinforcements from the United States. Troops had to be drained from the combat forces and posted at the airports to keep order. The military and commercial airliners that had begun bringing in the troops were loaded with civilians on the return flights, but there were obviously not enough planes. Nor were there enough planes to evacuate the wounded from overcrowded military hospitals in Europe.

There were no realistic plans to handle the flood of refugees pouring into the United States. Domestic transportation systems and domestic ports were already straining under the burden of military mobilization. The previous mobilizations had been run in a neat little military world without civilians. When Nifty Nugget realists added the civilian world, they brought in not only the panicky tourists but also the American home front.

Military movements were slowed down by an overloaded transportation system. Civil defense officials in the exercise added problems to remind military planners who they were fighting for. "We believe that the military must never forget that the civilian sector has a claim on the resources also," a civilian veteran of the exercise told me. "And they cannot run a realistic exercise unless there is input from the civilian side."

Tanks, trucks, and other military vehicles were supposed to be transported on railroad flat cars from various bases and depots to disembarkation ports. In Nifty Nugget, people checked the minor details of this seemingly routine bit of logistics. The mobilization plan's railroad schedules were out of date, but that was only the beginning. There were not enough railroad cars available where and when they were needed. (It takes 2,400 rail cars to move an armored division's equipment and basic supplies.)

Researchers, trying to follow the route of the phantom tanks, went to the military bases and the ports. They found that in many places railroad tracks had been moved or removed, and tanks could not be carried to docks for loading onto ships. And, even if the tracks were still there at dockside, in the years since the original mobilization plans had been drawn up, some tanks had become so hefty that the old railroad cars and tracks would not be able to bear them.

Ports with special facilities for the loading of ammunition could not handle the surge of munitions dangerously piling up on the docks. Planes in the Civilian Air Reserve Fleet—passenger airliners pressed into military service—were called up, but there were no provisions for quickly converting them to military use. Only 7 percent of the commercial pilots and flight engineers were in the Air Force Reserve or Air National Guard. Laws were vague concerning the status of airline employees and other civilians whose skills were needed during a mobilization.

The Pentagon had three separate transportation agencies for handling the movement of men and matériel. Each operated under a schedule uncoordinated with the others. If a surprise were thrown into an exercise—sabotage of a rail line, the destruction of an ocean terminal—the shipments could not be diverted. They had to stay on their treadmills to nowhere.

This was the Nifty Nugget horror story that senior people lived through for three long and frustrating weeks. Never before had there been such an all-star cast for a mobilization. High officials ran Control, which included the newcomer "Congress" alongside its traditional wraiths of weather, time, and other countries. Stanley Rezor, the Under Secretary of Defense for policy, played the Secretary of Defense, and Lieutenant General John A. Wickham, Jr., a future Army Chief of Staff, played the Chairman of the Joint Chiefs. Each of the chiefs sent a three-star deputy as a stand-in. "That was considerably more horsepower than we had ever had before," Burcham said.

Gary E. Lindquist, a retired Army officer who, like Burcham, went from the Pentagon to Brehm's think tank, had headed the JCS Joint Exercise Division. (One of his tasks on that job is a reminder of an exercise's proximity to real war in the wargaming spectrum: Lindquist had to get White House approval for any exercises that were "significant and sensitive.") Nifty Nugget, Lindquist remembered, was carried out "when there was a lot of discussion about whether war would remain conventional or would go nuclear. Is mobilization a feasible thing? Is conventional war feasible? Is there any chance of the U. S. mobilizing and fighting another conventional war as we did in World War II?"

One answer came in the wake of Nifty Nugget when officials at the Industrial College of the Armed Forces, which trains officers in mobilization and high-level logistics, dug out of the files the mobilization plans that had worked in World War I and World War II. They had to be better than what had been spewed out of the Nifty Nugget computers. Under that burden many of the computers broke down, as did many elements of the worldwide communication system. I was told by a mobilization specialist in 1986 that if a callup were ordered that day he would dust off World War II mobilization plans and improvise with them.

The army that finally made it to Europe in the Nifty Nugget war suffered from shortages of food, ammunition, and fuel. Within a few weeks the 400,000 men were all but wiped out. "The Army was simply attrited to death," a veteran of Nifty Nugget later said.

Those doomed soldiers had little to back them up, because one reality that was tested in Nifty Nugget was the reluctance of a President to mobilize the reserves. "In 1964–65, when the President decided to go to war in Vietnam," Brehm said, "I saw how the OSD's staff was mobilized, and as I reflected back on it, I saw how little they knew about what to do. You can imagine what that did to the plans, because the Army and the Air Force are built around the assumption that all the reserves will mobilize for any substantial, multidivision event.

"I've never been in a foxhole. I can't imagine what that is like. But I have worked twenty-four hours a day, sleeping in the office— or not sleeping at all—while the Secretary of Defense is shuttling between the Pentagon and the White House laying down alternatives for the President. So, you know, you pick that up."

Drawing from this experience, Brehm examined the exercise for its political dimensions and for problems that civilians would have to solve. Out of the chaos of Nifty Nugget came the realization that mobilization was impossible without draftees—delivered, as Brehm described it, by "a system that is capable of producing qualified inductees at the same rate that the Army training base can accept them and the services need them." This was translated into the law obliging eighteen-year-old males to register with Selective Service so that there could be a quick start-up of the draft at the beginning of a mobilization. This was the most sweeping result of Nifty Nugget and the first major piece of domestic legislation ever to stem directly from a war game.

One of the Nifty Nugget players was Robert (Robin) B. Pirie, Jr., who had been in systems analysis, was a former Navy submariner, had worked in Andy Marshall's Office of Net Assessment, and had served as Deputy Assistant Secretary of Defense. He went into the exercise with the real-life post of Assistant Secretary of Defense for the management of military and civilian manpower. "Nifty Nugget," he said in an interview, "was initially an effort to see if we had the right kind of procedure and systems to start a war. If somebody says, 'Start a war,' do you have procedures and systems to do it?

"We found there was no legal system for calling up reserves and retired personnel, for seizing ships. Refugees would all be coming back from Europe. HEW [Health, Education, and Welfare] was supposed to sort them out, but HEW didn't even know it had the responsibility to handle refugees.

"It was not in the perspective of the players in the Pentagon that they had roles they would have to play. The Joint Chiefs of

Staff didn't know their job would shift from complaining about not having enough resources to do the job to the role of running a war. The JCS was taken aback to have such participants—assistant secretaries of defense and so on. It had been difficult to tell [Secretary of Defense Harold] Brown [about exercises] in JCS briefing style. Hearing about the JCS exercises was not the most exciting part of his day."

Nifty Nugget was different. People listened and helped to produce what Pirie called the laid-away laws—contingency legislation that would be introduced as part of a mobilization. Federal laws and Pentagon procedures were changed. The President, who could call up 50,000 reservists without declaring a national emergency, got from Congress the power to call up 100,000. To prevent interservice competition for the 100,000, the Pentagon produced a series of "packages," each of whose contents would reflect the manpower needs for a specific emergency. One unexpected job group, now in the packages, consisted of ship loaders—men with strong backs and the seemingly unmilitary skills needed to operate forklifts and cranes at dockside here and overseas.

Because the mobilizers' computers could not find thousands of the nation's 450,000 inactive reservists, the Pentagon wanted them to report for duty one day a year, just to keep track of them and their skills. Planners worked on ways to establish which of two jobs a reservist may keep in an emergency. About half of the members of the Immigration and Naturalization Service Border Patrol, for instance, were reservists who would be called to active service. The Veterans Administration had no plan for replacing employees who were reservists (nor for delivering health care during a national emergency). The Pentagon came up with two choices: Reservists holding essential jobs would be discharged from the Ready Reserve or emergency replacements would be trained to replace the reservists so they could be called up.

Federal disaster specialists were rounded up and put into one new organization, the Federal Emergency Management Agency, which later organized the National Disaster Medical System. In a real war this organization would try to minimize the disruption of the civilian health-care system caused by the transport of evacuated military casualties to civilian hospitals in the United States. This last task stemmed directly from the discovery in Nifty Nugget that U.S. soldiers wounded in European fighting would have to be flown back to the United States for medical treatment because of the shortage of military medical facilities in Europe.

Emergency laws pertaining to mobilization were analyzed in terms of Nifty Nugget and similar exercises—and from real experiences going back to World War I. The analysis led to an exhaustive report that showed the far-reaching effects of make-believe, for from the exercises came the realization that the laws on the books were inadequate for a real mobilization. The report put on the shelf for future congressional consideration such issues as censorship. (The report said that the Pentagon's little-known Wartime Information Security Program [WISP] "is now no more than an unfunded, unstaffed, undeveloped acronym.")

The report also recommended amendments to the draft law that would allow the drafting of people not by chance but by needed skills, such as health workers (including women), pilots, computer programmers, merchant marine crewmen, and interpreters. The report speculated on the possibility of a "mobilization competition" between the United States and the Soviet Union, in which both countries escalate their mobilizations to disrupt each other's economies—but do not go to war.

The report, in its discussion of the need for a health-occupations registration plan, gave a glimpse of the immense, secret, and complex Defense Master Mobilization Plan, which feeds on exercises like Nifty Nugget. The plan's timeline for the conscription of health workers shows how detailed the plan must be. It anticipates that a law will be passed and signed by the President in eight days, the first health inductee will report for active duty on the fifty-fifth day, and more than 20,000 will have been processed by the eightieth day.

The Joint Deployment Agency, whose mission is the coordination of mobilization transportation plans, was created almost immediately after the exercise. The buildup of a sealift quickly began. Ships were stocked with the vehicles, munitions, and other supplies for a Marine Amphibious Brigade and sent to sea or kept in a state of readiness, able to leave port on two hours' notice. The Military Sealift Command acquired "roll-on, roll-off" ships that can rapidly load and unload tanks and trucks. In a NATO exercise in 1984 one of these ships carried 1,100 vehicles to Europe and unloaded them in thirty hours.

The Marines got storage ships that could sail from an East Coast port and reach the Middle East in fifteen days, with enough equipment and supplies to back up 16,500 Marines sent out separately. Moving such ships, according to the Marine Corps Commandant, General Paul X. Kelley, is itself a move in a real global game—a "signal in the realm of escalatory signals in a crisis. It is the most cost-effective means of crisis management we've ever had."

Nifty Nugget was not the turning point that the beans-and-bullets pushers had hoped for. Two years later an exercise named Proud Spirit took the same stage, with a slightly changed script (just mobilization, no war-fighting and thus no wounded soldiers), and the show was a flop.

The Worldwide Military Command and Control System broke down, blacking out some military commands for as long as twelve hours. Ammunition stockpiles were disastrously low. There were not enough U.S. troops in the United States even to bring U.S. forces in Europe up to their wartime strength. And this time the Pentagon horsepower was not there. "Among eight major agencies needed to carry out a mobilization for war," a professional military journal reported, "not one responded to an invitation to send the top executive officer—the agency secretary—to a two-hour briefing on Proud Spirit."

The worldwide communications system, which Congress had been investigating and condemning for years, "just fell flat on its ass," according to the journal, which also reported that, because of schedule mix-ups, Air Force transports landed at military bases two days before the troops they were supposed to carry got their orders to board the planes. And all this happened in an exercise that had been "simplified" by the amazing expedient of decreeing that there would be no war to fight.

Unofficial reports of ammunition shortages and inadequate medical facilities have continued to trickle out of the Pentagon in the years since Nifty Nugget, but secrecy so enwraps mobilization plans that outsiders get only peeks, granted by disgruntled leakers. The context of the mobilization exercises is also hard to bring into focus. As a former U.S. Ambassador to NATO has observed, "Military planning is caught between the short-war scenario, based on the early use of nuclear weapons, and the long-war scenario, involving the reinforcement and supply by sea of NATO's forces in Europe."

The Army is supposed to get ten divisions to Europe in ten days in an emergency. This is theoretically possible, at least on paper and in scenarios, but, according to a report in 1984, the Army does not have enough soldiers to meet its global responsibility and has only about half the equipment it needs to field the forces it has, "and much of what it has is outdated." The unreality that haunted war mobilizations in 1978 still stalked the exercises well into the next decade.

The Pentagon's political-military gaming and mobilization exercises are close enough to each other on the wargaming spectrum for an observer to say they are only different shadings of the same unlikely vision. To wonder whether ten divisions really can reach a NATO battlefield in ten days is the same as wondering whether a President really would fire nuclear missiles at the Soviet Union. It has not happened. We are looking at make-believe and hoping it will stay that way.

But terrorism has happened, and the gaming for terrorism has no spectrum. Today's game is a rehearsal for tomorrow's crisis. Today's players will take their same places at the actual controls tomorrow. Today's unseen enemies will be transformed on some terrible tomorrow into real actors acting out a drama that the world will be watching.

CHAPTER 14

In the Theater of Terrorism

On the morning of Tuesday, July 17, 1979, a fictional TWA Flight 847 took off from Athens en route to Boston and was hijacked by wargaming terrorists. On the morning of Friday, June 14, 1985, a real TWA Flight 847 took off from Athens en route to Boston and was hijacked by real terrorists. In an incredible paralleling of fact and fiction, the two flights would follow similar scenarios.

In both dramas attempts to end a hijacking confronted similar realities. And when the moves in the hijacking game are compared to the moves in the real hijacking, the fiction is like a pattern for the reality that came six years later.

Locked in the psychological combat of a long and grueling game, terrorist experts took on their roles so realistically that one of the playacting "terrorists" said after the game, "I thought I wouldn't be able to identify with terrorists. But after two hours I think that I was one of the hard-liners on our team, and I must confess that I really enjoyed killing those two people. I think that this was not only because you don't usually get such an easy chance to execute people, so it was, for a change, something new to me, but it was also that we really identified with our jobs."

The fictional Flight 847 began a scenario prepared for experts on terrorism from several nations, including the United States, West Germany, Israel, and the Republic of South Africa, gathered at Tel Aviv University's Center for Strategic Studies for an anti-terrorist war game code-named Kingfisher. The game was part of an International Seminar on Problems in Political Terrorism, sponsored by the center and the Office of the Prime Minister's Adviser on Combatting Terrorism. Delegates to the conference included U.S. Ambassador Anthony Quainton, director of the State Department's Office for Combatting Terrorism, and members of the U.S. Delta Force antiterrorist military unit.

The game was far more complex than a typical Blue-Red-Control war game. There was an Israeli emergency crisis team, a U.S. emergency crisis team, a terrorist team, and seven Control teams, each representing some aspect of what would be an intense and prophetic exercise. The clock was real—each tick was a real tick in a synthetic crisis whose pressures were real, not only because game time was real time, but also because outside the game rooms a real crisis was building in Iran. Five months before, in February 1979, an Iranian mob had temporarily seized the U.S. Embassy in Teheran. Four months after the Tel Aviv game dealing with hostages in an airliner at Teheran airport, some of the American players would be dealing with a real hostage crisis at the U.S. Embassy in Teheran.

At eight o'clock in the morning (0800 by the game's timekeeping system, which will be used in this narrative) the Kingfisher game's plane took off from Athens. Ten minutes after takeoff the captain of Flight 847 called the Athens airport control tower and said that the flight had been skyjacked by men identifying themselves as members of the Squad of the Martyr Zuhair Ukasha, that the flight had been diverted toward Gibraltar, and that one of the terrorists was about to make a broadcast over the airliner's radio.

The control tower recorded these words:

> The Martyr Zuhair Ukasha Squad of Sons of the Occupied Lands Organization announced that it took over a TWA 707 airliner. From now on all communication with the aircraft must use the code "Mount Carmel." Calls which will not use this code will not be answered.
>
> Our heroic fighters, the sons of Jaffa, Ramle, Nablus, and Al-Kuds, with the aid of our freedom fighter brothers from Germany, are carrying out this operation in order to arouse public opinion and make the world aware of the oppression of our people by American imperialism and its lackey, the oppressive Zionist state, and against the traitor Sadat and the exploiting capitalist regime of West Germany.

The world has to know that U.S.A. airplanes transport weapons to the warehouses of the Zionist enemy in order to oppress our people in Palestine and Gaza. The Zionist passengers are not civilians but foreign conquerors of Palestine who rob the Palestinian people of their land.

We fight against the betrayal plot to exterminate the Palestinian people and deny their just national rights. Under the protection of Sadat, who deserted his Arabism and betrayed his people, and with American bombs the airplanes of the Zionist enemy attack refugee camps in Lebanon.

Through her agents Begin and Sadat, America wants to take over the land of the Arab people in order to exploit its oil. European countries which support this exploitation and colonial policy have to know that they will not be spared either.

The statement ended with the cry "War until victory!"

The next message from the aircraft came at 1711, when the captain asked permission to land at Damascus International Airport. Permission was not granted. Next he tried Tripoli International Airport. Permission was denied and trucks were driven onto the runways. A few minutes later the captain desperately radioed Algeria's Dar-El-Beida International Airport and said that if he could not land there he would have to crash-land at sea.

The Algerian government said that for humanitarian reasons the plane would be allowed to refuel and take on food and water and then had to take off. The plane landed, but the skyjackers would allow only refueling. The plane quickly took off again and, after obtaining permission for flying through Syrian and Iraqi air space, landed at Teheran's Mehrabad Airport.

The pilot taxied the plane to a side runway and shut off its engines. Iranian troops surrounded the plane. At 1807 the terrorists' leader transmitted another message. The terrorists' "noble demands" included the release of more than fifty prisoners held in Israel and five terrorists held in West Germany; they were to be flown, via Switzerland, to Algeria. They also demanded that the United States release Sirhan Bishara Sirhan, the assassin of Robert Kennedy, and arrange his free passage to Algeria.

In addition, U.S. Secretary of State Cyrus R. Vance "must visit Palestinian refugee camps in the Tzor area which were hit by Zionist bombers using American bombs." Vance was then to pay

$100 million to the refugees for rehabilitation of their homes. Israel was to pay $30 million as "compensation," with the money to be funneled through a United Nations refugee organization. West Germany was to pay 10 million marks to a Swiss lawyer "who will transfer this sum to our hero brothers, members of the Red Army Faction in Germany." Belgium was expected to release two terrorists and "publicly declare its support of the Palestinian people's struggle."

Warning the governments that any delay in fulfilling all our demands would "result in severe repercussions to the people and the airline that we hold," the terrorists set 1818 as the deadline for publicly acknowledged acceptance of the demand.

"We warn," the message concluded, "that any delay in carrying out our demands will result in the immediate execution of one hostage every thirty minutes, starting at 1818. After 1908 any further delay will result in the execution of *two* hostages every thirty minutes."

The game began with the distribution of the scenario to the players. It was 0900 in real time—and in scenario time (0900 the morning after the hijacking).

There were three major teams: the Israeli emergency crisis team, the U.S. emergency crisis team, and the terrorist team. Control encompassed a wide array of principals: the Israeli, German, and U.S. governments, all Arab countries, including those who support terrorism; pressure groups (families of the kidnapped passengers and U.S. oil companies); U.S., Israeli, and international media; the U.N. secretary general. Control decided whether decisions made by the other teams could be carried out.

Control gave the teams four tasks: Suggest a coordinated U.S.-Israeli-German proposal; suggest ways to exert political pressure on the terrorists and Iran; suggest ways to put pressure on the terrorists through "indirect military action"; prepare "suggestions for courses of action should negotiations fail."

Control also had the U.S government send a "nongovernmental representative" to Beirut to establish a contact with the Palestinian Liberation Organization, which has sent an official to Teheran to work with Iran mediators.

Pressure is building on the governments. American families be-gin contacting their congressmen, who press the administration for information. Reporters and the families of Israeli hostages aboard the plane demand a statement from Israeli officials. (The media later turned down a terrorist's offer to come aboard for a live in-terview, fearing the interview could hamper negotiations. But when an Arab spokesman, off the record, puts out the false rumor that the hostages would be exchanged for the Shah of Iran, the media, though sensing that the information is false, spreads the report.)

At 1035 U.S. crisis managers ask for a meeting with the crisis teams from Israel, West Germany, and Belgium to coordinate con-tingency plans and responses to the media. Israel begins looking for ways to influence the Iranian government, including, through diplomatic channels, a request to the President of France to "ex-ercise his personal relations with Khomeini." Soon the United States and Israel form a joint team for possible military planning. Later German antiterrorist forces are included in the planning.

Meanwhile, an ill female passenger is released from the plane. She reports that the terrorists include four males and a female, some speaking Arabic, some German. Intelligence services advise their governments that the terrorists are members of the Red Army Faction of West Germany and the Popular Front for the Liberation of Palestine (PFLP). A mob of pro-terrorist Iranians surges onto the runway and confronts the Iranian Army units around the plane.

Diplomatic probes are being sent out. Algeria offers to mediate and gets permission without any public announcement. The Vat-ican is asked if the Pope can be persuaded to appeal to Khomeini for help. A secret U.S.-Israeli-Belgian-German military planning conference begins at a German Army base in Stuttgart. A U.S.-Israeli combat team is proposed. Israeli intelligence specialists comb the Koran looking for passages that can "divide Khomeini and terrorists on Islamic principles." The specialists also analyze pre-vious Palestinian terrorist raids to see how moral strictures of the Koran were violated.

At 1430 the plane's captain reports that the terrorists are seg-regating the American, Israeli, German, Belgian, and British pas-sengers in the rear of the plane. At the same time the Belgian government hints at an interest in accepting the terrorists' demands. Pressure is put on Iran to induce the hijackers to refuel and take off, the idea being that this will buy time for the victims and the multinational crisis managers.

But at 1500 the Iranian government stuns the crisis teams with a communiqué: "Iran is neither Uganda nor Somalia. We shall not allow foreign forces to operate on Iranian soil. Imperialism has been expelled from Iran together with the Shah. We do not support skyjacking, but the Palestinians' plight must not be forgotten. There is no military solution to the problem at hand. We shall not take part in any reckless attempt to release the hostages by force. The hostages' fate is in the hands of the U.S.A., Germany, Belgium, and Israel." The Iranian Army now has moved nearly two divisions to positions in and around the airport, eliminating the chance for a quick, compact rescue mission.

The United States and Israel step up their military planning. Israel orders a "dummy mobilization," which sounds extensive but actually amounts only to the movement of one hundred tanks toward northern borders with a maximum of publicity. At the same time real plans are made for surgical bombing strikes against PLO sites and an extensive military strike into southern Lebanon. The United States examines—but does not carry out—its own military options: Move the Sixth Fleet closer to the Lebanon coast; put U.S. forces in the Mediterranean and Europe on a higher state of alert; buzz PLO camps with U.S. Navy carrier planes; launch surgical strikes against nonpetroleum Iranian targets.

The Belgian Foreign Minister announces Belgium's unilateral decision to accept whatever terrorist demands it can. The terrorists release seven hostages—a Belgian man and wife, three passengers with Arabic-sounding names, and two who could be French or Belgian.

The released hostages say that the five terrorists, one of them a woman, are armed with handguns and grenades. They have placed what appear to be explosive devices around the cabin and have connected them with wires. The terrorists are treating the passengers harshly and have beaten two Americans.

At 1645 the terrorists release a communiqué: "We . . . have proven our goodwill by releasing the Belgian hostages; however, the promises made to us have not been kept. At exactly 1800 hours we will execute the first hostage, Mr. Ben Chaim from Israel, who is standing here next to me. Revolution till victory!"

As U.S., Israeli, and German military officers escalate plans for a major air-ground assault on the airport, President Carter, Prime Minister Begin, and President Schmidt confer and rule out military action. Israel toys with the idea of leaking to a reporter the news that terrorist prisoners in scattered Israeli jails are being moved to a central spot.

The terrorists release Ben Chaim's wife at 1750. Twenty-five minutes later they throw his body from the plane. They say that "as a sign of good will" toward Iran, they are postponing the next execution until 2000, instead of carrying it out at 1900.

In a previous announcement the terrorists had indicated that they would alternately kill Israeli and American passengers. At 1915 the terrorists name the next victim, an American identified as Mr. Black. Fifteen minutes later they release Mrs. Black and warn that the plane "will be blown up automatically" if a rescue attempt is made. At 2000 Black's body is tossed out of the plane.

The crisis managers desperately turn toward the idea of opening negotiations with the terrorists, weighing the offer of safe passage in return for release of the remaining hostages. Day One of the crisis ends.

Day Two begins with a communiqué from the terrorists at 0930: "We have waited enough. The situation in the plane is critical. We have nothing to lose, and we are prepared to die. We will agree to negotiate after the governments announce their intention to accept our demands. If we have no positive reply till 1030, we will execute three more hostages. Stop playing for time. In case of a positive reply we will free all the twenty women and children."

Israel makes plans for air and ground strikes on PLO bases in Lebanon and in Arab countries that support PLO terrorists and an invasion of Lebanon, including Tyre. At the same time the Israelis work out a "bazaar bargaining mode" for negotiating with the terrorists. The idea is to bargain through a series of options, holding until the last the best item in the bargaining bazaar: a deal tying release of the hostages to payment of all demanded money and the safe conduct of all terrorists "not guilty of shedding blood."

At 1000 the terrorists issue another communiqué: "We accept

to negotiate with the U.S.A., FRG [West Germany], and Iranian representatives at 1100, with the condition that they are high-ranking officials and are authorized to make decisions. We are delaying the executions until after the meeting. In the interest of the hostages we intend to talk business, and no more words, words, words."

Israel independently decides that if negotiations fail, all contact with the terrorists will stop and the Iranians will be warned that they must give up aid to the terrorists and support U.S.-Israeli-German military action, with the implicit warning that failure to do this will put the Iranians in the line of fire. The Israeli armed forces will "use" this "terrible opportunity" to punish the PLO with massive military actions.

The three nations agree on the maximum concessions they will make: All hostages and plane to be released. Safe passage of terrorists—but no release of murderers among terrorists holding plane. No visit of Vance. Payment of money demanded. Release of no more than eighty terrorist prisoners (none of whom can be murderers).

Iran works out the protocol for a conference on board the plane. Participants are the terrorists, Iranians, PFLP negotiators, and PLO representatives, who urge the terrorists to stop the killing because they are hurting the PLO's international image. The terrorists agree to a U.S.-German proposal to release all women and children. Secretly the Iranians allow four armed PFLP men to stay aboard the plane.

King Khalid of Saudi Arabia privately informs the U.S. government that he will put pressure on the PLO "to save American lives," but the fate of the Israeli passengers cannot be part of any arrangement involving the Saudis. The King also tells the United States that in exchange for Saudi help the United States must be willing to make unspecified concessions to the PLO.

Israel, simultaneously urging the United States and Germany to push for Saudi intervention, starts a multifaceted public relations campaign against the PLO: condemnations from victims' families, religious leaders, and intellectuals in several countries; pressure at the United Nations to expel PLO observers from the UN; appeals for intervention to socialist leaders throughout the world; a special campaign among African contacts to influence the PLO and Iran;

an exploration of ways that airlines of several countries can threaten Iran economically.

The terrorists release the women and children at 1205 and demand an immediate meeting with the ambassadors of the United States and West Germany. The United States has developed two tracks for its negotiator. One is simple: Pay the ransom to save hostages of all nations. The other is complex: Try to get the release of a certain number of U.S. hostages, along with an equal number representing all other nations on board—providing that the move does not result in Israelis being the only passengers left on board.

The terrorists drop their demand for the release of Sirhan but maintain their demand for the release of the West German and Israeli prisoners. They also simplify their ransom demands to $50 million on the spot.

The United States agrees to pay the $50 million. Israel will release twenty nonmurderers and pay $10 million. Germany will free two nonmurderers. All offers are on the condition that the terrorists free all hostages immediately.

At this point the game ended.

The ending of the drama of the real Flight 847, six years later, was a real ending: The surviving hostages were released. Hijacked soon after takeoff from Athens, Flight 847, like its game counterpart, sought a place to land and, as in the script, once was thwarted by blocked runways. From the hijacking to the negotiations that ended the ordeal sixteen days later, Flight 847 followed several parts of the game scenario: The hostages were segregated. The hijackers sought out Americans and passengers with "Jewish-sounding" names. An American was singled out and killed. The terrorists demanded the release of prisoners held by Israel.

In the real hijacking, Israeli prisoners, most of them Shiite Moslems, were released (in what was supposed to be a coincidence). And, as in the script, negotiators sympathetic to the terrorists' cause became the principal negotiators, eclipsing the efforts of antiterrorist nations. In the real case the hostages were freed and the terrorists slipped away because that was the way the sympathetic negotiators played their real game.

After the Tel Aviv game players talked about both the game and real terrorism. Reading the transcript of the discussion I got the impression that, to many of the players, terrorism *is* a game. It has been played so often, in reality and in contingency plans, that players on both sides now know the rules and the real stakes. Like two sovereign nations at war, the terrorists and the hostage negotiators look beyond the victims of the conflict to the conflict's greater issues.

A member of the U.S. team, while admitting that "death concentrates the mind," said, "We have to be constantly aware, even if one or two people are killed, of the fact that there are longer-term objectives which one is trying to pursue. The whole strategy need not be transformed by the fact of one or two deaths."

As if trying to rationalize the fictional deaths from the killers' viewpoint, a member of the terrorist team blamed the killings on a sense of powerlessness. "Our first frustration came from not being taken seriously," he said. "When we decided to execute our first hostage, one of the reasons for it was that we had to show the people out there that we would really do something. Nobody seemed to believe us."

The family players said they shared the terrorists' frustration. Like the terrorists, the families were "forced to act in a more and more extreme way" to get officials' attention—and, like the terrorists, they turned to the media to get that attention.

The need to put human life above all other issues, said Ariel Merari, chairman of the psychology department at Tel Aviv University, "touches upon the basic philosophy of liberal democracies and is, therefore, very difficult to alter. It is quite paradoxical, but nevertheless true, that governments, which in wartime send thousands of soldiers to their death in the service of political goals, are willing in times of peace to stretch—to thin—their most essential political principles in order to save a few scores of civilians.

"It is time to recognize that for many nations the struggle against political terrorism amounts to a state of war, albeit one which is undeclared and fought according to different rules."

The blurring of reality at Kingfisher seemed more intense than in typical Blue-Red war games. As a leading member of the U.S. team said, the crisis conditions "under which we work in Washington" are "very similar to those in the game." Another player suggested that "relations with Iran . . . may be more important than solving the hijacking" and that the United States' action should give the oil crisis priority over passengers on an airplane.

These were men talking not about the game they had just played but about the bigger game of nations that encompassed what they did and what they planned to do if they ever stepped out of the game room and into a situation room. Kingfisher's records showed the reality hidden in what was supposed to be a game. I wondered how much potential reality also hovers around such high-level war games as Ivy League.

A U.S. team member noted "an obsessive search for a military option" and said it was driven at least in part by the existence of elite units, such as the Delta Force, some of whose anonymous members actually played in the Kingfisher game. There was, the U.S. player said, a "psychological need for retaliatory action."

The game and the hot washup took place in July 1979. On November 3, 1979, about ninety people, including sixty-three Americans, were taken hostage at the American Embassy in Teheran. On April 24, 1980, eight American servicemen were killed and five wounded in an accident that occurred when Delta Force led an ill-fated attempt to rescue the hostages.

The man who played the President of the United States on the Control team in the Kingfisher game was Dr. Robert H. Kupperman, executive director of the Georgetown Center for Strategic and International Studies, a foreign-policy research organization with ties to academe and to Washington's large colony of national security worker bees, who are ever ready to toil for the present or next administration. Kupperman was deputy assistant director and chief scientist of the U.S. Arms Control and Disarmament Agency and former assistant director for government preparedness in the President's Office of Emergency Preparedness. An internationally recognized authority on terrorism, he is also one of Washington's leading gamemasters, an alchemist who turns plausibility into synthetic reality.

Kupperman said Kingfisher had its origin in a game played the year before in Berlin, during an international conference on terrorism, which he had co-chaired. "We wanted to try something different and so we decided to run a game," he recalled. "In the case of terrorism, this had not been done very often." Among the

Israelis at the Berlin conference was Ariel Merari, who suggested that a game be played at Tel Aviv the following year.

For the Berlin game Kupperman invented a country called Ruritania—"a mixture of Switzerland and Germany and Austria. We had a Mid-Eastern group making demands. But what they did—and this followed on the heels of Legionnaires' Disease*—they took a chemical substance—a toxin known as ricin, a military agent that we made in World War II.

"It's an incredibly toxic substance that is derived from solvent extraction from the castor bean. Its lethality in aerosol form is a microgram per kilogram of body weight. It produces an allergic reaction. You die from suffocation. It follows all the symptoms of Legionnaire's Disease.

"What we were doing was hitting hotels and conventions in different parts of Germany and knocking off fifty to a hundred people at a time and creating real panic in the height of the tourist season. It was theater. In a way, the Kingfisher game, which utilized lesser technology, was nevertheless more theater."

In another one of Kupperman's dramas terrorists have hijacked a tanker off the East Coast and forced the crew to sail her into New York Harbor. The terrorists threaten to kill the thirty crewmen and blow up the tanker if their demands are not met. Full of oil, the tanker will explode with a force equivalent to a ton of TNT. But the terrorists threaten to pump out the oil, converting the tanker to a superbomb containing explosive fumes equivalent to one hundred tons of TNT. The explosion will ignite the oil slick, transforming it into a mammoth sheet of flame that the blast will hurl onto New York City. . . .

Kupperman rounded up former and current high-level government officials to play members of a special committee formed to handle the tanker hijack. As in the real crises that many of the officials had faced, they did not know what was to happen from one moment to the next. (This exercise in crisis management was telecast as a special edition of ABC's *20/20* show in 1981. Since then an expanded form has been used as a training film for prospective crisis managers in the CIA, the Federal Emergency

*The mysterious sickness killed twenty-nine persons who attended an American Legion convention in Philadelphia in July 1976. The bacterium that was the cause was discovered after a year-long investigation.

Management Agency, and the State Department, according to Kupperman.)

Members of the crisis committee formed an old and new boys' network of gamers and real-world crisis managers. Retired Admiral Thomas Moorer, former Chairman of the Joint Chiefs of Staff, took a demotion as simulated special adviser to the Chairman of the Joint Chiefs of Staff. Retired General John Vogt, former Commander of the U.S. Air Force in Europe, was promoted in the game to the Chairman of the Joint Chiefs of Staff. John F. Lehman, Jr., played the Secretary of the Navy, which is actually what he became between the time the exercise was taped and the broadcasting of the game. Richard N. Perle, who had been Lehman's partner in a defense think tank when the taping began, played the Deputy Secretary of Defense—and would become the Assistant Secretary of Defense for international security policy. Richard R. Burt, who joined the game as a *New York Times* reporter and played a hard-line deputy national security adviser, became a hard-line director of the State Department's Bureau of Politico-Military Affairs and then an Assistant Secretary of State and later Ambassador to West Germany. Ray Cline, former CIA Deputy Director for Intelligence, played the Director of Central Intelligence. Joseph Sisco, former Under Secretary of State, played the President's national security adviser. (The President, who was part of the backstage Control team, was Kupperman.)

In the scenario Francis A. Bolz, Jr., a real-life hostage negotiator for the New York Police Department, makes the first contact with the tanker terrorists—by approaching the tanker in a New York City police launch and speaking through a bullhorn. The terrorists demand to talk to national authorities.

Navy Seals and Army Delta Force teams are secretly deployed in Manhattan. The committee, working against a forty-eight-hour ultimatum, decides on a two-track approach: negotiating with the terrorists while simultaneously making secret preparations for military action.

A strategy is worked out: Put limpet mines on the hull of the tanker and sink her if the terrorists attempt to discharge the oil into the harbor. Sisco turns down Lehman's alternative plan: Precede the mining with a "riot-control disabling gas attack. . . . Probably we'd wind up with most of the hostages and the ship intact, whereas if we go the mining route, you end up with the ship on the bottom leaking oil and all the hostages dead."

The dialogue often sounds authentic. When Sisco suggests stalling for time, Seymour Weiss, a former State Department official and veteran gamester playing the White House chief of staff, cracks, "You'll have Walter Cronkite saying, 'It's the 241st day of the ship in the harbor.'" And when Cline wants a phone tapped, George Carver, a former CIA official playing Cline's deputy for CIA operations, says, "We will, Mr. Director, but as a member of your staff I would suggest that you get an instruction in writing from the Attorney General to protect us against future legal complications"—and even that, Carver says, "may not keep us out of a congressional committee."

Lehman—raising his voice, interrupting, acting angry—seems actually to lose patience with Sisco, who urges attempts to talk to the terrorists. "There is no reason to play fair with these people," Lehman says. "They are deceiving us and lulling us to continue the negotiating path while they create a superbomb. . . . This pursuing the negotiating road is slowly closing off our military options."

The TV version of the scenario ends without resolution of the crisis. In the training version the President gives the terrorists half an hour to give up the crew members held as hostages—or face the consequences. The President leaves the consequences to his advisers. "The presumption," Kupperman said, "was that we would sink the ship if they didn't release the hostages in half an hour."

Antiterrorist gaming, in Kupperman's opinion, shows "that we are institutionally victims of our own backgrounds"—military people think in military terms and diplomats in diplomatic terms. "There seems to be no intellectual capacity to join these thoughts in any meaningful way. Joe Sisco, for example, really wanted a diplomatic solution. John Lehman and John Vogt really wanted to sink that damn ship from the moment of day one."

Kupperman said that Moorer's experience in Joint Chiefs of Staff games did not prepare him for antiterrorist gaming. As Chief of Naval Operations, Kupperman said, Moorer played "the SIOP kind of stuff. Terrorism is a very different thing. The incidents are either nuisances—real political disruptions domestically—or they are precursors to an undeclared kind of warfare."

Kupperman seemed tired of playing war games, especially about terrorism. "I'm weary of what I have seen because I can kind of predict the outcome in every case. And yet I think they are very worthwhile as a learning experience, a sensitizing experience. I'm very interested in the international gaming, even conceivably with the Soviets. Crisis resolution. That's a different matter." He has talked to Soviet officials about playing joint games with the Soviets. "My own concern is really that we say it is a learning system and not kid ourselves about understanding each other's philosophy and motives. But it's not going to be on a government-to-government basis; it's going to be on quasi-government levels, such as the Soviet Academy of Science and maybe the CSIS or somebody else."

In the television game, he said, "I wanted to create a high degree of dynamics. The operational people hate it because it shows indecision and an unwillingness to provide clearer orders on a timely basis and to expeditiously obtain tactical assessments. What I'd love to do—because I don't know how it would work out—would be to link real policy types with real operational types. I'd like to take the innermost horrors and paranoia of each side and then integrate it, edit it, synthesize it.

"When SAGA plays games they create enforced neutrality. Nobody has a role. There's a team leader. There's a seminar reaction. Control maybe reacts three times in two days to anything. It's a near-academic setting. You sit down and write down what your objectives are and why God put you on earth. I've played in a number of SAGA games, which, because of their classification I can't discuss. But they are by contrast—and I'm not trying to advertise what I am doing—boring. Kingfisher had an element of screwup to it and so did the *20/20* game. SAGA games never have those elements. Because they are seminars.

"I really try to create the atmosphere of a White House: the confusion, the leaks, the turf problems. When I put a game together I'm looking for certain decisions to be made and I want to confront them with certain problems. And then I want to give them out to these human beings and I want them to get upset. Not *too* upset. It's not a highly refined art form. And yet you can kind of predict where it's all heading because you know what the options really are.

"I think the term art form is correct. I don't think there's a science in this. It's not inferentially based. Nor are there controlled conditions under which it's done enough to get neat statistical evidence."

The search for that neat statistical evidence has led military game designers away from human beings and toward the most rational of all players, the computer.

CHAPTER 15

Real Problems, Simulated Solutions

At one end of the war-game spectrum, right next to real war, are the major exercises that give U.S. forces a world stage to rehearse for war. Each year the Office of the Joint Chiefs of Staff runs as many as seventy of these operations, which involve tens of thousands of soldiers, sailors, and Marines in Europe, Asia, South America, and the Middle East. Several of the major exercises have a secondary simulation built in: The soldiers and sailors are casts of thousands, actors in a show of force designed to dramatize U.S. presence in a region.

U.S. statecraft productions, sometimes called acts of "perception management," are particularly visible in Latin America. In 1983 U.S. troops began exercising in Central America almost continuously. The exercises provided realistic "drills" for Navy SeaBees and Army engineers, who built at least six airfields, two of which can handle large transports; two radar stations; many miles of roads; and eleven miles of tank traps near the Honduran-Nicaraguan border. As part of one exercise near the border, Marines staged an elaborate hostage rescue from a make-believe embassy, sending a signal to anyone with hostage-taking ideas.

Every fall NATO stages REFORGER—for REturn of FORces to GERmany—to test the alliance's reinforcement plans. (One realistic result is that millions of dollars of warlike damage is done to streets and farms.) In an annual Pacific exercise called Team Spirit, U.S. and South Korean forces counter a mock invasion by North Korean forces. The 1985 edition, with nearly 200,000 U.S. and South Korean troops, was the largest exercise outside Soviet bloc countries that year.

Exercises can look so menacingly real that the United States and the Soviet Union have made agreements that NATO and Warsaw Pact headquarters notify the other in advance of major maneuvers. On the high seas the U.S. and Soviet Navy ships, playing what sailors called "chicken of the sea," used to pass the time by simulating attacks and harassing each other. In 1973 the two nations signed the Incidents at Sea Agreement, which, on a practical level, means that high-ranking officers from both navies get together regularly and work out rules of the road.

But realpolitik still rules the waves. Navy aircraft, surface ships, and submarines still use actual Soviet submarines as training targets in what both sides call anti-submarine *war*fare. "It's a semi-real world," a pilot of a Navy submarine hunting plane said. "We're doing something real—looking for Soviet subs and tracking them. But we just find them. In war, we'd find them and sink them."

The wargaming spectrum bursts at simulation, a vast and booming business. Military forces and NASA spent $1.6 billion on flight simulators, missile trainers, and other such training devices in fiscal year 1983. A large-scale example of where that money goes can be seen in the training of Navy fighter pilots at the Naval Air Station in Oceana, Virginia.

Aerial dogfights are replayed on a sixteen-square-foot television screen within an hour after the planes land. Ground instruments pick up flight, maneuver, and weapon-firing data automatically transmitted from the plane. Hits or misses are determined by computers working from the tracking information and from data supplied by instruments on the plane. No missiles are fired, but when a pilot presses the firing button, computers instantly figure out whether a real Sidewinder or Sparrow or Phoenix would have hit its target. Oceana simulators trained the crews of the F-14 Tomcats that shot down two Libyan Su-22 fighters over the Gulf of Sidra in 1981 and the crews of the F-14s that intercepted the Egyptian plane carrying the *Achille Lauro* hijackers in 1985.

Marines train under live ammunition during large-scale exercises, but, while this gives the players a vivid experience, they cannot stage realistic maneuvers because of safety precautions. The Army solves this problem by using laser "bullets." A soldier goes into mock combat with a detector array strapped to his helmet and chest. If he is fired at, he hears two short tones—*near miss*—from his detector and takes evasive action. If he hears a continuous tone—*killed*—he takes himself out of action by removing a key from his rifle transmitter to disable it and inserting the key into his detector to silence it. A similar system is used on tanks and other vehicles. The Marines also use this laser system.

The increasing use of simulation hardware is spurring new interest in just how human beings act and react in a strange world where the unreal imitates the real. The manufacturer of an electronics warfare training system boasts, for example, "Trainees cannot distinguish between simulated signal displays and genuine signals seen and heard on tactical equipment." Simulators aboard submarines, using the ship's own sonar as part of the simulation, can realistically synthesize any one of forty-eight targets, including whales and Soviet submarines. As the simulator's manufacturer advertises, "The realism is so great that if the exercise is not announced, the crew doesn't know it's only a drill."

Some specialists worry about the way people and their senses treat the simulated world. Pilots used to "fly" their simulators over a checkerboard display as they learned to handle fast-flying planes on low-altitude missions. Observers noticed that the pilots used the varying sizes of the squares in the checkerboard to determine whether they were flying dangerously low. The pilots could get passing grades—they rarely "crashed" into surprise checkerboard hills. But checkerboard topography does not exist in the real world, and the visual cues learned in a simulator did the pilots little good in actual low-level flight.

Now, incredibly realistic, computer-generated images are replacing the checkerboard. Trees and shadows, rocks and pebbles are seen as they would appear through an altitude range of three to 100 meters, rippling past the pilot's field of vision at high speeds. Simulators also display such special effects as storms, bright sun, and darkness; air-to-air missiles fired by the simulated plane or at it; bomb explosions on the ground; tracers and flak in the air.

The Naval Tactical Game (NAVTAG), which bristles with computers, is played on a much grander scale, for it tests the fleet-level tactical skills of surface warfare officers. Up to three hundred surface ships, submarines, and aircraft can be deployed in any one of several possible scenarios. Stationed at a computer terminal instead of a bridge, the officer is confronted with fast-moving battle situations. Issuing orders and acting on information given him by the computer, he deploys fleets against an enemy at another computer.

Players see graphic images of real ships. The Red fleet includes the guided-missile helicopter carrier *Moskva*, guided-missile cruisers, and an intelligence ship. The Blue player deploys the aircraft carriers *Forrestal* and *Nimitz*, guided-missile cruisers, and other ships, all bearing actual names.

Some of the scenarios are based on real recent battles, such as those in the Falklands War. Much of the data—on weapons, the Soviet Navy, weather patterns—is supposed to be realistic. But in some Navy games the reality of the opposition (often still the old unspecified foe, Orange) has been questioned. A 1984 report on the Navy's Interim Battle Group Tactical Trainer noted "invalid" Orange plays and said, "Blue play is irrelevant and possibly counterproductive. . . ."

Information gathered from the Navy's own fleet exercises has also been questioned by officers who have access to this highly secret information. The problem of honesty in exercising goes back to the late 1960s and early 1970s when electronic and counter-electronic devices and systems added new and bewildering elements to tactics little changed since World War II. At the same time the Navy had been concentrating its tactical interest on HVUs—high-value units, particularly aircraft carriers.

After years of laboratory experiments and wargaming dry runs ashore, the two issues were joined in Uptide,* a major, meticulously planned at-sea test of electronics warfare and carrier protection. The exercise began in the first six days of October 1973, about four hundred miles southwest of San Diego. "The measure of BLUE mission success," said a secret Uptide report, "is *HVU*

---

*An acronym for "Unified Pacific Fleet Project for Tactical Improvements and Data Extraction."

survival time." A retired Navy strategist who had worked on Uptide directed me to the recently declassified reports on the exercise because, he said, "Uptide was the last honest exercise the Navy ran."

In the Uptide exercise the U.S. Navy is, as usual, Blue. The enemy is known as Orange (not Red, though Soviet hunter-killer tactics are assumed). During "a state of non-nuclear hot war . . . between BLUE and ORANGE nations," the Blue strike force, built around the aircraft carrier *Ticonderoga*,* is to make five transits across a rectangle three hundred miles wide and four hundred miles long. Somewhere in it are Orange surface raiders and two Orange submarines.

In a "break with traditional exercise philosophy," on three of the transits Blue, rather than seek out the foe and try for a kill, must "frustrate" the enemy, through tactics based on the use of electronic deception. The other two transits are to be essentially old-fashioned. The electronic devices include acoustic transmitters that produce deceptive and confusing underwater sounds. One, which will be carried on Blue decoy ships far from the *Ticonderoga*, makes noises that will sound like a carrier to an enemy submarine sonar operator. Another decoy device, puckishly named "Nymph Voices," simulates radio traffic on phantom ships.

Orange has not precisely located the Blue strike force but is within two hundred miles. Two Navy destroyers, playing Orange missile-armed cruisers, attempt to disguise themselves as merchants by sailing merchant routes and rigging merchant lighting. The two Orange submarines are often fooled by the deceptive devices, but the submarines manage to attack the *Ticonderoga* on three runs. Referees rule out one of the attacks. But officially the carrier is twice clobbered by missiles and put "out of action."

The word *sunk* never appears in the Uptide reports. The reality is there, however, and that is what the retired officer meant about the last honest exercise. The Uptide reports, he said, "are interesting evidence of a rare period of tactical innovation in the Navy where war games and fleet exercises were used to explore new ideas in the execution of war at sea. The effects of these experiments

*The carrier has been retired; the name now belongs to a cruiser.

reverberate even today and some of the once bold innovations are now widely accepted." He said this very precisely in a letter. In a conversation he lamented what he simply called cheating.

Eventually, in ways that do not involve memos or orders, the understanding grew that aircraft carriers were not supposed to get sunk in Navy exercises anymore. A hint of this dirty little secret of modern gaming surfaced in 1983 when a commander who had served on four submarines and commanded a fifth, writing about ASW (antisubmarine warfare) exercises, said, "Operating a submarine against a carrier is too easy; the carrier's ASW protection often resembles Swiss cheese. . . . There aren't many carrier battle groups. In a war with the Soviets, there will be fewer unless fleet ASW is top-notch."

Revelations about cheating in Navy exercises have been few because the Navy has kept such information secret. But, because of an attempt to publish information about one exercise, Ocean Venture 81, the smokescreen temporarily lifted. From August 1 to October 16, 1981, the Atlantic Fleet ranged the high seas in Ocean Venture, one of the largest series of exercises ever staged. The fleet sailed first to the South Atlantic for operations with Spanish, Brazilian, and Argentinean navies, continued operations in the Caribbean, then headed up the East Coast for joint practice with the Canadian Navy, and finally crossed the Atlantic for NATO exercises that began with a sweep to the Norwegian Sea through what strategists know as the GIUK Gap, the narrow passages commanded by Greenland, Iceland, and the United Kingdom.

Some three hundred warships, including two carrier battle groups—one led by the *Eisenhower* and the other by the *Forrestal*—took part in Ocean Venture. In NATO exercises planes flown from the carriers gave Navy air support to a mock NATO-Warsaw Pact battle in Europe. The fleet also tried out various tactics for a vital wartime mission: the protection of merchant ships running the submarine gauntlet across the Atlantic. A few "Orange" submarines dogged the carrier battle group, as did numerous real Reds—Soviet aircraft and more than twenty Soviet surface ships and submarines.

Lieutenant Commander Dean L. Knuth was the chief analyst of Ocean Venture. In December 1981 Knuth, by then a reservist no longer on active duty, submitted an article on Ocean Venture to the U.S. Naval Institute *Proceedings*, the highly respected professional journal of the Navy. Because the potential article contained operational information, the editors sent the manuscript to the Navy for a routine security review.

The Navy kept the Knuth manuscript bottled up until April 1982, when an inquiry about the proposed article was made to the Navy by the Senate Armed Services Committee, which was holding hearings on a defense budget that included requests for two aircraft carriers. The day after the inquiry the Navy notified the managing editor of *Proceedings* that the manuscript had been classified secret.

Knuth, in an interview with *The Washington Post*, later said, "It is hard for me to accept the proposition that the Eisenhower and the Forrestal reached the Norwegian Sea from the North Atlantic without being attacked successfully, even though the Orange threat was very low. . . . The fact is, our aircraft carriers were attacked by torpedoes or missiles from submarines in our major exercises. And yet the Soviet submarine force is many times larger than the handful of Orange submarines." (The Soviet Union by early 1985 had 380 submarines, with 200 of them deployed in the Atlantic and the Norwegian Sea.)

The Navy has had no comment on Knuth's claims. But the Navy did launch an apparently inconclusive investigation into the staff of *Proceedings*. The proposed article remains classified as secret.

Simulations are the next best thing to testing weapons and defenses against real Soviet submarines or tanks. And here the spectrum gets cloudy. Testing is a fine art that balances the reality of battle against the abstraction of a would-be weapon. Testing is an art like other arts, for it attracts creators, clients, patrons—and critics. In recent years the usual critics, found primarily in Congress, have been joined by people from the Project on Military Procurement, an advocacy group set up to transmit revelations from Pentagon whistle-blowers to the media.

The Project operates out of a couple of rooms, full of stacks of xeroxed reports and whistle-blown memos, in a run-down building near Capitol Hill in Washington. I talked there with Paul Hovin, who had a special interest in tanks. The Project's first campaign had been against the Abrams M-1 tank, which, among other problems, sometimes gave off so much heat that it set fire to the bushes around it. "When the Air Force was in the process of selling the A-10 [attack aircraft], designed to kill tanks," he said, "their idea was to fill a tank with *water*, put dummies in it, then do their measuring. We say, 'Why don't you put fuel and ammunition in it, shoot at it, and see if it blows up?' "

He paced around the battered desks and stacks of paper and began talking rapidly about weapon after weapon in the U.S. arsenal. I plucked one weapon assessment from his stream of acronyms and cuss words: "The Air Force said the Sidewinder [an air-to-air heat-seeking missile] had a pk of .98 [a probability of killing the target 98 percent of the time]. In the Falklands, the experience was pk .73."

In games such pk numbers are part of the calculations that determine hits and misses, losses and gains. If the pk of .98 stays in the calculations, players will get false views of the Sidewinder's effectiveness. And, if those players should eventually find themselves in a real war, their tactics will be based on that false information.

But sometimes, Hovin said, gaming will expose the reality about a weapons system. "From our experience in wargaming, you occasionally get caught up in your own lies." He paused and laughed. "But those [hobbyist] board games are often better than the real war games."

Most of the criticism about weapons testing centers on the way manufacturers trumpet scores racked up in *developmental* tests, which look at an evolving weapon in an experimental situation, and then withhold or mute information about scores made during *operational* testing, in which a weapon is tried out in a realistic situation, a war being the ultimate realism. In the weapons-testing game manufacturers' specifications get to be considered performance specifications—and these find their way into wargaming. One need look no further than Nifty Nugget to see that realism often is one of the first victims of a war game.

Models of warfare are not as accessible as games. Numbers get locked into the black boxes at the heart of the models, and model users often have trouble getting at the numbers or learning their origins. But model-building, like simulation, has been booming. One of the most popular models has delivered "attrition" numbers. The attrition model, with its tables of dead and wounded, has been constantly rising in the ranks, from theoretical "duels" between individual soldiers or weapon-against-weapon to attrition in small-unit combat, battalion-sized battles, division-against-division engagements, and, finally, theater-level warfare.

Modelers have faced their own kind of nuclear horror. Tactical nuclear weapons in a battlefield model blot out traditional concepts of front lines and rear echelons. One model attempted to solve the problem with a nuclear snap-on accessory. "You can stop the conventional war at any time you like," the model's booster said, "plug in the nuclear portion, look at the resulting forces, and then continue to play the conventional game."

Firepower scores, the traditional basis for measuring attrition, once came from simple historical combat records: how many guns were fired and how many soldiers were shot. Around the late 1960s complex firepower "models" began getting popular. The old firepower score was linked to new, numerical ways of measuring combat effectiveness, and FPS/ICE (firepower scores/index of combat effectiveness) numbers were applied as tags showing the value of every kind of ammunition, from pistol to howitzer. Next came WEI/WUV (Weapon Effectiveness Index/Weighted Unit Value), and some real soldiers began asking the modelers what the numbers meant. Do they say anything about shooting at an enemy and killing him? The answer was that this is a model, something on a higher plane than a mere exchange of rifle shots.

When the modelers multiplied weapon effectiveness by manpower, they got a force ratio. The killing and wounding of soldiers was calculated in the form of "casualty curves" and incorporated with force ratios. From such battlefield mathematics emerged a model called ATLAS (A Tactical, Logistical and Air Simulation), designed for theater-level force planning. ATLAS signified a grinding war of attrition that Lanchester and generals of World War I would recognize. To post-Vietnam Army strategists who advocated fast-moving maneuver warfare, ATLAS was a monster from the past.

But attrition models produce numbers, and in many Army-sponsored models there seems to be a compulsion for numbers. In model after model designers cite attrition as a feature. One touted a "killer/victim scoreboard." Another, with focus on individual soldiers, "provides a complete coroners [*sic*] report." A "casualty stratification model" of a "wartime theater scenario" told not only how many soldiers would be killed or wounded but also what their jobs and ranks would be.

Attrition drives most models, for as one side loses men, it falls back or drops out of battle, causing changes in the forward edge of the battle area—FEBA in modelers' shorthand. As the FEBA moves, territory is lost or gained, and modelers get another thing to measure. As one modeler told me, "It comes down to how many casualties are you willing to pay for how many klicks [kilometers]." This simple concept of kills-for-klicks is the essence of most NATO-Warsaw Pact models.

The FEBA, however, is dangerously simple, in the opinion of Raymond Macedonia, an Army officer who helped to popularize the FEBA—and later regretted it. To understand Macedonia, the FEBA, and the war that was slowly building up against models, it is necessary to follow Macedonia around the board a bit.

Dr. Raymond M. Macedonia, who was working on ideas for robot soldiers when I talked to him, first became interested in wargaming in 1965 at West Point, where he taught leadership and psychology. "It's very difficult to teach leadership in the classroom," he said. "I had students falling asleep and hitting their heads on the chair and getting hurt. I decided I had to do something about this." He started getting his students involved in classroom wargaming and they stopped napping in class. He developed enough of a reputation for his gaming so that it set up his next move several years later. The Pentagon, looking for an officer who knew something about wargaming and had a Ph.D., found Macedonia. He was assigned to SAGA, not for political-military games but to work on simulations that would aid U.S. negotiators in talks with the Soviets on Mutual and Balanced Force Reduction.

"We used simulations to measure the differences between options," Macedonia said. "If you took so many forces out, what impact did that have on a theoretical battle of the future?" Macedonia had two basic models, one that "was good on the ground but not very good in the air"—it was ATLAS—and another that "was rather good in the air but rather gross on the ground. So we had to use both. And these were very large analytical models. The [computer program] code was getting into thousands of lines. It took a very large computer and a very long time to run. Everything was in a black box. No one really understood what was there."

Macedonia and his team, though in SAGA, worked for the National Security Council. "We would meet sometimes in the basement of the White House and sometimes in the Executive Office Building, and there would be representatives from the State Department, the Arms Control Agency, the intelligence community—CIA and DIA—and a few other agencies, and we would present results and they would attack them. For every person doing analysis, there were one thousand critics.

"We used gaming to look at, say, warning time. What did the Soviets have to do in order to attack? How long can he keep his missiles uncovered? There are certain things people have to do in preparing for a major war. How long does it take to move forces from one point to another?

"It was the same problem: large models, the black box, not really understanding all the interactions, the long time it took to train someone to use these models, the tendency to believe that the models really represented reality and were not an abstraction of reality.

"I made a major, major mistake early on. I wanted to make the thing so it would seem a little sexier than it was. I used as the criterion the movement of the FEBA—the Forward Edge of the Battle Area—comparing scenario to scenario on the movement of the FEBA. I put a caveat in the study and in my briefings: 'Who knows what a day in a computer is? These are just relative figures. All these things are sensitive to assumptions.'

"But, lo and behold, the figures got out: It took so many days to get to the Rhine, and what have you. And I lived with that ever after. That became the system and it became used over and over again. The movement of the FEBA. People would say they knew the numbers were relative, but they always asked me, 'How many days would that equate to FEBA?'

"I learned that when an analyst becomes an advocate, you are going to have problems, and that is what happened. I had become an advocate. When people talk about running one war game or two reiterations of a war game, and they say, 'That's the way the war is going to go,' that is as ridiculous as one can be. It takes

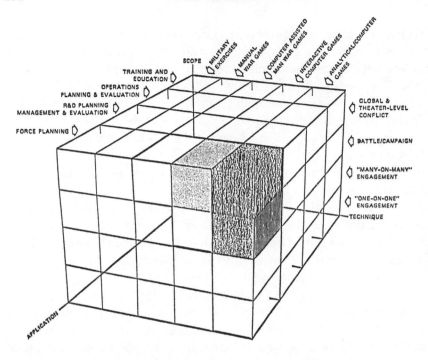

A wargaming cube—its sides labeled *application, technique,* and *scope*—symbolizes the complexity of modern gaming. *Application* ranges from leadership training to decisions on the size, composition, and arms of military units. *Techniques* span military exercises and analysis based on purely computer games. *Scope* begins with man-to-man duels and soars to theater-level and global warfare. (Credit: L. J. Low, *Proceedings on the Workshop on Modeling and Simulation of Land Combat.* Shaded area shows topics covered at workshop.)

hundreds—thousands—of runs, just to have some feel for your basis of uncertainty. If there is one thing about a war game or a simulation about war, it's not going to be the way you have it in your simulation. *That* is the certainty."

Models not only were growing more complex—they were making promises they could not keep. Some modelers were especially concerned about the evolution of the wargaming spectrum; it had become the wargaming cube, whose far edges now encompassed force planning, pure computer games, and theater-level warfare. Gaming along those edges seemed especially risky. Modelers argued over definitions of "V&V"—*validation* to determine whether a model is internally consistent, *verification* to see if a model does what it is supposed to do. Cynics growing rich on selling models to the Pentagon shrugged and said, "Validation is a satisfied customer" and "ultimate validation is a follow-up contract."

By 1977 there was something in the wargaming community mildly akin to the atmosphere among theologians on the eve of the Reformation. The Wittenberg would be Leesburg, a little Virginia town about forty miles up the Potomac from the Pentagon. It stretches the metaphor to make Andrew Marshall, the incorruptible abbot of U.S.-Soviet net assessment, the Martin Luther of wargaming, but he did make his stand at Leesburg, and the effect was a contest for the soul of wargaming.

One path to the Leesburg gathering of the wargaming hierarchy begins in the McNamara Pentagon, where the Systems Analysis Office hatched the idea of a "one-move war," a U.S.-U.S.S.R. nuclear exchange whose catastrophic results could be easily calculated by constructing a mathematical model. This is the model primordial, spawned by what a retired Army officer described to me as "a body of analytical theory based on zero-sum, two-sided games."

This "one-move war" as the origin of modern military modeling is the theory of Leslie G. Callahan, Jr., acting director of the Georgia Tech Institute of Military Education. There he is fulfilling his longtime wish to make a science of war (which he calls *polemology*, from the Greek word for *war*).

A blend of retired colonel and very active Ph.D., Callahan has been living in both the civilian and military worlds ever since he left the Army in 1969. As a thesis adviser to both graduate students heading for civilian engineering and graduate student-officers advancing their military careers, he might oversee "Perishable Seasonal Inventory Control: An Industrial Dynamics Analysis" one day and on the next read "Methodology for Probability of Kill Against a Moving Target on Air-to-Ground Gunnery."

In the interview Callahan did not get right to the point in the crisply assured way that analysts usually do. He graduated from West Point in 1944 in the last class that produced coast artillery officers, and he began by talking about artillery. "In the old coast artillery," he said in a soft, dawdling drawl, "they had two base stations in towers where guys did optical plotting of the ships going by. Down inside the operating room there was a big map, a high-resolution map, and you had a master gunner and every six seconds you had a bell—*bung* . . . six seconds . . . *bung*. And they would predict where that damn ship was going to be. And they'd take a sixteen-inch gun and train it on her. Now this was innovation.

"Our air defense notions came from coast artillery doctrines about shooting at moving targets. As the targets got faster, the system had to get faster. Point defense, you sit around the target. Area defense, you try to cover an area. Air Force is area defense. Army is point defense.

"Lanchester said one machine gun was worth sixteen rifles. Now most of these Lanchester models are used in ecology and population models. When I say Lanchester, I mean for attrition. Take the battle of Iwo Jima. The Japanese had no new supplies, and we know what the casualty rates were over a thirty-five-day period. We can make a correlation between Lanchester and an actual model. It's been validated. It's a rare thing, a validated model." (Modelers agree, but some say that the Iwo Jima model is *not* validated. Modelers even argue over the meaning of validation.)

Callahan estimated that 80 to 90 percent of all basic military decisions have become institutionalized or frozen in technology and stuffed into computers. "When 'so much is institutionalized," Callahan said, "you give up something. A commander goes aboard the *Ticonderoga* or some ship like that, and he can't change much. The same thing is happening in the Army." (The cruiser *Ticonderoga*, commissioned in 1983, is the lead ship in a class that carries the most complex radar-weapons control system ever to go to sea: The Aegis fleet air defense system, which can track and fire at numerous targets at the same time. In the combat information center there are two large video displays that show *Ticonderoga* crewmen electronic images of the outside world. Those displays are descendants of the huge, glittering war-games screen that was first lighted at the Naval War College in 1958.)

"In the United States the roles of the professional military man and the scientist once were very cleanly separated," Callahan went on. "The scientist's role was that the scientist helped out, sold things to the Army. It was always to the man in the field. 'Come with me to Macedonia'—that expressed the classic military belief beautifully."

I knew from the tone of his voice that he was quoting, but I was not familiar with the quote. I also knew that Ray Macedonia shared many of Callahan's beliefs, and this crossed up communications for a moment. Before I could ask him about Ray Macedonia, Callahan added, "If a guy wants to tell me what to do, let him come out with me in the field and I'll give him a tent. But don't tell me what to do."

Later I looked up what Callahan believed a military officer's response should be to civilian advisers, including, presumably, modelers. The words were spoken by Lucius Aemilius Paulus, a Roman general and consul, as he was about to go off to decisively defeat the Macedonians in 168 B.C. "Commanders," he said, "should be counseled chiefly by persons of known talent, by those who have made the art of war their particular study and whose knowledge is derived from experience, by those who are present at the scene of action who see the enemy. . . . If therefore anyone thinks himself qualified to give advice respecting the war which I conduct . . . let him come with me into Macedonia."

Where they all went—Callahan, Trevor Dupuy, Jim Dunnigan, Andy Marshall, and many more who had seen the computer and analytical techniques enter wargaming—was Leesburg, to what had been billed simply as the 1977 Leesburg Workshop on Theater-Level Gaming. The workshop would become the arena for an unabashed and unprecedented dissection of gaming by professional gamers.

The scene setter at Leesburg was Jack A. Stockfisch, an economist and sober critic of modeling, who, in a Rand-sponsored study two years before, had warned that "unverified findings of modeling" were being "taken as fact." His previous criticisms had been couched in polite words, but in Leesburg he was wildly subjective. He said, by way of military introduction, that he spoke as a former airman who had been "shot at on about thirty different occasions in air combat missions." He said that, looking about at the "eighteenth-century version of modeling," he felt like Prince André in *War and Peace* saying, 'What are these people talking about? Does it make any difference?'

"Suppose a war should break out in NATO in the central front, say a year from now. Will anything that any people in this group have done in following their trade of modeling or researching or

whatever it might be have made any difference on the outcome?
. . . I submit that, on the whole, much of this that most of us have
been participating in probably won't make any real difference. Had
we all been, say, harvesting corn or designing corn-picking machines or looking for oil, I suspect the overall human condition
would be somewhat better."

Turning to WEI and WUV and firepower scores, he said that
"much of this so-called data, as illustrated by the firepower score,
by my conception of the word is simply not data at all. A number,
yes, but by and large, the kindest word I could say for it is that
it's garbage. . . . I'm suggesting that what we are dealing with when
we use the word data is really non-data; it is mainly the outputs
of other models, most of which have not been validated. . . .

"Indeed, the answer is to try and get all the goddamned data
you can, whether it's junk or not" and "throw the information into
a computer.

"The chance of getting honest numbers in this kind of setting,
where the real war is between the Army, Navy, and Air Force over
dollars, over budget (and if that war is going on within the Army,
it's between the tankers and the artillery men and the infantry),
the chances of getting detached honest analysis, of supporting empirical work in the field, of getting those field trials done honestly
and in a hard-nosed way are very slim. . . .

"The modelers say, 'Yeah, we know the data's crappy, but it's
not our fault.' The people who produce the data say, 'Yeah, we
didn't produce the data for the modeler. We produced it for something else.' But the numbers producers change those numbers from
time to time, and they can do it often, in very good conscience,
because there is no ongoing, systematic experimentation and rigorous testing.

"But we don't want too much testing; we don't want hard information. Because when you get it too hard, the options available to
the decisionmakers . . . are not available. If the evidence suggests
clearly that you should get this type of airplane, or this type of
tank, then you lose some degree of freedom with respect to where
the contract goes, whether it's in this country, or some other country, or whether it's in Detroit, or Cleveland, or what have you. So
you don't want it to be too rigorous. . . ."

The meeting began to sound like one of those religious revivals where sin is trumpeted and repentance vowed. Years later gamers would still be telling tales about the confessions and the accusations made at Leesburg.

Robert Schneider, from the Office of the Secretary of Defense's Planning Analysis and Evaluation Group, spoke up: "Of course you and I know that when the chips are down, models often are used for political reasons. If it comes out with a right answer, no matter how it got the answer, that's the right answer, and so it is used."

John Shewmaker of the Congressional Budget office, who had just retired from the Navy, talked about Navy models he had known. He especially mentioned aircraft carriers, about whose fate the Navy is so protective. A study of carrier effectiveness, he said, had shown that "four carrier groups could survive in the Norwegian Sea and land an amphibious task force. Well, if you think about that, the Soviets have probably half their naval power in that area. One might scratch one's head until one read the study, and found out that it was done in a SEA MIX scenario in which the first war took place in the Eastern Med and the Soviets had transferred most of the Northern Fleet bombers to the Black Sea Fleet to attack in the Mediterranean. Their submarines were all at sea sinking convoys, and the only thing the Soviets had in the Norwegian Sea were a few bombers that didn't work too well and some patrol boats. So the carriers survived very nicely and landed the amphibious force."

Marine Major Raymond Bednarsky worked at the Pentagon's Command and Control Technical Center, which operated twenty-two theater-level models for SAGA. It took six to nine months to feed information to crucial data bases—those supporting models of force planning and weapon characteristics and performance—and the job was "a nightmare."

A typical operational run for a hundred-day war, he said, called for 200,000 to 300,000 bits of data. Besides this, there were huge data bases for NATO and Warsaw Pact forces, and data files on U.S. forces and equipment, on ammunition, on scenario plots, and on such arcane bits of logistical data as the Dutch railway system. A SAGA colleague backed what Bednarsky said and added that there was little solid information on three factors that have a great effect on games—and war: rates of advance, casualty rates, and the density of forces engaged in combat.

During a discussion of the problem of inserting nuclear and chemical weapons into models of conventional warfare, Dr. Edward Kerlin, of the Institute for Defense Analyses, talked of "stimuli" and similar matters detached from battlefield reality. Then someone asked him, "What about the morale effects of gas? No one has seen anything of this nature for over sixty years. This might be the biggest variable. Has any work been done on that?"

"We certainly don't consider it," Kerlin replied. "And I don't believe the people that I've talked to at Aberdeen [Proving Grounds] and Edgewood [a chemical-warfare center] have done any work on it either."

"But the effects of the first effective use of gas in World War I were catastrophic," his questioner persisted, "and it was primarily a morale effect."

"You mean in terms of forces just getting up and leaving the foxholes?"

"Yes."

Kerlin responded that troops can get used to anything, such as the awkward protective garments they don against chemical weapons. And he ended up by saying of such questions, "I don't know how to measure them."

The problems of the unmeasurables in nuclear weapons surfaced again and again. There was, for example, the report on "Techniques for Modeling Tactical Nuclear Warfare," presented by Stanley Spauding of Vector Research, Inc. The doctrine of tactical nuclear weapons use calls for their release in a "time frame" of twelve to twenty-four hours at a "pulse" rate of firings every forty-five to ninety minutes. To "promote perception" by the enemy that the weapons are being carefully controlled, the National Command Authority specifies such constraints as the number of weapons to be used, the timing of use, and the damage expected.

"We tried to model some of the behavior aspects," Spauding said. "One point I wanted to make is that I think the behavioral aspects of the response of the units that are hit and the units in the vicinity of units that are hit is one of the things we really don't have a handle on. . . .

"Our friends from West Germany probably have a very active interest in that particular area. What's going to happen to all those people if one side or the other starts using the nuclear weapons?

Probably they're going to try to evacuate, clog up all the roads, you know, and create a tremendous impasse. We don't really know what's going to happen there. . . ."

The model he described allows a setting for "collateral damage"* and does not allow the use of nuclear weapons in a place where more civilians would be killed than the number authorized by the collateral damage ceiling. Vector ran about three hundred analyses of the model, including simulations in which the United States was the first to use tactical nuclear weapons.

Although the results of the analyses were classified, the general outline was revealed at the conference. The model, for example, used the "value" of destroyed targets as the driving force of the model. The model also examined the recovery rate of Blue battalions or Red regiments hit by tactical nuclear weapons.

In response to a question from Trevor Dupuy about radiation effects, Spauding said that "in the games we were playing, which essentially called for a pulse on one side followed by a pulse on the other side, I don't really know whether considerations of those people who'd gotten enough radiation to make themselves sick would have been useful. How useful are those people going to be anyway? Basically, our model considered that even though you didn't kill all of the unit, you hit 30 or 50 percent; they were destroyed and they weren't in the game anymore."

Even seemingly routine logistics models did not look very good when candled in the light of reality. An elaborate model of the European transportation network showed how combat divisions would be supplied. Some questions were asked about the model. An Army logistics specialist responded about reputed ammunition shortages—which perennially show up in exercises like Nifty Nugget. "We don't really have any ammunition problem," he said, "because we'll run out of POL [petroleum, oil, and lubricants] much sooner than we'll run out of ammunition."

Philip Louer, of the Army's Office of Deputy Chief of Staff for operations and plans, brought up a serious problem lurking inside an Army model. "The Army has looked at some alternative forces against the Red threat in Europe and they played the game out

---

*Secretary of Defense Weinberger once defined collateral damage as "the euphemism for how many women and children you kill" when you are attempting to do something else.

for 180 days," he said. "Starting the game with one force config-
uration, they find that it loses badly. It is swept off the continent
even before 180 days expire.

"Another force, then, is given some additional help, and at the
end of 180 days, it is in great shape. This force is sitting with very
little area lost, and the Red force depleted, but Blue is in good
shape. However, if you look carefully—and the unfortunate thing
is that the decision is made that favors this second force—but if
you go back and look carefully at the history of the battle, you'll
find that back around D + 1 [the first day after start of battle] or
D + 15, Blue force has maybe only 10 or 20 percent of his tanks
surviving at that time. His personnel are down 30 to 40 percent
in both games. Yet, at the end of the war, in the second case, Blue
is in good shape.

"But the thing that is important that we have failed to look at
[is] the fact that if you really put yourself in that D + 15 situation,
you don't have many tanks, your personnel are low, the Red force
is continuing to build up. You would have never continued the
war." The model, Louer said, had "somehow failed to get distilled
into the decision" the hopelessness that a human general would
have seen.

Jim Dunnigan looked around him and, speaking as a wargamer
not a modeler, suggested that generals and decisionmakers get
terminals and get computerized versions of the models so they can
be played as games. "I mean quite frankly," he said of his suggestion, "the fellow might sit down and just sort of stare at it—like
you'd stare at the seven o'clock news. He may not understand
everything that's coming across, but once in a while he'll say, 'Well,
look at that! I just wiped out half of East Germany!' or something
like that."

Martin Shubik, a vigorous critic of gaming from inside the community, said modelers should be "trying to design games with
malice aforethought. . . . The question of foul-up gaming is something we haven't spent enough time with." He also made an observation about the modelers rather than their products: "When
one draws a graph of the age structure and you look at the number
of individuals between, say, twenty-six and thirty-four or thirty-
five, they're getting pretty few and far between, and when the
industry was healthy, it was the other way around."

Of all the voices, the one that reverberated the most was Andy Marshall's. A friend of many at the conference, an important customer and contact for some, he was, as everyone there well realized, the *totally* informed insider. They all knew secrets, but he knew more than anyone else did, and now he told them one that few, if any, of them knew. "When I first got into this business in 1972," he said, "one of the first things I did was to call in some people and try to survey the models that existed to see whether there really was anything that was going to be of much use to me. The answer was, there wasn't."

Here was the Director of Net Assessment, the man they assumed to be the custodian of the best they could make, and he was saying this—and more: "In none of the work that I have done have we run a single model."

But what, they asked each other, did Andy Marshall want? They answer was that he wanted something new. To get something new he was willing to ask for a game from a toy-store game designer, which he would do three years after Leesburg when he met with Jim Dunnigan. And he was planning a way to force something new from the major-league players of gaming. That, too, would come.

Leesburg had inspired a search for the new. Leesburg had also sanctified an important new dogma: It was all right to criticize the orthodoxy of gaming. Previously, people like Dunnigan were clucked at for being irreverent because they said harsh things about gaming, which, after all, as part of the nation's defense effort, was not to be criticized. Now it would not be unpatriotic to speak frankly about the shortcomings of gaming. In fact, some of the sharpest criticism would come from people in Marshall's own shop.

Formal reaction to Leesburg was slow in coming. By the time Lawrence J. Low, a respected executive of SRI, a California think tank, analyzed the meeting in a 1981 report, a critical assessment of gaming had already been issued by the General Accounting Office. In it a modeler had wondered if by asking computer models for "answers to complex social, political, or behavioral problems . . . we may be returning full circle—back to the shaman and the oracle, asking for a magical mechanism beyond the range of human consciousness and understanding."

Low seemed to aim his report at the community itself, a loose assembly of officers, military and civilian government officials, defense contractors, and academic modelers and players. "Models have retained a strong 'engineering' orientation," he wrote. "They are notably weak in their treatment of human behavioral characteristics and the impact of such characteristics on combat processes."

Low analytically described the central problem, and in doing so, intensified it by displaying the analyst's insistence on making human activity sound and look consistently sensible and systematic. "If one recognizes the fact that combat involves a life and death struggle between arrays of various man/machine systems," he wrote, "it follows that the effectiveness of these systems in performing their function depends on the characteristics of machine or equipment, which are generally measurable, and the characteristics of human behavior under stress, which generally are not."

He focused on a model named Carmonette, which the Army had been using since 1958. Carmonette's human beings were "groups of 'doctrinal' soldiers with identical behavior patterns in executing orders." The model, used primarily as a tool for evaluating weapons and tactics, can simulate a battle over an area ten by fifteen kilometers with as many as five hundred "maneuver units" on either side.

Various kinds of terrain and weather conditions can be injected into Carmonette, whose "doctrinal" soldiers follow computer orders—move, stay, fire—and whose weapons follow doctrines about accuracy and lethality. The fog of war rarely grazes this make-believe battlefield, where orders are always understood and obeyed, where there is no shortage of ammunition (because there is no computer loop for logistics), and where no nuclear weapons can be introduced by either side because the model does not know how to handle nuclear warfare.

As Low looks deeper into the limits of gaming, he becomes philosophical, saying that while a model could probably pick a good antitank weapon or help to determine defensive tactics against land mines, there was little chance that a model could answer such an enormous question as "What NATO force levels are necessary in the 1985–95 time frame to deal with projections of Warsaw Pact capabilities at that time?" Yet he points out that at Leesburg no less than three theater models were devoted to that question. Low politely scoffed at such use of modeling and said he could only hope that there were not very many analysts who let "their make-believe world of modeling take on a real world aspect."

By the time Low's critical report started to circulate through the gaming community, Andy Marshall had begun to establish the rules for a superbowl game that would change the make-believe world, but not in a way that Low or anyone else expected.

CHAPTER 16

The Search for the Ultimate War Game

With his brief Leesburg speech, Andy Marshall made the first move in setting up what would become his own wargaming game—a contest to find, in his words, "more systematic methods to analyze the state of the military balance" between the United States and the Soviet Union. The second recorded move came in August 1978, when the Defense Science Board met at the Woods Hole Oceanographic Institution on Cape Cod. During a discussion of national strategic problems, Marshall suggested a search for a new style of analysis through wargaming. The moves would lead to an unchronicled but crucial competition that no one called a competition, a contest that no one called a contest, an ultimate game played by computer programs named Sam and Ivan.

The story of the wargaming game loops through time, ideas spiraling, players making their moves and countermoves. To replay the game chronologically is impossible, for, as in a chess match played by mail, it is sometimes a long time between documented moves, and, besides, the postcards are not what the game is about. In this game, time goes back to the 1950s and into the mid-1980s. But many of the players do not change. As Shubik had observed at Leesburg, the modelers were growing old, and the unspoken crisis may have reflected the reality that modeling was feeling its age.

In the late 1960s and early 1970s political science journals frequently carried articles about model-making, simulations, and gaming. There was a hiatus in the late 1970s, and then the number of
articles began increasing again in the 1980s. Gaming seemed to be
coming back, trying to find again the academic home it had when
Harvard and MIT first fostered political-military games back in the
1950s. I wondered if changes were welling up, perhaps because of
Leesburg, perhaps for other reasons. I asked a couple of younger
gamers what they thought was happening.

"My reading," William Martel of Rand said, "is that political
science in particular went through a tremendous quantitative revolution in the late 1950s, early 1960s. What they did was throw
out many of the traditional approaches to politics—political science
research—in favor of heavily quantitative exercises: regressions,
game theory, factor analysis.

"Then I think the methodological revolution hit them—the realization of hubris, that they could not really measure the world.
People slowly backed away from this, slowly became disillusioned.
People started saying you couldn't game or measure value processes. If you don't understand why people make certain decisions
in a game, what is the reason for the game? Political science went
through a pretty bloody period. The new guard—statisticians,
mathematicians, modelers—pushed out the old guard, in my opinion in a pretty brutal fashion."

Martel takes the view that war games are "a training exercise."
Games, he believes, "teach you how to think in a crisis mode: What
is the background noise I can screen out?" Before he and Paul
Savage ran the nuclear war game at Harvard (page 173), they tried
it out at St. Anselms College, a small Catholic college where Savage
taught and Martel had got his bachelor's degree. Martel said he
discovered something in that game. "We threw in a crisis in the
Middle East and coupled it with a Libyan nuclear terrorist attack
on Boston. It pushed the teams over the edge. We didn't do that
at Harvard."

The crush of so many crises all but wiped out the gamers and
the game. In such an environment no one can learn the dimensions
of decisionmaking. And learning how to make sense of policy is
what Martel believed gaming was for. In their book on nuclear war
Martel and Savage say that the United States has no real strategy.

What "is actually produced is an incoherent strategic 'philosophy' that has emerged from the infighting and competition among the oligarchic defense bureaucracies. What we have is the illusion of strategy—not the reality." In quest of that reality, Martel became one of Rand's players in Andy Marshall's ultimate game.

Sherman Greenstein, an analyst and gaming enthusiast at Systems Research and Applications Corporation, a Washington think tank (page 165), talked of academe's grand plan "to combine the best of political-military simulation—high interest, exciting scenarios—with a research design that would have a more systematic evaluation of a problem." In his view the merger did not happen, and he admitted some disillusion.

"I use to think," he said, "that there was a lot of insight provided by simulated games. But one of the disadvantages of the academic game comes when people try to put probabilities on something occurring. For example, I was involved in some games dealing with the Middle East. What we would introduce was an Israeli nuclear capability to see whether that would be an impetus for peace."

The result of each of more than one thousand games was the same: Israeli nuclear weapons did not change anything. The results of the games, which Greenstein had thought to be significant, meant nothing to policymakers—probably because none of them had played the game. They were too busy with the real Middle East to bother with the results of games, no matter how fascinating those results might have been.

For some policymakers, however, the introduction of computers into war games has given gaming the appealing illusion of precision. "The use of computers in games," Greenstein said, "sometimes is like looking under the lamp for the quarter you lost, even though you lost it in the dark park, because the light is better under the light of the lamp. Computer-supported gaming has gotten more attention, more money than policy gaming because there's a pseudo-precision involved."

The computer had become a war gamer even before Marshall had begun setting up the arena for his wargaming game. The other contender was only partially computerized. Its handlers called it the people-in-the-loop concept.

After Leesburg the models went on evolving and, as a gamer

put it to me, "there were two schools, the nothing-is-good school and the everything-is-good school." But some of the players got together and produced changes that can be tracked through the great cube of gaming. Ray Macedonia, assigned to the Army War College, reintroduced wargaming there and, with the unpaid assistance of Jim Dunnigan and some of his hobby-shop board games, started developing a new military center for gaming and model-making.

"What was pushing wargaming at this time was Nifty Nugget," Macedonia said. General Edward C. Meyer, Army Chief of Staff, fresh from the discovered disasters of the mobilization exercise, "came to the war college, actually grabbed me by the tie, pulled me into a conference room, and sent someone to get the deputy commandant. He said he wanted to improve contingency planning in the Army and the military."

The way contingency planning then worked, Meyer told Macedonia, "it might take six months to set up a game on a contingency, and by that time the contingency was over. He said we needed to have something that we can quickly set up—something that could be set up anywhere in the world and be able to play within twenty-four hours, with the data base ready to go."

Macedonia had been computerizing Dunnigan's board games for use in the college with the aid of three programmers, one of whom was Fred McClintic. He went to work on what Meyer had so urgently ordered and, using new principles of computer architecture, created what became known as the MTM—the McClintic Theater Model.

By 1981, inside the wargaming community, the MTM was already famous. It was cranking away at SAGA and at the Army Concepts Analysis Agency, the source of many Army games and models, and it had been put to highly classified use by the Readiness Command and Rapid Deployment Force. When Leslie Callahan, who had helped set up the Leesburg conference, ran the next big modeling workshop, at Callaway Gardens, Georgia, in 1982, the MTM was one of the stars.

Modelers peered into the MTM's structure and saw that it was not a black box. The MTM gave a theater commander the ability to "look down" through what modelers like McClintic called a transparent model. To them transparency meant a kind of com-

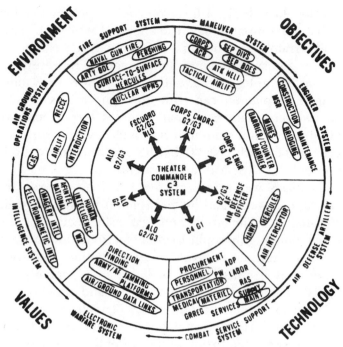

The theater commander's overview of war is simulated in the McClintic Theater Model, named after Fred McClintic, who designed the model at the Army War College in 1980. The model was developed around this diagram, which shows how a theater commander's C₃ system—command, control, communications—links him to the intricate modern battlefield. Circled items were in the original model; the other items were to evolve. The Office of the Joint Chiefs of Staff has used this "high-level planning and decision aid" for real-world contingencies. (Credit: McClintic Theater Model's *War Game Director's Manual.*)

puterized honesty, a simplicity that enabled someone planning for a contingency to understand that the model was serving him, even though he might not have enough knowledge about models to know exactly how he was being helped.

To the nongamer a diagram of the MTM looks like a wheel of chance, with many arrows radiating from the theater commander or contingency planner at the center and aiming toward outer sectors of the wheel. Each sector is filled with all he needs to know, from the status of his bridge-building equipment and medical supplies to the range of the electronic warfare and intelligence-gathering systems available to him.

The battlefield is built of hexagons, with each hex coded by a formula that differentiates, say, the effects of artillery fire on a hex in forested terrain from a hex in a clear area. A hex can be as small as ten meters across, with an individual soldier represented on it, or as large as ten miles across, with a division or corps unit on it.

The gamer or planner can riffle through various possibilities, using an awesome data base that includes terrain and weather information, each unit's ammunition and gasoline supplies, and even accounts of World War II battles contributed by the history department of the U.S. Military Academy.

When he plays war games on the MTM he can replay those battles in various ways, and he can remind himself of realities as old as warfare itself. The intelligence-gathering sector, for example, besides including highly sophisticated electronic intelligence, also dispenses very humble but valuable information, because when MTM Blue or Red units "enter or leave a city, members of the local population inform the enemy of its size and direction of movement." The gamer gets constant doses of realism, such as communications snags. He does not get a certain percentage of the messages he is supposed to get, and some of the messages he sends are lost or intercepted. He begins to feel he is in a real army in a real world.

MTM soon left the war college for the Lawrence Livermore National Laboratory, a nuclear weapons facility, and, in a process Macedonia bitterly described as "taking off the Army label," MTM was modified, given a stronger nuclear identity, and named Janus. Although MTM still lives in the Army Concepts Analysis Agency, it is Janus, not MTM, that now runs the major games at the Army War College. Janus is highly secret. I was not allowed to see any aspect of Janus. Reportedly, during a Janus game an officer ordered the use of tactical nuclear weapons and wiped out his own forces. I was told that there could be no comment on that.

Even before the Livermore adaptations MTM already had an arsenal of nuclear, biological, and chemical weapons. But gamers could use them only with the permission of the wargaming controllers. In an unusually frank revelation of U.S. doctrine on the use of chemical and biological weapons, a description of MTM said:

"If a force calls for a nuclear or biological and chemical attack on a specified coordinate, and permission has been granted by the controllers, the attack execution time is a function of the delivery means (aircraft, artillery, missile, etc.). Use of dirty or persistent nuclear or chemical rounds contaminates that hex for a certain length of time; however, noncontaminating rounds may be specified in the order, in which case no contamination results. Both forces are notified that a nuclear or chemical and biological weapon has been detonated in the specified hex. Any unit entering a contaminated hex will suffer a predetermined percent loss and a significant time delay en route through the hex. Every hour that a unit remains in a contaminated hex will result in additional losses."

MTM's computerized data and unconventional weapons symbolized Marshall's principal preoccupation as Director of Net Assessment. At this point in his career—which covered much of modern wargaming—he could look back some thirty years. But he could not look ahead as much as he believed the nation needed to look. He also believed that wargaming, which could so help the nation in the future, still had not reached its potential. As long ago as the mid-1960s he had been arguing for the use of wargaming as a tool for strategic analysis. "However," he wrote in 1982, "wargaming has been used far more for training purposes than for analytic purposes and wargaming methods have been used less for strategic forces and general war situations than they have for conventional forces and theater warfare."

National strategy had become steadily more complicated since the early 1950s, when he entered Rand as a young economist, changed professions by becoming an analyst of nuclear war, and helped to found modern wargaming. His mentor at Rand, Herbert Goldhamer, had made Marshall a player in what is the first recorded series of political-military war games in the United States (page 148ff). At Rand, with Goldhamer and another young Rand analyst, Nathan Leites, Marshall had invented the model of a nuclear Pearl Harbor, and many games and models since then had built on this premise.

Some of the models had barely evolved beyond what Leslie Callahan labeled the one-move war: the Soviets strike; the United States, stabbed in the back, manages to retaliate massively and, though gravely hurt, survives. End of war. As a bonus, the Pearl

Harbor scenario inspired a valuable patriotic image, equating Red with deception and Blue with a vital need for vigilance—and a Strategic Air Command on high alert.

Now, with tactical nuclear weapons, with missiles of ever-sharpening accuracy, with constantly shifting strategic doctrines, the Pearl Harbor scenario was out of date. Marshall, along with other architects of American nuclear strategy, knew that a Soviet surprise missile attack—what the people in the strategic community all call the "bolt-from-the-blue" war—was no longer looked upon as a likely event.

Strategists, aware of the propaganda value of that sneaky Red image, do not usually talk about the change, but Roger C. Molander did. "Through SALT or sanity—or whatever you think has led over the last fifteen-odd years to the situation that prevails today," he said in an interview, "there is pretty high confidence that there will be no bolt from the blue, except among real nuts who think that the Soviet threshold of pain is so high that I must have a situation where the Russians know there *will be* a five-thousand-and-first warhead that lands on that machine shop a hundred miles outside of Novosibirsk."

Molander earned a Ph.D. in nuclear engineering at Berkeley and got a job in 1967 at the Institute for Defense Analyses, where, as he put it, "I entered the nuclear war, nuclear weapons business." He went from there to the National Security Council and spent seven years, in the administrations of Nixon, Ford, and Carter, working on National Security Council plans for nuclear war. After leaving the Carter Administration he founded Ground Zero, a movement inspired by his fear that the world was on "the slippery slope to nuclear war."

Molander has since become president of the Roosevelt Center for American Policy Studies, a public policy research institute. There he has developed a war game, Wildfire, which examines what he sees as the successor to the bolt-from-the-blue route to nuclear war: Nuclear proliferation—the possibility that the war "might start with an accident, or it might start with unauthorized use, or it might start with a terrorist attack."

In Wildfire players take the parts of five nations. Pakistan and India are at war, and Pakistan is threatening to use nuclear weapons. As Pakistan plans to fire a "demonstration shot," the United States, China, and the Soviet Union try to get India and Pakistan to negotiate an end to what is now a potential nuclear war. The Roosevelt Center has sponsored sessions of Wildfire in cities and towns throughout the United States as a way to teach people about the perils of nuclear proliferation.

In setting up a Pentagon-style war game, Molander says, the "emotional component to escalation is very, very hard to model." At those games players are always extremely reluctant to call for nuclear weapons. "If you are playing along in a game—whether you start in the Middle East and somebody shoots Hussein, or somebody in Iran assassinates Khomeini, or something like that— you can play the game right up to the nuclear threshold and then you've got to find some artificial way of stepping across the nuclear threshold," according to Molander.

Hesitancy at the nuclear threshold was a bit of lore well known inside the wargaming community, and Marshall realized he had to face that reality because it was no longer a secret. In 1981, for example, a military journal's survey of about twenty-five military and civilian players showed that nuclear confrontation had to be forced in high-level games. One player, Air Force Lieutenant General Kelly H. Burke, told of a game in which both the Blue and Red teams kept making moves that avoided a nuclear attack. Control, according to Burke, "threw both sides ever more implausible situations until, finally, it was able to force one side to launch its nuclear force."

After the game an officer on the attacking side said to the game director, "If your control group had just left us alone, we could have negotiated a lasting peace." Similarly, in a NATO game, a player striving to avoid the release of nuclear weapons was told by Control, "You can't *do* this! You'll ruin the entire exercise."

Yet, those who did launch nuclear attacks often were derided for their foolishness or their ignorance of military options. The former head of the State Department's Bureau of Politico-Military Affairs, Seymour Weiss (who played in the televised terrorist game described in Chapter 14), said that doves turn hawkish in games, "especially in their willingness to employ nuclear weapons." Weiss, reputed to have played in more games than almost anyone else in

Washington, said that unless the scenario takes players across the threshold, it will not be crossed. "When we got up to that point—with really *top* people playing—the game usually ended."

Robert W. Komer, whose career encompassed Rand, the White House, the CIA, and the position of Under Secretary of Defense for policy, said he had played in about a dozen games that had crossed the threshold. "It's been my experience," he said, "that, almost invariably, it's the civilians who are the most bloody-minded of all and the military who are the most prudent. And the *most* bloody-minded of them all have been the State Department types."

Former Secretary of Defense James Schlesinger added other reasons for the different nuclear attitudes of military and civilian players. "One of the *reasons* for organizing such games," he said, "is to demonstrate the inadequacy of our deployments and force structure. Once the military is able to demonstrate that in these crisis scenarios, they can relax and may not care that much how the game ends. . . . The military—particularly the Air Force types—have been deeply saturated in the belief that if you use *one* nuke you can't stop from going all the way. The civilians, on the other hand, tend to believe more that the demonstrative use of a nuclear weapon may help create the 'pause that refreshes.'

"For the most part, the reason these games are played is to *educate* the civilian adviser and decisionmaker and make clear the *consequences* of the options they have to decide. Thus, the games are often intentionally structured to *impel* the civilians to 'go nuke' early on: then the military controllers can sit back and watch us either back off or squirm."

Thomas Schelling, the Harvard professor who had led games in the Kennedy and Johnson administrations, said in a speech in 1983 that "it's terribly hard even in a game with fairly bloodthirsty players on both sides, terribly hard to get anybody to contemplate any use of nuclear weapons except a diffident, shy, cautious use."

In Marshall's Pentagon cavern the dirty little secret about nuclear timidity was not publicly discussed because of the Net Assessment Office's homage to secrecy. But the office yielded for some reason in 1985, when, at a little-publicized National Defense University conference, Air Force Lieutenant Colonel Barry D. Watts, one of Marshall's military assistants, delivered a broad and devastating critique of gaming.

Watts, who had flown F-4s in combat in the Vietnam War, was a specialist on the Soviet Air Force and a war-game veteran. He had played, usually as the Red air commander, in many games— three global war games at the Naval War College, in a major global war game at the National Defense University, and in several games run by Rand.

In his paper for a "thinking Red" conference at the Defense University, he described himself as "a user of theater-level war games." He also knew his way around some other gaming models, such as the Air Force's Advanced Penetration Model, which, he said, had a phenomenal defect that was not discovered for several years. The model was supposed to fly a bomber into enemy air defenses and then determine whether the bomber would be detected. For the model to do this, it first checked the bomber's location. "But in the process of doing so," Watts said, "the model would inadvertently change the bomber's latitude or longitude, thus frequently transporting the bomber magically outside of hostile surface-to-air missile coverage (much like Captain Kirk beaming up to the Starship Enterprise)."

In games he played in during 1983 and 1984, Watts saw a "Red" order of battle that gave Warsaw Pact forces 1,800 fewer fixed-wing warplanes than they actually had. On the second day of a full-scale NATO-Pact war, a game had "NATO flying over 2,200 sorties into the teeth of Pact air defenses without suffering any losses." These make-believe wars had battleships steaming at seventy knots (exactly twice the actual top speed), NATO destroying in four days 50 percent more Soviet Backfire bombers than the Soviets started with, and "Austrian border guards stopping an entire Soviet combined-arms army in its tracks."

Watts mentioned a game—"in which, for a change, I played on the U.S. side"—designed to look at U.S. and Soviet thinking about how strategic nuclear forces "might be affected if both sides possessed some ballistic-missile defense, stealth aircraft," and other anticipated high-technology weapons. But the game controllers concentrated on conventional forces and virtually ignored strategic nuclear forces because the players were more comfortable dealing with information they knew and understood.

In "all too many games," he said, "the Control Team (or game-master) forces Red to initiate combat" in situations in which most

Red experts believe the Soviet General Staff would not recommend combat. Once, when Red did go nuclear, an indignant Blue sent Red a message: YOUR BARBARITY IS UNPRECEDENTED IN HUMAN HISTORY. MAY YOU BURN IN HELL AS YOU WILL BURN HERE ON EARTH.

"I vividly recall a global game from 1983 in which the Red 'General Staff' complained bitterly, right from the outset, about being denied even the most elementary planning information necessary for planning and conducting operations in accordance with Soviet methods of troop control. Not only did those in charge refuse to give Red rudimentary information on Blue forces, but Red was frequently denied similar information on Red's own force dispositions.

"Yet when, in the second week of the game, the Red 'Politburo' began considering nuclear escalation in light of Red's apparent lack of success on all fronts, one of the game masters immediately revealed that the 'correlation of forces in Europe was still quite favorable for Red by about 3-to-1.' Based on this revelation, the Red Team quickly shelved all plans to escalate. Obviously this sort of intervention, whether intentional or not, can so taint the contents of a game that few (if any) meaningful 'lessons learned' can be drawn from what occurred during the course of play." Two of the "lessons" he learned were "Bad numbers are better than no numbers at all" and "We're doing 'God's work'—believe me."

Another kind of intervention, Watts said, is "rheostat twiddling." He told of a game in which players questioned what they believed to be excessive aircraft losses. The protests "led to the rheostats being lowered to the point where both Red and Blue could fly enormous numbers of sorties with fairly low losses. The catch was that a stated game objective was to ascertain air power's impact on the ground war in central Europe." But, since the twiddling had made it so that the opposing air forces had little effect on each other, "gauging air power's impact on the ground war became impossible."*

There is extensive evidence from World War II and Vietnam that in air-to-air combat upwards of 90 percent of the pilots who

---

*Watts credited game designer and Dunnigan protegé Mark Herman with the concept of a game rheostat.

are shot down never see what or who hit them. But this fact, which Watts called situation awareness, is not included among the air-war factors in typical theater-level games.

Red and Blue forces are often mirror images. In the concept design paper for TACOPS, the air-war adjunct model to the McClintic Theater Model, the phrase "mirror image" is actually used to describe the similarity of Red and Blue air forces. Both "mount similar, opposing groups of missions each day in a conflict," but the opposing missions are so perfectly alike that each air commander merely has to decide what percentage of his planes should be allotted to what missions. Watts said that when he told this to a former air intelligence officer now with CIA, he responded with a phrase "a bit more colorful" than "absolute balderdash."

Watts also cited a mirror-image problem that appeared when a combined force of soldiers and Marines fought more than 150 battles against a "Soviet" force in 1980 in an exercise in California. During the first two months the Soviet force repeatedly defeated the Marines. At first a computer error was suspected, but then an analysis of the tank guns' videotapes and other computerized data showed that the problem was with the U.S. battlefield techniques: American fighting men were not able to respond to tactics that were drastically different from U.S. tactics.

As the director of the exercise later wrote, the Soviet approach to tactics is "simple, straightforward, brutal. . . . It moves straight at its opponent at high speeds and when contact is made, one element fixes while the other maneuvers." The opposing Marines, who thought they were fighting a "dumb enemy," were whipped until they developed tactics designed to thwart, not mirror, Soviet tactics. About halfway through the six-month test, the Marines, their lesson learned, began to dominate the battlefield.

"Kriegsspiel against a mirror image of yourself is a form of self-delusion," Watts said. ". . . You could even say that a professional military establishment that does not take interactive, 'people in the loop' wargaming seriously, probably does not take war seriously."

The tendency to mirror adversary play (as in the opening move in chess) has made wargaming suspect at the policymaking level, and this added still another inspiration to Andy Marshall's quest for a better way. In his paper Watts repeated what had become an anthem among critics of gaming: the statement by Paul Davis of Rand that he had "never heard a senior or mid-level official of the U.S. government base an argument on the results of combat simulations."

Davis had echoed the assertion that Marshall had made at Leesburg—that Marshall had never used a single model in net assessment. There was another echo in Watts's words about mirror images, an echo from Marshall's own boss of long ago.

Herbert Goldhamer, Marshall's mentor at Rand, had been writing a book that was to have been titled *Reality and Belief in Military Affairs*. He had already written down some of his ideas—that, for instance, "military-political behavior is often a function of images of the enemy and of the self"—when he died in August 1977, a month before Leesburg.

Marshall wrote the foreword for Goldhamer's last work—the drafts of the first three chapters were found—and poignantly made it a "report for the Director of Net Assessment." Goldhamer apparently had intended to stress the mirror-image warning. He had written, "U.S. scenarios certainly take account of differences in the order of battle of the two antagonists and their weaponry, and in some doctrinal and tactical characteristics stemming largely from an assumed Soviet offensive posture, but otherwise tend to treat Red and Blue as mirror images of each other."

He had also written of war: "Strategic nuclear war, by becoming, relatively speaking, a war with one big battle, reduces war again to the point where a nation's fate rests on the uncertainties of a single event."

Marshall wanted to get rid of as many of those uncertainties as possible, and the way he proposed to do it was to make wargaming as important as Goldhamer had once envisioned it. Marshall saw wargaming as a new way to analyze the U.S.-Soviet strategic balance from three perspectives:

- An "assessment structured as the Soviets would structure it, using those scenarios they see as most likely and their criteria and ways of measuring outcomes."

- A "wider range of contingencies than is currently feasible." Noting that "standard calculations tend to focus on surprise attacks," he said "analysts need to look at the performance of forces in a wide range of situations, from crisis to conventional theater war under the threat of strategic systems, escalation from theater conventional to theater nuclear war to all-encompassing general nuclear war, with a protracted period of warfare that may ensue beyond the large exchanges."

- An "assessment of the perceptions of allies and other major third parties" to "promote alliance cohesion and the continued development of adequate overall military forces in the West."

He also asked for the breaking of "new ground" by the integration of certain aspects of conflict—such as space warfare, command and control, antisubmarine warfare—not usually included in policy-level analysis and assessment. To get what he wanted, he issued invitations for formal proposals. Two major defense think tanks were selected to give what were politely called demonstrations. And the wargaming game was on.

CHAPTER 17

When Ivan Plays Sam

Marshall's wargaming game would be between Rand, which had been his own first stadium for playing nuclear war, and Science Applications, Incorporated,* another California-based firm with strong Washington connections. Each contracted to produce a new kind of war game that answered Marshall's demands for a better gaming system.

The formal invitation from the Pentagon sought "a new analytic methodology" for force designers, operations planners, and assessors of the U.S.-Soviet strategic balance. "It appears to us," the invitation said, "that to permit the exploration of complex scenarios, a wargaming style of analysis which blends human judgement with computerbased models and bookkeeping routines will be required. But whatever approach is chosen, it must provide the opportunity to systematically work through many different branches of complex scenarios which involve the use of nuclear forces at levels of crisis and conflict ranging from peacetime to the aftermath of major nuclear conflict."

The competition was technically not secret, but details of it were tightly held inside the strategic community. The two firms were the only ones to get research contracts to launch a search for Andy Marshall's ultimate war game. And only one of them would get a chance to transform the revolutionary new theory into an operational center—a futuristic command post where modern wargam-

---

*Now SAIC, Science Applications International Corporation.

ing would be linked directly to operational planning and policymaking at the highest levels of government.

A bemused Pentagon spectator impiously called the idea "a new global sandbox," but it was an idea that seemed destined to be realized—an operational assessment center. Eventually, as Rand foresaw it, the center would be part of the Office of the Secretary of Defense or the Joint Chiefs of Staff. SAI was not that specific, but people working on the SAI entry also talked about a center.

The center would serve as a storehouse for strategic data, for strategic concepts, and for continuity of strategy. One President's administration would pass this strategic treasure on to the next administration, which could then test or experiment with strategy inside the center instead of at scattered bureaucratic outposts or at competing think tanks. Eventually the center would probably become a place where contingency plans would be worked on, possibly during crises. Rand warned, however, that the one thing the strategic assessment center could not be was "an answer machine" by political leaders.

The match for developing the center pitted Rand, the fountainhead of nuclear-warfare wisdom and the nation's most influential defense research firm, against SAI, a younger, less famous, more diversified firm. Anyone touting the match would bet that Rand, the wary old competitor for Pentagon contracts, would reach back to its traditions, born with the Goldhamer political-military games, and SAI would thrust its entry into the future, propelled on buzz words like *artificial intelligence* and *executive system*.

The "people-in-the-loop" idea that Watts had endorsed in his paper was also a good bet. The "people" were from the political-military gaming tradition, and the "loop" consisted of computers supplying data bases and graphic displays. This would be safely in the middle of the Pentagon road, not particularly innovative, certainly not revolutionary.

The sandbox remark was typical of the response from critics who had been watching gaming now for decades and still had not seen much interest or acceptance on the policy level. Watts told of a meeting with some Americans and West Germans in the *Bundeswehr* Office of Operational Analysis and Exercises. A West German admiral asked who played command positions in U.S. war games. The admiral was surprised to learn that in U.S. war games majors and colonels routinely played as stand-ins for their star-shouldered superiors (who sometimes made "guest appearances" at games).

"The Germans," Watts recalled, "were frankly astonished by our answer. The admiral in particular made no bones about his feelings on the matter: War games are too important to be left to majors and colonels. I think, moreover, that the admiral is right." Like many other advocates of serious gaming, Watts also pointed to the fact that high-ranking Soviet military officers saw the importance of wargaming, which he said "appears to be a growth industry in the Soviet Union," where "senior civilian and military leaders take the time to participate substantively in war games."

Political scientists and defense analysts who had studied political-military games often found that the games were valuable for raising questions but not for producing answers. In a typical game players would grope for solutions, find some, and then quit, leaving sparse accounts of what had happened. When Robert J. Murray, the first director of the Naval War College's Center for Naval Warfare Studies, took over his post in 1981, he found relatively few records in the nation's oldest citadel of wargaming. "The mice had eaten them or something since Nimitz's time," he said.

People are not very orderly, and they do not keep very good records. On those two counts at least, computers won out over people, and both Rand and SAI turned toward computers for planning the new generation of games. Two favorite words of the planners were *sophistication* and *rigor*, the latter having for analysts the special meaning of standing up to validation. The idea was to develop techniques that would make wargaming more scientific by building into it records that later could be rigorously examined.

Christopher Makins, who had been a foreign-affairs expert on the SAI-aided *Nightline* television game (page 244, 245), worked on the Marshall project for SAI. Much of the data recorded in a game, he said, "is unbelievably boring." SAI "had three teams working away six or eight hours a day, and all that stuff was put on video. We had something like six days of solid video viewing to do if one wanted to be sure that one had not missed any of the nuggets or misunderstood any of the comments.

"Players never assimilate the material to equal degrees. Different teams go off on different tacks. You have to really look and see the

internal games the teams are playing, as well as the contributions
they are making to the game you are trying to get them to play,
and it is just very, very tough to draw conclusions from that kind
of thing that will stand up to any analytical rigor."

"One of the things we were trying to see was whether there was
any science that one could use in terms of choosing the players.
One of the big problems we tried to address in a rigorous fashion
was, 'How do you know what you've got?' "

The SAI system, according to a report by SAI researchers, would
use "the human intellect as an integral part of the architecture,
through 'people-in-the-loop' gaming." Other human beings would
train and support the players, whose computer help would include
a large data base and a "simulation library" continually updated to
give players the benefit of real-world situations and information
from past games.

SAI and Rand also faced the usual problem of creating a modern
Red whose sophistication carried Red beyond the mirror-image
level. Although house experts could provide voluminous information on Soviet crisis behavior and Soviet military and political
organization, the trick was to translate that knowledge into lore
useful for gaming.

Little appears in open literature about the games Soviets play.
But MIT's Lincoln Bloomfield has had the extremely rare experience of playing real Soviets in a war game. Bloomfield was lecturing at the Institute of U.S.A. and Canada in Moscow in 1970
when his hosts began urging him to run the kind of political-military
game he had been directing in the Pentagon. "I didn't want to run
that game," he told me. "I was euchred into it, and I made it very
tough for them: a Mid-East crisis.

"At that point they had a passion for American social science.
They thought we had some strong magic—sort of the way they
had thought about the Ford Motor Company in the 1920s—and
the word was, 'Find out all you can about this.' "

Bloomfield and the Soviets haggled over the setup of the game,
which he wanted to center around the breakdown of a Middle East
cease-fire. "I said, 'You've got to have an Israeli team.'

" 'Some of us have to play Israelis?' they said.

" 'You're goddamn right,' I told them. 'You've got to play Israelis
if you want a game.'

" 'Well, can't we have a game on some other subject?'

"I said, 'No Middle East, no game.'

"I was tired. I was being exploited. I said, 'Come on. I understand that the Soviet Union *invented* political-military games.'

"Anatole Gromyko [son of Foreign Minister Andrei Gromyko] said, 'I'll go and ask Daddy.' And he went, and he came back and said, 'No. It's always military. We may have some diplomatic statements in the scenarios but we've never done them.' "

Nevertheless, the game began, with Anatole Gromyko heading the U.S. team. There were also Soviet, Arab, and Israeli teams. "And *they* suggested that there be an Arab extremist team. There were the usual anonymous cop types posing as scholars and sitting around the room, and there was a lot of nervousness about this.

"Within an hour the teams were all into the dynamics of role-playing, despite the nervousness. By the end of the day the Israeli team was doing its best to screw the Soviet Union, and the Arab extremists were giving the Russians a terrible time. The Soviet team was worthless. They just prattled propaganda.

"The U.S. team showed a tough American posture. In order to enforce the cease-fire, they had jets from the American Sixth Fleet overfly the Sinai Peninsula, and at that point I said, 'Wait. Come on. You've got Americans saber-rattling here, and we're much more quiet and cautious during a crisis.'

"Three days after I left Moscow, Sixth Fleet jets overflew the Sinai. [A shaky Israeli-Egyptian cease-fire and other events had sparked another Middle East crisis.] I was asked not to publish the results [of the game], but I gave quite a few deep briefings in Washington, and there was a general feeling that it was a very valuable thing for the Soviets to get into our shoes."

SAI planned to get Americans into Soviet shoes by, among other things, providing Red players with a handbook containing decisionmaking, signaling, and command and control from a Soviet viewpoint. "What we tried to do was a little different from Rand," Makins recalled. "What we said was, 'If you are playing a Red team, can you calibrate those people in whatever scale or scales you want to—as to how Red they are?'

"The answer is that there are various ways you can do that, various techniques for self-calibration: you give them a lot of statements to comment on, which is very political-science, but does not strike one as impressive. Then you can do it in insightful or impressionistic ways. You know that Richard Pipes [a Russian émigré professor of history at Harvard and an adviser to President Reagan] is sort of a hard-liner and Marshall Shulman [a Soviet scholar who pursues a softer line] is whatever you think Marshall Shulman is.

"The next methodological question is, what do you want your Red team to consist of? Do you want a Red team that consists entirely of Richard Pipes and another one, maybe, that consists of Marshall Shulmans, or do you want a Red team that reflects what you assess to be the different Ivans that there are in Ivanland at any given time?"

Rand went completely automatic. Rand's Red would be a computer program. So would Rand's Blue. There would be no people in Rand's loop. There would only be computers. Human players would be replaced by "agents" whose behavior "is rule-programmed through extensive use of computers." And the agents would have characters, a variety of Ivans on Red's side, several kinds of Sam on Blue's side. Rand would enter Andy Marshall's contest by demonstrating a computerized war game in which "machine-controlled players" could do what human players had been so loath to do: cross the nuclear threshold.

"Conceptually, the notion of automating game play, though provocative, was neither unique nor new," members of Rand's development team pointed out in a report. "Computerized games are widely available, and automated players are now acknowledged as competent for games like chess and backgammon." As to the human beings failing to cross the nuclear threshold, "While heartening, that reticence does not lend itself to exploration of the circumstances and problems that may attend the transition to (or away from) nuclear conflict. If gaming (as opposed to invoking) the initiation of nuclear conflict holds any analytical value, it might only be possible with a programmed agent."

SAI and Rand demonstrated their war games to the Department of Defense in January 1981, Rand at its headquarters near the beaches of Santa Monica, California, and SAI in a building in suburban Virginia, hard by the Washington Beltway, the landmark for countless defense think tanks big and small. A number of Pentagon officials, including representatives from Marshall's office, attended the demonstration. There is no publicly available guest list or roster of players. Information about the demonstrations is sketchy. Participants and observers talk warily about what went on. Marshall's attendance at the demonstrations is reported by some people and not by others.*

SAI's demonstration lasted five days, beginning with a day-long briefing. SAI's predemonstration research had focused on a game involving the use of tactical nuclear weapons in Europe and a game in which surviving leaders of a "large-scale U.S.-Soviet strategic nuclear exchange" decide to fight a "protracted general war." Presumably, the demonstration included some nuclear war-fighting, which had been much in fashion in recent years.

The demonstration's wargaming began with a political-military game that focused on the Middle East and was played in the seminar-style, Red-Blue-Control tradition. Two moves were made. On the third day SAI shifted to a board game with rigid rules; the two moves of the previous day related to the board game, although the locale had shifted to NATO countries. On the fourth day two more moves were made in the political-military game. On Friday, the fifth day, everybody got together—SAI game directors, players, sponsors—to discuss the demonstration.

In Santa Monica, Rand demonstrated a game that had "control over the variables rather than being subject to the vagaries and inconsistencies of human teams." But, apparently due to the vagaries of time, Rand had not been able to get all its "automata," as it called them, ready by the show's deadline. So Rand settled on just developing the "Red Agent" automaton and letting "Blue Agent" be a group of human beings. The other agents—*Scenario, Monitor,* and *Force*—were programmed in time for the demonstration. *Scenario* contained enough political information to make a "world model." *Monitor* kept track of moves, time, and game rules. *Force* was full of data bases on the projecting, supplying, and reinforcing of military forces throughout the world.

For demonstration purposes the automated Red and the human Blue moved in what Rand said was "a Ping-Pong mode as opposed to having them both move at the same time." Rand did not reveal where or what Red and Blue fought over, though a report did frankly admit that even with a robot in the loop, escalation can be

*Marshall declined to be interviewed for this book.

balky. "We were asked to produce a scenario that followed a particular escalatory path. But we had built a machine that developed its own scenario from the ingredients fed into it, not a machine that followed a prescribed path. The result was frustration. Repeatedly, our automated war games would fail to escalate or would skip a certain escalation level. Much 'fiddling' with the BLUE moves and one override of the SCENARIO control program were necessary before we finally arrived at the kind of scenario required by the client. . . ."

Rand said at the time of the demonstration "our primitive automated agents" were not very sophisticated. They were, however, sophisticated enough for Rand to have won the wargaming playoff. If *won* is the right word. People at SAI do not say Rand won and they lost. As in a prizefight or professional tennis match, both opponents got purses. But Rand went on to develop its entry, which it named RSAC, Rand's Strategy Assessment Center.

There was an exquisite philosophical contrast in the Rand-SAI competition, whatever it was called. On one side were robots capable (with a little fiddling) of mindlessly going to nuclear war, and on the other side were human beings who usually could not. On one side was a world encapsulated in a computer, and on the other side was a world drawn from human history and the personal experiences of the players.

"SAI was looking at an intuitive tool like gaming and trying to make it rigorous," Joel Resnick of SAI told me, looking back at the demonstration he had helped to run. "Rand probably felt that maybe the art was in need of improvement. Even if the demonstration offered only anecdotes, it will be useful. SAI felt we had to take the ragged thing called gaming and model it.

"By setting such a high standard, you did something good. But in setting the high standards, the sponsors were disappointed. There are things that you can't analyze—troop response to nuclear weapons, morale. [General] Andy Goodpaster [former Supreme Allied Commander, Europe] once said, 'I think I can identify some generals who could keep their troops going *once* after a nuclear shot. Not twice.' "

Resnick had been in the midst of nuclear warfare studies since he began his career in the late 1950s at MIT's Lincoln Laboratory. Before joining SAI he developed analyses for SALT at the Arms

Control and Disarmament Agency and, as an analyst working for the Director of Central Intelligence, worked on what became PD-58, the Carter presidential directive that validated the idea of a protracted nuclear war.

When he talked about the demonstration he sounded more philosophical than analytical. And instead of directly describing SAI's entry in what he said had not been a contest, he broadened the topic to a discourse about the ideas behind people-in-the-loop war-gaming.

"What makes a problem a tough problem—a problem we can't solve?" he asked—and answered, "The problems that you can't handle are the tough ones.

"The two-sided exchange"—the bolt from the blue—"was a difficult problem. We have machines, and now we have worked out all of the nuances of the exchange.

"Suppose a solution to a problem was within the grasp of one smart person. You turn to that one person to get that answer. This happens in medical diagnosis. There are doctors who can solve a problem, be the one smart person.

"If there isn't one smart person for your problem, then you can get ten smart persons, but this is a problem in itself. You need to structure their interaction. Say you need experts—economic, cultural, political, military, morals. But they don't have a common language. So you say, 'I'll speak to each, one at a time.' Meetings. There is a limit to how long you can keep a meeting going. You can go to each smart person and interview each one. Or you can have decisionmakers do it. Or you can have a meeting and let everyone interact. All of this takes a lot of time.

"If neither of these ways is a satisfactory solution, if the problem bothers us, I must look for another tool. I must find another way for people to interact in a way that doesn't have the problem of time. Well, we have summer studies. We take the Defense Science Board to Woods Hole for a week or two. A pleasant environment and it keeps people together.

"Gaming is an artificial environment, but the juices flow. People can play for day after day. A game brings out the human emotions. I can bring a lot of smart people together. We get at least the hope of getting into some very good ideas."

Resnick coined the word *outliers* to explain unexpected results from a game. "Suppose you were to ask ten people what the stock market would do a year from now. One might say the Dow Jones would hit one thousand, eight might say between eighteen hundred and nineteen hundred, and one might say thirty-three hundred. Those two guys with the thousand and the thirty-three hundred would be outliers.

"Suppose you play a game ten times, you would expect it would be predictable, as in a chess game between a novice and a master. But every once in a while unlikely things happen. Outliers. That's the real world. It helps you understand how you can use the normal results. But how can you *use* outliers—if you find one that you like?

"In a series of Central European war games, say, nine times out of ten the Warsaw Pact forces make it to the Channel in fifteen days. But in one game, in three months they still haven't made it. Then you discover that this happened because a probability went that way. Suppose somebody threw sixes three times in a row and some division's artillery was much more effective. The results of the campaign resulted from those three sixes—and maybe you start thinking about whether you should make the artillery more effective. Or the weather works against you because of probability. You don't get as many air sorties as you want. You say, 'I don't want that to happen. Maybe I should do things to make these unlikely events *happen* in the real world.' The game is letting you see how these events produce a different result. It's not saying it's predictive."

Soon after Rand got the contract for the assessment center, Paul K. Davis, who had become notorious among gamers for pointing out that he had seldom heard a senior U.S. official base policy on games, set out to make automated wargaming a major policymaking tool. He was the director of the development of RSAC, and he took over the task of getting RSAC into the Pentagon. A former acting Deputy Secretary of Defense for program analysis and evaluation, he, like Resnick, had worked in the Arms Control and Disarmament Agency and had learned about nuclear warfare from deep inside the strategic community.

He has written extensively on RSAC, which he has always seen as a true center—a *place*, not a concept. In that place, connected with the Joint Chiefs of Staff and the Office of the Secretary of Defense, wargaming automata would be asked such questions as these:

- What if the Soviets attack Washington on Inauguration Day? Would we be "decapitated"? What does "decapitation" mean on an operational level?

- What if the Secretary of Education becomes the National Command Authority (NCA)? Will he know enough to make timely strategic decisions?

"Once in ten or twenty years something like RSAC happens. Until now we have had no analytical models to cover political ideas," Davis said in an interview. "If one has purely political games, the military issues don't work. The political people will blithely think that the armies can be moved with the speed of light." He told of an Assistant Secretary of State "who nuked Aden," and he added, "When people get sense, they don't do the stupid things they do in novels.

"In the RSAC system the politics and the military are put together. We use models very similar to what would be used at the Pentagon. For instance, the escalation model. No one—Kissinger or Paul Davis—can predict with any degree of confidence what the U.S. or the Soviets will do in any crisis. So we don't build escalation models to see what will happen. But we put down what strategists worry about. We acquaint people with issues and interrelationships rather than predictions.

"In the human war game serious people are less concerned with force levels. They are wondering, 'Is my command and control system working? How robust is my alliance? Is there a way out of this?' They are less interested in a math game than in averting catastrophe. RSAC consists of languages and a number of rules intended to look at the world the way senior leaders do.

"Political scientists don't know about quantitative issues—such as how fast can armies move. They write essays. We are trying to get them to write down *rules* in a programming language anyone can understand."

RSAC once spoke ROSIE (for *R*ule-*O*riented *S*ystem for *I*mplementing *E*xpertise), an English-like computer language. This evolved

into the presumably better, though sterner-sounding, RAND-ABEL.
The stilted computerese language is familiar to any personal computer user who has ever tried to read the instructions ("documentation") accompanying software. RAND-speak tries to be friendly
to the user and useful to the computer. The effect is a language
that hovers between pidgin English and a weird analytic poetry.
Here is a sampler of RSAC's language:

```
If the actor is a conflict location,
  let the actor's threat be grave and
   record grave [threat] as ''being a
   conflict location''.
If the actor's Ally = [is] USSR and the
   actor's superpower-presence & [is] Red-
   major,
  let the actor's threat be grave and record
   grave [threat] as ''major Red force in
   its territory''.
If the actor's Ally ~ [isn't] US and the
   actor's superpower-presence = [is] Blue-
   major,
  let the actor's threat be grave and record
   grave [threat] as ''major Blue force in
   its territory''.
If the actor is a follower of (some leader
   such that that leader's threat = [is]
   grave),
  let the actor's threat be indirectly-grave*
   and record indirectly-grave as the string
   {''grave threat to'', that leader.}
```

"The basic idea is simple," Davis and another Rand researcher
wrote about RSAC. "We seek to maintain the basic structure of a
political-military game while replacing some or all of the players
with computer programs serving as automatons. This permits (but

---

*An "indirectly-grave threat," when "viewed broadly or in terms of its long-term conse-
quences, is equivalent to actual or imminent bombardment or invasion." A "grave threat"
is the same situation "viewed narrowly or in terms of its near-term consequences."

does not guarantee) efficient documentation, replicability, and transparency. It certainly has the virtues of depersonalizing issues and rendering them amenable to analysis." The biggest problem in automating wargaming, they say, is getting "political scientists to spell out what they believe about the behavior of individual countries" and getting into a computer "what we believe are the processes of war."

One political scientist who lacked faith in RSAC was Lincoln Bloomfield. "The notion that you can substitute for judgment a computer program to do typical Soviet responses—I think is extremely dangerous," he said. "I want senior American officials not to look at Rand Corporation to tell them how Ivan behaves but to do some homework of their own.

"And that's why Ivan is not a substitute for the kind of game that will go on in the Situation Room when you say, 'Jesus! This is what has happened. What do you think Gorbachev is thinking? And when he looks at us what does he see as our options?' You can have a conversation like that in a game. A game is a way of doing that, and using computers for everything you want, data, maps, displays.

"I've spent some time in the Situation Room and it's not like some of the things we research. There's a different order of things. Someone might not say, 'What do you think Gorbachev is thinking?' Someone might say, 'Let's show that son of a bitch that he can't get away with this' or something like that. Fine. Okay. But someone then ought to say, 'And here's what the reaction is likely to be.' "

In developing RSAC, Rand came full circle to one of the oldest forms of wargaming, chess. Chess-playing computer programs, a Rand report on RSAC said, "embody the concept of a competition between two intelligent, but not omniscient, entities." The players in the RSAC game are "agents."

They are *Red Agent*, a "computerized model of Soviet behavior"; *Blue Agent*, a "computerized model of U.S. behavior"; *Scenario Agent*, a "computerized model of nonsuperpower behavior, as well as a 'bookkeeping' model describing political aspects of the world situation"; *Force Agent*, consisting of computer models and data bases "used to keep track of military forces worldwide and to determine the results of combat." Also tucked into Force Agent is "Campaign," which watches over military forces and operations, mobilizations, and the outcome of naval and ground battles. *Systems Monitor* compiles game records and has "housekeeping programs" for "determining the times of the various automaton moves, formatting information for use by the different agents."

Red Agent and Blue Agent are "essentially identical" on a technical level and can make "realistic mistakes," but they have "drastically different behaviors." These are arranged in models labeled Ivan and Sam, with numerals following their names to indicate the particular mood they are in during a particular game.

Ivan 1 is "somewhat adventurous, risk-taking, and contemptuous of the United States." Ivan 2 "is generally more cautious, conservative, and worried about U.S. responses and capabilities." Suppose, for example, the United States had its forces on worldwide alert. Ivan 1 "might regard that as a minimum measure, characteristic of timidity," but Ivan 2 "might be alarmed." Davis and his colleagues write quite a bit about Ivan, but Sam remains somewhat of a mystery automaton whose personality will probably vary as much as presidential administrations do. By measure of how much Rand writes about the two agents, Sam seems to be more of a secret agent than Ivan.

Red Agent "is sensitive (to greater or lesser degree depending on which Ivan is used) to indicators of U.S. will and intentions." RSAC's handlers program Red Agent's game behavior by feeding the computer "descriptors," each of which has possible "values." For instance:

Descriptors	Values
Expansionist ambitions.	Adventuristic, opportunistic, conservative.
Willingness to take risks.	Low, moderate, high.
Assessments of adversary intentions.	Optimistic, neutral, alarmist.
Insistence on preserving imperial controls.	Moderate, adamant.
Patience and optimism about historical determinism.	Low, moderate, high.
Flexibility of objectives once committed.	Low, moderate, high.

| Willingness to accept major losses to achieve objectives. | Low, moderate, high. |
| Look-ahead tendencies. | Simplistic one-move modeling, optimistic and narrow gaming, conservative and broad gaming. |

At the demonstration Red Agent was Mark I, a "breadboard proof-of-principle" automaton, and Blue Agent was not even on a breadboard. By 1983, when RSAC was already an up-and-coming candidate for serious Pentagon jobs, Mark III Red and Blue Agents existed. With far less human help than their ancestors needed, they could play games from complex scripts that contained several "idealized start-to-finish strategic level" Analytic War Plans and "some aspects of real-world confusion."

The agents also had some knowledge of each other. "Red has a model of Blue, assumptions about third-country behaviors, and his own combat models"—but Red's model of Blue may be wrong. If Red wanted to invade Europe or merely seize Iran, Red looked into "a family" of plans, each of which branched off with more plans and more decisions: Blitzkrieg? (High risk, but potentially high payoff.) Cautious campaign? (A chance to tailor moves to the vigor of Blue's reaction.) Guessing each other's behavior, the Mark III Red and Blue could shift from one plan to another as they fought through a scenario that began, say, with a Soviet invasion in Southwest Asia, escalated into war in Europe, and ended with a prolonged, intercontinental nuclear war.

Scenario Agent was also evolving. At first Scenario was "expected to reflect the behavior of third countries well enough to create a realistic context for the superpower confrontation." Scenario was told, "It is not important how the nonsuperpowers reach their conclusions (for instance, we certainly do not wish to model the politics of the French Cabinet), nor how well they achieve their independent objectives in the conflict."

But Scenario seemed to be writing itself a bigger part, for soon Scenario was letting a third country join with a superpower for a while and then use that relationship as a lever for invading a neighboring country, giving Red and Blue new problems. Scenario's champion was William Schwabe, a Rand graduate student who looked into what he called "Strategic Analysis as Though Nonsuperpowers Matter." His ideas were developed and put into the scripts, and soon playing got as complicated as a gin rummy game in which the kibitzers now and then snatch cards from the table and otherwise bother the gin players. Kibitzer countries are labeled *Blue* (sides with the United States), *Red* (sides with the Soviet Union), and *White* (sides with neither superpower in the current conflict), and given variables:

Firm:–very unlikely to change its side.

Moderate:–fairly unlikely to change its side.

Soft:–fairly likely to change its side.

Noncoordinate:–does not grant transit rights to the military forces of either superpower.

Coordinate:–grants logistics access to its ally.

Cobelligerent:–grants combat access to its ally.

Nuclear Releasor:–fully cooperates with its ally, including agreement on use of nuclear weapons.

Each country gets a temperament rating—reliable, reluctant, initially reliable, initially reluctant, neutral—and is rated as a "leader" or a "follower" that is "opportunistic" or "assertive," which has a special meaning: "If nuclear capable, exercises independent nuclear deterrent. If gravely threatened, requests allied nuclear strike against opponent superpower homeland. If abandoned by superpower ally, becomes noncombatant. If aided by superpower ally, becomes reliable."

RSAC research concentrated on Red Agent, especially models of Ivan. An automaton referred to as *it* in early descriptions, Ivan soon began appearing in documents as *he*. Davis, RSAC's master, along with a colleague, wrote that before they began building National Command Authority versions of Ivan, they had to "form a strong image of the particular Ivan—his personality, his grand strategy, his 'temperament.' " Sometimes Ivan's creators "write little essays" like this one: "Ivan K* is somewhat aggressive, risk-taking, and contemptuous of the United States, but is ambivalent—

*K for " 'complex'—neither dovish nor hawkish, and neither simplemindedly doctrinal nor softheadedly optimistic about prospects for intrawar bargaining."

believing that the United States can be aggressive and irrational sometimes. Ivan K believes Russian military doctrine is essentially correct, although not always applicable."

Ivan K, "contemplating how to invade Southwest Asia, hopes for a quick and straightforward takeover of Iran because of his tentative belief that the United States will not intervene. He recognizes, however, that he could be misassessing U.S. resolve, so he prepares to raise the stakes by blockading Berlin if he sees the United States preparing to send CENTCOM [Central Command] forces. The reasoning here is that the fear of war in Europe would divert U.S. forces that otherwise would go to Southwest Asia, and might well divide the Western alliance."

Red Agent operates on three levels—national command, area command, and tactical command. While the National Command Authority is ordering up war plans from the levels below, Red Agent is also calculating what Blue Agent is probably going to do. Red Agent issues commands with an *if-then backup default* formula: "If roads are open in the Transcaucasus and Iranian forces are below two divisions in Khorasan," then deploy three motorized rifle divisions or else deploy five motorized rifle divisions (backup), with a "default" of three divisions instead of five.

A script that calls for certain action writes finis to itself when something happens that had not been expected in that script. Suppose, for example, Ivan makes what seems like a relatively unprovocative move while working from a scenario on an area command level. Unexpectedly, Blue reacts strongly and NATO mobilizes. That abruptly puts The End on that level's script, and the Red forces at the area command level must turn to the national command for further instructions. This hands another script to Ivan.

The war game then continues in whatever direction Ivan and Sam decide. They are, after all, running the war.

CHAPTER 18

Across the Threshold:
RSAC Goes to War

If the Persian Gulf game I played at the Naval War College had been played by RSAC's Ivans and Sams instead of by human beings, RSAC would have begun by swiftly moving through a baseline scenario with several "phases" that could have branched off into other scenarios with more phases. RSAC does have to thrash around wondering what is going on in the Persian Gulf. RSAC briskly looks through its data bases, peers ahead, and, deciding on a more interesting game than the one we had played at Newport, quickly settles on *Starting Conditions*:

It is August 22, 1990. A Soviet client government controls Iran, which is near civil war. Iranian Army units have defected to "the Iranian white faction." Soviet advisers, but no forces, are in Iran. Most nations in the region are uncommitted to either superpower, but the U.S. has basing and overflight agreements with Egypt and some of the Gulf Cooperative Confederation (GCC) states.* Quicker than anyone could write *assumptions* on a blackboard, Scenario dumps an "assumptions about nonsuperpower behavior" into the scenario. The table begins with Afghanistan—*no nuclear capacity, average military strength, Red Major superpower presence, Red orientation, reliable temperament*—and ends a litany of nations

*GCC members are Bahrain, Kuwait, Oman, Qatar, Saudi Arabia, and the United Arab Emirates.

341

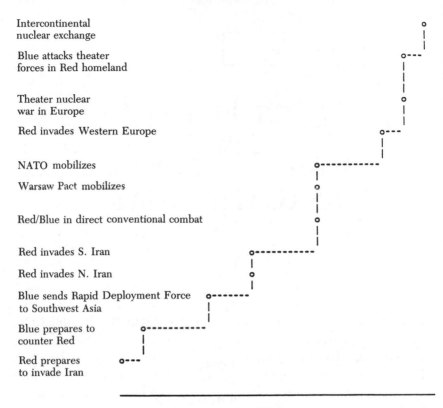

Intercontinental
nuclear exchange

Blue attacks theater
forces in Red homeland

Theater nuclear
war in Europe

Red invades Western Europe

NATO mobilizes

Warsaw Pact mobilizes

Red/Blue in direct conventional combat

Red invades S. Iran

Red invades N. Iran

Blue sends Rapid Deployment Force
to Southwest Asia

Blue prepares to
counter Red

Red prepares
to invade Iran

Time

with United Kingdom—*has nuclear capacity, average military strength, Blue Major superpower presence, Blue orientation, initial-reluctant temperament.*

RSAC then clicks in the nuclear escalatory ladder (above), a scaffold for the series of branching scenarios that will drive the crisis.

RSAC sorts through "cases" that would affect the scenario and produce variations in the game. *Case Oil Consumers*: Several European countries are dependent on Persian Gulf oil and perceive a threat as Red prepares to invade Iran. Does this make Blue move more quickly? No. Blue is not dependent on these countries (though they are NATO allies). Case not important. *Case Oil Consumers and Suppliers*: Persian Gulf oil suppliers feel threatened by Red. Blue can deploy faster because of cooperation. This could limit Red to northern Iran. *Case Egypt*: If Egypt provides forces to augment Blue's Rapid Deployment Force, Red could be deterred. Game

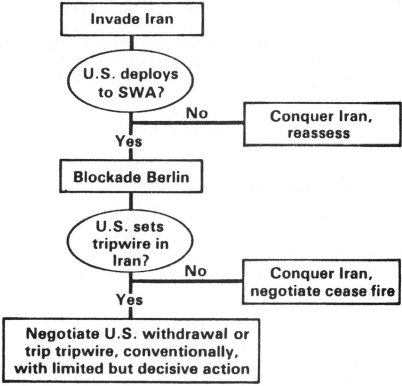

Rand's computerized war pits an automaton named Sam against an automaton named Ivan. In this branch of Ivan's "grand-strategy tree," Ivan sees no problem if Sam does not deploy forces to Southwest Asia in response to an invasion of Iran. If Sam does, Ivan raises the stakes by blockading Berlin. If Sam moves on Ivan, Ivan may trip the wire—drawing Sam into combat. Ivan would react "conventionally," meaning without nuclear weapons, though Ivan and Sam, unlike human game players, have no inhibitions about crossing the nuclear threshold. (Credit: Rand Strategy Assessment Center.)

could end here. Case folded in, but game goes on. *Case Libya*: Will Libya interfere with Blue forces to aid Red? No. Libya perceives direct threat from Blue; Red constrained not to attack Blue before Red D-day for invasion of Iran. Case folded into baseline.

Case Pipeline: European countries served by Soviet Siberian gas pipeline susceptible to Red blackmail. Neutral temperament assumed for Belgium, France, West Germany, Italy, and Netherlands. Lack of alliance cohesion will undermine Blue efforts to defend Europe and affect game. Case folded into baseline. *Case*

Saudi-Israeli: Saudi Arabia and Israel may not cooperate on same side to aid Blue. But Saudi may be offended by fact that GCC states are cooperating. Run this through leader-follower rules. *Case Turkey*: Turkey has early threat perception because it sees Red mobilization on Iranian border, also Red mobilization on Turkey border.

If RSAC slowed down and showed what is going on inside itself, this is how the scenario would look.

Phase 1: Early Blue Efforts to Gain Support

MOVE 1.1 RED REGIONAL STRATEGY DECISION

Situation:—Red receives a request for military assistance from its client government in Iran and must select an intervention plan that fits the circumstances and Red objectives.

Response:—Red sets the present day as D-day for an invasion of Iran, begins mobilization and deploys additional forces to the Indian Ocean.

Rationale:—Red is inclined to provide assistance unless military balance considerations suggest an unsatisfactory outcome or high risk or undesired escalation. Red's intelligence estimate is that combat outcome prospects in the event of escalation in Europe are satisfactory. Intelligence estimates for Southwest Asia suggest Iran could be occupied by Red successfully because of the slowness of Blue's anticipated response.

Red Agent asks Force and Scenario Agents for intelligence estimate for several future contingencies: "Conduct a Force Operations look-ahead in Europe, assuming the execution of Red plan/script D2[5] [a Red plan for invasion of Europe] following mobilization and deployment of the WP [Warsaw Pact] forces starting today, with NATO assumed to initiate mobilization and deployment of reinforcements after a few days' delay and further assuming that conventional weapons only will be used by both sides.

"If Red is projected as occupying and controlling Western European territory as far as France and less than X* days of combat operation (penetration along two or more major axes of Red attack), then initiate the process of selecting which of Red Southwest Asia plans/scripts (A, B, C) is to be executed starting today—

"*Else if* European look-ahead projects Red control objectives as being achieved in more than X days but less than X days, then initiate process of selecting which of Red Southwest Asia plans/scripts . . . is to be executed starting today—

"*Else if* European look-ahead projects Red control objective as *not* being attainable in X days or less and no successful Blue defense within X miles of the West German eastern border on X or more major Red attack axes, then initiate process of deciding whether to execute Red plan/script A starting today or to take no action in Southwest Asia—

"*Else if* European look-ahead projects a successful Blue defense within X miles of the West German border on X or more major Red attack axes—

"Then take no Red military action in Southwest Asia."

But look-ahead—which is a quick game-within-the-game—in effect plays out a few future moves. The look-ahead foresees Red victory in X to X days, and so Red can decide whether to proceed with Plan A, an occupation of northern Iran only, or Plan B, a takeover of Iran. Red decides on Plan B after being told that Blue will not be able to get the equivalent of an armored division to Iran within three days of Red D-day.

MOVE 1.2: EARLY BLUE RESPONSE: Blue begins negotiating with European and Middle East allies for bases because Blue cannot project power until it has a way to put forces in the area. This is a complex move. (The Middle East is so complicated that it can baffle even RSAC, which resorted to using human beings for Saudi Arabia and Israel in one game. The human beings were asked to fill out a form that ordered and explained the move; explanations included "oppose communism or Soviet hegemony" and "hedge bets." In game without human beings, RSAC simply gave up and told Scenario Agent, "Deny Israel is a player.")

---

*Rand's examples of RSAC rules in unclassified documents often do not include actual numbers for times, forces, and distances.

As Blue sounds out other states, RSAC examines other games and finds that there had been disputes over what Egypt and other Middle Eastern countries would do to help Blue. RSAC decides to have Egypt not only grant Blue access but also offers its forces, which Blue begins airlifting closer to potential combat. West Germany, France, and the United Kingdom back out of any commitment, saying they see no threat to their own interests. Spain and Portugal allow Blue logistical access, which Turkey denies. (Even though a case had Turkey perceiving a threat, a "rule review" has resulted in a decision that Turkey is trying to be neutral.) Italy and Greece also deny access to Blue.

Egypt's decision makes Red think through its plans again. Then Blue mobilizes the Rapid Deployment Force, strategic mobility forces, the Military Sealift Command, and the civil reserve air fleet (in effect, commandeering civilian airliners). That does it. Red decides to limit the invasion to northern Iran. The Gulf states change their status from reluctant to reliable and allow Blue logistical access.

Blue sets DefCon III and sends the Rapid Deployment Force to Southwest Asia with the intent to land in Iran if Red invades. Now Egypt and the Gulf states place their forces on call to Blue. As Red launches the invasion of Iran, Turkey grants Blue combat access. Red attacks Blue forces in Egypt, Bahrain, Kuwait, Oman, Qatar, Saudi Arabia, the United Arab Emirates, and in the Indian Ocean. Red bombs the Suez Canal, which shuts down.

Blue puts the Strategic Air Command and strategic warning networks on high alert. Blue lands forces in Iran and steps up naval forces.

Red attacks Blue naval forces and merchant ships in the South Atlantic, Indian Ocean, and South China Sea.

Blue, whose forces have been in combat against Red in Iran, prepares to withdraw from Iran and simultaneously launches air strikes on Red forces in Afghanistan.

Red mobilizes the Warsaw Pact to bleed off Blue resources from Southwest Asia and prepare for possible war in Europe. Red launches air strikes against Turkish military air bases.

Blue calls for NATO mobilization. The Soviet natural-gas pipeline, source of much European energy needs, causes some problems (and "a departure from the baseline scenario"). Canada,

Denmark, Greece, Iceland, Luxembourg, Norway, Portugal, Spain, Turkey, and the United Kingdom mobilize. Belgium, France, Italy, and the Netherlands, fearing the pipeline will be turned off, will not cooperate with Blue. West Germany independently mobilizes its forces to protect its own borders.

Blue withdraws forces from Iran and begins mining the Baltic Sea and other waters that Red ships must transit.

Red decides to invade Western Europe and, aware of the split in NATO, sends Rand-speak warnings to Belgium, France, Italy, and the Netherlands—*demand you remain white noncoordinate noncombatant*—and promises to respect their neutrality.

Red invades through West Germany. Red, with Czech, East German, and Polish forces, fights Blue, which is joined by Canadian, British, and West German forces. Blue predicts that Red will reach the western border of West Germany in a matter of days.

RSAC has reached a moment it was created for: the nuclear threshold.

A "case" offers a way out: Blue believes that it cannot win even if it uses tactical nuclear weapons. Red and Blue agree to a cease-fire. Blue will remove troops from West Germany and will begin negotiations for reunification of Germany.

RSAC skips that case. The baseline script moves relentlessly forward. The war is in a final phase.

Phase 5: War in Europe

MOVE 5.1: BLUE RESPONSE TO EUROPEAN INVASION

Situation:–Red strikes France and European NATO with conventional weapons.

Response:–Blue defends with conventional weapons, but requests NATO nuclear release authority. Blue strikes Warsaw Pact countries, except for the Red homeland, with conventional weapons.

MOVE 5.2: NATO AND FRENCH NUCLEAR RELEASE DECISION

Situation:–NATO members and France have received requests from Blue for nuclear release authority over nuclear weapons located in those countries.

Response:–They comply with Blue request.

MOVE 5.3: RED THEATER NUCLEAR PREEMPTION

Situation:–Red has started to penetrate Blue defenses. Red learns that Blue has requested nuclear release authority.

Response:–Red preempts with theater and battlefield nuclear weapons in Europe and Southwest Asia, initiates civil defense, initiates strategic antisubmarine warfare with nuclear weapons, and neutralizes Blue satellites.

MOVE 5.4: BLUE THEATER NUCLEAR RESPONSE

Situation:–Blue and NATO forces in all combat theaters are under nuclear attack.

Response:–Blue strikes Red theater nuclear forces, including those in Red homeland. The rationale: "Blue considers this a militarily necessary response."

RSAC passed into the custody of the Office of Net Assessment, where one of Andy Marshall's military assistants, Navy Commander James J. Tritten, watched over RSAC's transition from demonstration model to "operational delivery to prototype users." RSAC's name was changed as soon as it went to work. It is now officially known as the Rand Strategy Assessment *System*. "Center" was dropped along with the idea of the automata having their own place to play games. RSAS can exist wherever its masters set up the necessary hardware and software; RSAS exists essentially only as lines of computer code. R-for-Rand still advertises the system's origins, but RSAS's public life has ended. Sam and Ivan have entered the secret world of nuclear realities.

Tritten, who received his doctorate in international relations from the University of Southern California, smiled when he talked about RSAS. He smiled the way naval officers a generation ago would have smiled about a sleek new warship. "RSAS captures the imagination," he said. "People will say, 'I have five days to focus on a problem. I have a chance to interact with a lot of people—

and I will be exposed to all they know. And the sponsor learns by skimming the cream of all those bright people.

"When you say to someone, go play a game, he doesn't have to prepare a paper and get it cleared. He comes in with ideas and he stimulates others. And they leave the game with their ideas challenged. He'll say, 'We ran the European war—a sixty-day war in thirty minutes—250 times and X always happened, regardless of the variables.' "

I asked him what he thought this new kind of unpeopled gaming might do. He paused, and then answered in the measured way that Frank Kapper used when he told me, "Specifics are classified."

Tritten said, "It might figure out what Sweden will do, what Finland will do. If someone says, 'Let's play the war with Finland defending against Blue *and* Red,' then you can play it that way."

Sweden and Finland. Perhaps. But RSAS has in fact taken on one of the biggest problems in wargaming: the NATO-Warsaw Pact battle. Sam and Ivan have already raised questions about just how well NATO can expect to do on that great battlefield of games. A mention of the RSAS gaming was contained in a report that said, "The bottom line judgments about the feasibility of a successful defense by NATO . . . appears to be substantially at odds with those that have begun to emerge from preliminary work on Europe contingencies using the RSAC [as it was still named]." That was the last dispatch we will probably hear from the RSAS wars.

RSAS is reproducing itself—and its Ivans and Sams—and is at work. Its first group of masters are the Army's Concepts Analysis Agency, the National Defense University, and the Joint Chiefs of Staff, which has begun using it for wargaming and contingency planning. I wonder if the chiefs will put it where the old SAGA sign pointed, down in the basement, where wargaming was always played by people who had trouble fighting a nuclear war.

Sam and Ivan can detonate nuclear weapons because they lack what keeps players of flesh and blood from crossing the nuclear threshold, even in a game. Sam and Ivan loiter at that threshold not because they are robots responding to the wizards of nuclear war. They loiter there because, I think, they have a message, as much about war games as about war. They can cross that threshold because neither one of them has a conscience.

We learn very early in life that games have rules. The value of a war game is that it shows the results of a war when it is played as a game, with rules. Ever since the days of the Battle of Maldon the warriors of Western civilization have tried to wage war by rules. When nuclear war became the new kind of war, we tried to stick with the old rules. If Ivan and Sam have taught us anything it is this: The nuclear threshold is the place where war by the rules ends. Beyond that threshold, no war game can go, for beyond that threshold there are no rules.

Notes

The Spectrum of War: An Introduction and Scenario

Game on the future of President Marcos: I obtained an undated, partial tape of the game. I verified the tape with sources knowledgeable about the Philippines game. I was also able to identify two of the voices. In reporting on the contents of the tape, I decided not to identify any voices.

SINBAC: Background briefing by a Pentagon official who desired to remain anonymous; quotation about the Joint Chiefs is from this official.

Some forty universities and think tanks: "CSC Report, War Gaming," published by Computer Sciences Corporation, El Segundo, California, 1986.

BDM Iraqi game: *The Washington Post*, November 18, 1986; November 24, 1986.

"which had best be forgotten": Copeland, *The Game of Nations*.

Extent of wargaming in Department of Defense: Hoffman, "Defense War Gaming"; Mandel, "Political Gaming . . ."

Iklé quotation: Brewer and Shubik, *The War Game*.

Simulation and modeling replace "gaming": Although the change was supposed to have taken place some time in the early 1980s, *The United States Government Manual 1984/85* still listed the name of the agency as Studies, Analysis and Gaming Agency. My briefing officer, who had access to classified records, said gaming started in the Pentagon in May 1962. Haurath (*Venture Simulation*) says the games date to 1961, when a Joint Chiefs of Staff Joint War Games Coordinating Group evolved into the Joint War Gaming Agency with three divisions: General War, Cold War, and Limited War. The Cold War Division ran the games. Another source— Mandel, ibid.—gives 1961 as the founding date of *political* gaming by the Cold War Division of the Joint War Games Agency (JWGA, later named SAGA). Most game records remain secret. I was able to obtain some records in the National Security Files of the Lyndon B. Johnson

Library through the Freedom of Information Act. One of those documents says JWGA gaming started in 1961.

Young computer hackers.: A UCLA student was arrested on November 2, 1983, for breaking into more than two hundred computer systems, including those of the Naval Research Laboratory in Washington, the Rand Corporation in Santa Monica, California, and the Advanced Research Projects Agency Network, which contains Department of Defense information restricted to authorized users of the network. *The Milwaukee Sentinel* reported on August 11, 1983, that a group of young hackers had entered about sixty computer systems, including the Los Alamos National Laboratory, a nuclear weapons research center. The young men, who called themselves the 414s, after the Milwaukee telephone area code, all said they had seen *WarGames*, but they had started entering computer systems before they had seen someone do it in a movie.

Naval War College history: Hattendorf et al., *Sailors and Scholars*; Vlahos, *The Blue Sword*

"Let's write off Germany": confidential source.

Unclassified assessment of gaming: Mandel, "Political Gaming . . ."

Persian Gulf game: I played the game described in this chapter during the Third May Conference on International Security, May 9–12, 1984, at the Center for Naval Warfare Studies, Naval War College, Newport, Rhode Island.

Newspaper quotations: *Washington Post* May 17, 1984; June 8, 1984.

Chapter 1: War Stadiums: Then and Now

National Defense University gaming: Faculty members who wish to remain anonymous; Lawrence, "Playing the Game." (Quotes from Lawrence are also from this source.) The game on Poland seems to have been based on *Poland 1980–81*, which was published in March 1985 by the National Defense University Press. Like the game, the book looked at the relationship between loyalty to the government and loyalty to organizations. The book was written by Nicholas G. Andrews, a senior fellow at the university on detail from the State Department. He was at the American Embassy in Warsaw in 1968 to 1971 and was deputy chief of mission at the embassy in the crucial months bracketed by the strikes at the Gdańsk shipyards in the summer of 1980 and the imposing of martial law in December 1981.

Game that looked at undeclared war: Browning, "Rehearsing the Critical Decisions."

Estimate of game community: Several war-game players gave me the same estimate. Manufacturers of hobbyist war games claim annual sales of more than 2 million games. (Dunnigan, *The Complete Wargames Handbook*.)

Kapper remarks: Interview, Kapper. I also obtained a great deal of background from Kapper's articles on gaming (see Bibliography).

Pentagon gaming described in 1967: "General Information Political-Military Games," undated; distributed to players at a game in 1967. (Lyndon B. Johnson Library, National Security Files.)

Lack of changes: Background briefing, Pentagon Joint Analysis Division, February 13, 1985. The briefing officer asked that he not be identified.

Security restrictions: Records of the Joint War Games Agency, Cold War Division, Politico Military Branch, Quarterly Activities Bulletin No. 4, April 15, 1966. (Lyndon B. Johnson Library, National Security Files.)

Information on Epsilon I-65: Letter from Earle G. Wheeler, chairman of the Joint Chiefs of Staff, to participants in Epsilon I-65, November 4, 1965, and other documents. (Lyndon B. Johnson Library, National Security Files.)

Gaming disclosure in a comic strip: MIT Professor Lincoln P. Bloomfield, who was once deeply involved in Pentagon gaming, gave me a copy of the "Steve Canyon" strip and the story of how it came about.

Kerosene for jet fuel in an emergency: Wilson, *War Gaming*.

Ungerleider's trip to Israel: Interview, Ungerleider.

Exchange of words in 1964 game: Senior-level Critique of Sigma I-64, April 9, 1964. (More about this game in Chapter 10.)

Officer called "a wimp": Confidential source.

Lese remarks: Interview, Lese.

On "playing Red": At Naval War College, confidential source; students at Naval Academy, Associate Professor Jane Good, *Washington Post*, October 7, 1984; naval analyst as Soviet submarine commander, confidential source; American as convincing Red, interview, Bloomfield; how to diddle the U.S. government, Copeland, *The Game of Nations*. The "Thinking Red" wargaming workshop was held at the National Defense University, April 23–25, 1985.

Chapter 2: "It's Hard to Start a War"

India-China-Pakistan game: "Final Report," Summary of NU I-66, Joint

War Games Agency, Washington, D.C. Comments are from extracts from the transcript of the Action Level Review of NU I-66, February 2, 1966, contained in the "Final Report." Remarks of Game Director leading the Senior Review: "Senior Discussion," February 8, 1966. The filming of games is mentioned in "Nu I and Nu II-66, Final Report," Joint War Games Agency, January 11–25, 1966, Washington, D.C. (Lyndon B. Johnson Library, National Security Files.)

History of games: Game records, Joint War Games Agency, Cold War Division, Politico Military Branch; Quarterly Activities Bulletin No. 4, April 15, 1966. (Lyndon B. Johnson Library, National Security Files.)

"damnedest salesman's conventions": Jones in conversation with Shubik; see *The War Game* by Brewer and Shubik.

Antiballistic missile game: The portions of the game I received through a Freedom of Information Act request were "Memorandum for Agency Official Coordinating Beta I and II-67," March 10, 1967, signed by Army Colonel Thomas J. McDonald, chief of the Cold War Division; "First Scenario Projection, Beta II-67" from Control to Blue and Red; "Second Scenario Projection, Beta II-67" from Control to Blue and Red; "Move No. II" from Control to Red II; "Move No. III" from Control to Red II. All documents are stamped *Secret Noforn*. "Dear Walt" letter from Earle G. Wheeler to Walt W. Rostow, March 15, 1967. (All from Lyndon B. Johnson Library, National Security Files.)

ABM debate: See Herksen, *Counsels of War*, particularly Chapter 20, "The Balance of Errors."

Chapter 3: Measuring War

Lese remarks: Interview, Lese.

Dupuy published remarks: Dupuy, "Criticism of Combat Models . . ."

"undisciplined nature" of analytical community: "Anonymous," "In Pursuit of the Essence of War," *Army*. Retired Major General Edward B. Atkeson, who later identified himself as the author, wrote several pieces for the magazine, which is published by the Association of the U.S. Army, a private organization that is an unofficial, though influential, Army supporter. According to an editor, the magazine had a circulation of 167,500 as of December 1985. A total of 102,963 subscribers were on active duty in the Army.

Atkeson remarks: Interview, Atkeson.

Dupuy remarks: Interview, Dupuy.

A Kapper model definiton: Kapper, *Defense 81*.

Ambiguous English to unambiguous mathematics: Callahan, "Need for Multidisciplinary Modeling Language . . ." He showed the span of models on a chart, "Hierarchical Characteristics of Systems Engineering Models."

Variables in war game simplified: Kapper, ibid.

Aircraft carriers: "An Introduction to Wargaming and Its Uses," Peter P. Perla and Raymond T. Barrett, Wargaming Applications Project, Naval Warfare Operations Division, Center for Naval Analyses. Working Paper (Preliminary Version) 85-0463.09, March 21, 1985.

Standardized definitions: Brewer and Shubik, *The War Game*. The book, published by Harvard University Press, was copyrighted by Rand.

Questionable validity of mathematically based models: Dupuy, ibid.

Lanchester publications: "The Principle of Concentration," *Engineering*, October 2, 1914, and other articles were incorporated into *Aircraft in Warfare: The Dawn of the Fourth Arm* (London: Constable, 1916). For a modern analysis of Lanchester's equations, complete with mathematical formulas, see Taylor, "Tactical Concentration . . ."

Quotation from Clausewitz: *On War*, Book One, Chapter 7, as quoted in Dupuy's "Clausewitz's Deterministic, Predictive Theory of Combat," obtained from the author as an unpublished paper. In it Dupuy says Clausewitz "argues with himself."

Soviet citing of QJM: Dupuy mentions the Soviet acknowledgment (Soviet *Foreign Military Review* articles) in Vol. I, No. 1 (April 1984) of "QJM UPDATE," a quarterly newsletter "for friends and users" of the QJM. Dupuy said one article implied "that the Pentagon believes the QJM is the best US model" and another article concluded that "the QJM is now the accepted American model."

Reference to Bourgeois Gentleman: Dupuy, *Numbers, Predictions & War*.

Power Potentials: Ibid. One of Dupuy's most complex formulas is that for Strength (S).

$$S = (W_s + W_{mg} + W_{hw}) \times r_n + W_{gi} \times r_n$$
$$+ (W_g + W_{gy}) (r_{wg} \times h_{wg} \times Z_{wg} \times W_{wg})$$
$$+ (W_i \times r_{wi} \times h_{wi}) + (W_y \times r_{wy} \times h_{wy} \times Z_{wy} \times W_{yy})$$

Force strength (S) = overall weapons inventory value of a combat force, as modified by environmental variables. W = weapons effectiveness or firepower inventories of a force, s = summation of the operational lethality index values of all small arms (W_s), machine guns (W_{mg}), heavy weapons (W_{hw}), antitank weapons (W_{gi}), artillery (W_g), air defense weapons (W_{gy}), armor (W_i), or close air support (W_y). r_n = terrain factor, related to infantry weapons; r_{wg} = terrain factor, related to artillery; h_{wg} = weather factor, related to artillery; Z_{wg} = season factor, related to artillery; w_{wg} = air superiority factor, related to artillery; r_{wi} = terrain factor, related to armor; h_{wi} = weather factor, related to armor; r_{wy} = terrain factor, related to air support; h_{wy} = weather factor, related to air support; Z_{wy} = season factor, related to air support; W_{yy} = air superiority factor related to air support.

Dupuy on strength always wins: One of Dupuy's "verities," presented in a talk given to the U.S. Marine Corps Command and Staff College in March 1985. A summary of the talk was published in the *Marine Corps Gazette*, August 1985.

Description of battle on computer: Hardy allowed me to watch him in action in the course of my interview.

Haynes remarks: Interview, Haynes.

Behavioral scientist questions value of games: Pilisuk, "Games Strategists Play."

Pentagon report: *Models, Data, and War: A Critique of the Foundation for Defense Analyses*, U.S. General Accounting Office, March 12, 1980.

Chapter 4: World War III

Resnick on "squishy": Interview, Resnick.

Scenarios: One scenario evolved into the best-selling *The Third World War*, August 1985.

GAO report on gaming: *Models, Data, and War . . .*

twenty-two allied ships sunk: Brigadier General Michael K. Sheridan, U.S. Marine Corps, speaking on "The Norwegian Naval Campaign," 1986 Sea Power Forum sponsored by the Center for Naval Analyses, Washington, D.C., November 19, 1986.

U.S.-Soviet Union confrontation: Reed, *NATO's Theater Nuclear Forces*. . . . The first scenario and Reed's account of the "pulse" negotiations are from this source.

U.S. Military Forces: The basic U.S. Army unit is the squad, usually ten soldiers under a sergeant. Each higher unit consists of three or four of the smaller units below it, plus, in certain cases, support units. Three or four squads make a *platoon*, and so up through *company*, *brigade*, *battalion*, and *division*. There are about 18,300 men in an *armored division*. A *corps* consists of two or more divisions with corps troops; an *army* is built around two or more corps.

The Army has eighteen active divisions and ten National Guard Divisions.

The Marines have three active divisions and one reserve division; a Marine division usually consists of three regiments, each of which has three battalions (each usually with 824 Marines) and a headquarters company.

The Navy has 566 ships, including 14 aircraft carriers and 140 submarines.

Air Force has tactical aircraft deployed in 24.5 active combat wings, and 12 reserve wings. A typical wing consists of three squadrons, each of which has 14 aircraft.

The strategic bomber force consists of 315 bombers. The U.S. also deploys 998 land-based intercontinental ballistic missiles. Missile submarines carry a total of 496 missiles. (Source: Annual Report to the Congress Fiscal Year 1987. Caspar W. Weinberger, Secretary of Defense. February 5, 1986.)

U.S. active and reserve Special Operations Forces include special forces groups, PSYOP (psychological warfare) groups and battalion, a Ranger regiment and three Ranger battalions and a civil affairs battalion; two Navy SEAL (Sea, Air, and Land) groups, whose equipment includes at least one specially equipped submarine; an Air Force Special Operations wing, five active squadrons and three reserve squadrons, a helicopter detachment, and two reserve special operations groups.

EMP scenario: Sollinger, *Improving US Theater Nuclear Doctrine*.

Gas scenario: General Frederick J. Kroesen (Ret.) et al., "Chemical Warfare Study: Summary Report," Alexandria, Virginia, Institute for Defense Analyses, prepared for the Office of the Under Secretary of Defense for Research and Engineering, February 1985. (The institute used a subcontractor, Burdeshaw Associates.) General Kroesen had retired a short time before as commander in chief of U.S. Army forces in Europe.

Warheads "militarily useless": McNamara, interviewed on "Hiroshima Plus 40 Years . . . And Still Counting," *CBS Reports*, July 31, 1985.

Davis quotation: Davis, "Game-structured Analysis . . ." based on December 3, 1984, speech.

Secret NATO war game: Wilson, *War Gaming*.

Million civilian casualties: Gayler in *The Nuclear Crisis Reader*, edited by Gwyn Prins. (Many of the essays in the book grew out of a seminar on alternative strategies for NATO at Cambridge University in September 1983.)

Spectrum of NATO warheads: Interview, Callahan.

Atkeson on NATO: Interview, Atkeson and Atkeson, "The 'Fault Line' in the Warsaw Pact."

Greek-Turkish argument: The Greek delegation walked out of the NATO Defense College in Rome in January 1985 because of a perceived insult in a scenario for an exercise. *The New York Times*, January 25, 1985.

Collins on nuclear-war casualties: Collins in *The Nuclear Crisis Reader*. The manual he cited was the *Department of the Army Field Manual 100-30* (Test) Tactical Nuclear Operations (HQDA, August 1971).

Chapter 5: War as a Game

Dunnigan meets Marshall: "The Experience with Commercial Wargames," presented by Dunnigan at a workshop on Modeling and Simulation of Land Combat, Atlanta, Georgia, March 28–31, 1982. He also mentioned the meeting during an interview.

SPI: Dunnigan, *Complete Wargames Handbook*.

Definition of SAS: Lawrence, "Playing the Game."

The Army's use of "Firefight": Poole, "Simulations Business," and interview, Dunnigan.

Hexagonal grids: Dupuy criticizes them in an unpublished paper ("The Hexagon in Combat Simulations").

Tactical nuclear warfare games: The quotation is from the abstract of "Use of War Games in Analysis of Tactical Nuclear Warfare Doctrines," presented by Richard I. Wiles at the workshop mentioned above.

Definitions of victory: Dunnigan, *Complete Wargames Handbook*.

"Würzburg" affair: Poole, ibid.

"Nuclear Escalation" card game: "News Briefs," *Space Gamer*, March–April 1984.

Herman remarks: Interview, Herman.

Davis quotation: Davis, "Game-structured Analysis . . ."

"Cohesion" analysis: Henderson, *Cohesion The Human Element in Combat*.

Playing "Gulf Strike": My instructors were Dick Higgins and Bob Nedwich.

Avalon Hill history: Poole, ibid; "Silver Jubilee," a company history privately published in 1983.

Review of games: *Proceedings* (July 1984, pp. 116–117) reviewed "Atlantic '86" (Strategic Simulations, Inc.) and "Grey Seas, Grey Skies" (Simulations Canada). The premise of "North Atlantic '86" is that the Soviet Union has conquered all of Germany and Norway and is attempting "complete domination of the North Atlantic though the isolation of Great Britain."

Scale models of NATO and Warsaw Pact tanks: Fulton, "Wargaming in the United States Army."

Chapter 6: Playing Pearl Harbor and Other Games

Military chess, toy soldiers: Young, *A Survey of Historical Developments in War Games*.

Famous players: Morschauser, *How to Play War Games in Miniature*.

Wells games: Wells, *Little Wars*.

American Kriegsspiel: W. R. Livermore, *American* Kriegsspiel: *A Game for Practicing the Art of War Upon a Topographical Map*, H. O. Houghton, Boston, 1879. This reference is from Young. (A separate volume, published in 1882, contained tables and plates.)

Criticism of American Kriegsspiel: Major Farrand Sayre in *Map Maneuvers and Tactical Rides*, Springfield Printing and Binding Co., Springfield, Mass., 1908, 1910, 1911. In Sayre's war game, which resembled Livermore's, opposing forces were labeled Red and Blue, by then the inevitable color choices. (A tactical ride was an exercise in which officers, moving on horseback, went through the motions of fighting an imaginary battle over real terrain.)

Livermore at Naval War College: Hattendorf et al., *Sailors and Scholars*.

McCarty Little: Nicolosi, "The Spirit of McCarty Little."

Theodore Roosevelt and games: From letter to college president Caspar Goodrich, May 28, 1897, quoted in *Professors of War*, by Ronald Spector, U.S. Naval War College Historical Monograph Series No. 3, Naval War College Press, Newport, R.I,. 1977. On June 16 Roosevelt wrote Goodrich, "When I come on to Newport, I want to time my visit so as to see one of your big strategic games."

Plan still accepted doctrine on December 7: Vlahos, *The Blue Sword*.

136 games: Vlahos, "Wargaming, an Enforcer of Strategic Realism: 1919–1942." This excellent overview of Naval War College gaming shows the evolution of modern maritime strategy through war games. Vlahos, incidentally, believes that the continual use of the Royal Navy as a Blue adversary had a "subtle" purpose: It "helped to cement an aggressive, antidefeatest, anticomplacent combat ethos among naval officers by placing them consistently in adverse battle environments."

German gaming; British gaming German invasion of Belgium; effect of "rigid" war games: Wilson, *War Gaming*.

Marines gamed a landing at Grenada: Colonel Brooke Nihart, U.S.M.C. (Ret.), director of Marine Corps Museums and a fellow member of the Washington Naval & Maritime Correspondents Circle, told me this after the U.S. invasion of Grenada in 1983. Nihart had seen an old British chart of Grenada that appeared to have been used in a Marine gaming exercise.

Marines reenacted battles: Clifford, *Progress and Purpose*.

Pratt games: The creator of the game jauntily describes its origin in *Fletcher Pratt's Naval War Game*, an account of the game first published in book form in 1940. (Z & M Enterprises of Milwaukee, Wisc., published the original text plus updated material in 1976.) Pratt was born in 1897 and died in 1956.

References to Graf Spee: Cambareri, "Scuttle the *Graf Spee!*"

The Pratt formula: The formula, which is figured out for each ship in the game, is:

Value = $(Gc^2 \times Gn + Gc'^2 \times Gn' + 10TT + 10A^2 + 10A'^2 + 10A''^2 + 5Ap + M)Sf; +T$
Gc is the caliber of the main battery guns, Gn the number of guns. The next Gc and Gn set refers to the secondary battery guns. TT refers to torpedo tubes, A to the thickness of the ship's belt armor, A' to the ship's turret armor, A'' to the ship's deck armor, Ap to the number of airplanes carried, M to the number of mines carried, and Sf is the speed factor. The final T is the ship's displacement. The large number this formula produces is the number from which is subtracted numbers derived from "hit" factors. A ship does not sink; it is subtracted out of existence. As in the real-world case of the *Graf Spee*, it takes many hits to sink or even cripple a major warship.

Japanese Midway game: Wohlstetter, *Pearl Harbor* . . .

Japanese gaming: Robert D. Specht, "War Games," P-1041, Rand Corporation, March 1957; Wohlstetter.

Pearl Harbor game: Specht, Wohlstetter; Wilson, *War Gaming*.

Admiral Kidd and gaming: Conversations with Admiral Isaac C. Kidd, Jr., and others.

Effects of Japanese wargaming: "History of War Games," E. A. Raymond and Harry W. Beer, *Reserve Officer*, October and November 1938. Quoted by Young.

Chester Nimitz quotation: "Naval War College Meeting the Challenge," an information booklet distributed by the college. (The booklet is a reprint of a *Navy Times* article published on June 20, 1983.)

All but one wartime admiral: The exception was Randall Jacobs, chief of the Bureau of Personnel, 1941–1945. Hattendorf et al., *Sailors and Scholars*

Guderian on gamed campaigns: Guderian, *Panzer Leader*

German invasion games: Wilson, ibid.

Midway and games: Polmar, *Aircraft Carriers*; Fuchida and Okumiya, *Midway*.

German gaming during battle: The account comes from *General der Infanterie* (Rudolf Hofmann), translated by P. Leutzkendorf (Washington, D.C., U.S. Army Europe Historical Division, 1952), Manuscript No. P-094, according to a reference in Watts, "Diagnostic Observations. . . ." Another source, according to Watts, is "Survey of the Development, Concepts, and Use of War Games," prepared by Major (*Bundeswehr*) Herbert H. Mauerer for the OSD/Net Assessment and Institute for Operational Analyses and Exercises Conference, June 1981. I am indebted to Colonel Watts for the use of his paper.

Battle of the Bulge replayed: The updated Battle of the Bulge was played at the School of Advanced Military Studies at the U.S. Army and General Staff College, Fort Leavenworth, Kansas. The playing of the game was confirmed, in a letter, from Colonel Louis D. F. Frasché, director of the Army's Combat Studies Institute at Fort Leavenworth. He did not provide details or say which side won; the quotation is from a retired general officer familiar with the game.

Birth of operational research: Callahan, "Communication Between the Military and the Scientist."

How to evade kamikazes: Tidman, *The Operations Evaluation Group*.

Operational research in World War II: Much of the information on OR comes from Wilson, ibid. Wilson, military correspondent of the London *Observer*, knew war as reality. He was wounded in World War II. Also, Tidman, ibid.

Hopkins ORO: Hausrath, *Venture Simulation* . . . Hausrath was the chief of the military gaming of the Research Analysis Corporation, which emerged from the Hopkins ORO.

Systems analysis and Strategic Air Command: Kaplan, *The Wizards of Armageddon*.

Douglas Threat Analysis Model: Hausrath, ibid.

Marine-MacArthur incident: Krulak, *First to Fight*.

Marine permanent war-game organization: Interview, Simmons.

Navy Electronic Warfare System: Hattendorf et al., *Sailors and Scholars*. This source puts the cost of the system at $7.5 million. Brewer and Shubik (*The War Game*) say "NEWS represents an investment of at least $10 million" and was "a victim of technological changes" by 1966. NEWS was replaced by WARS (Warfare Analysis and Research System) and, in the 1980s, by the Naval Warfare Gaming System. The description of NEWS in action comes from anonymous reminiscences and from Young's *A Survey of Historical Developments in War Games*.

Bloomfield on Navy game: Interview, Bloomfield.

Simulations' "quasi-religious overtones": Brewer and Shubik, *The War Game*.

Enthoven remarks: Address at the Operations Evaluation Group Conference, May 16, 1962.

"It is better to be roughly right than exactly wrong": Enthoven, address at the Naval War College, June 6, 1963.

Models of missile exchanges influenced strategic thinking: This is the conclusion of J. J. Martin, an executive of Science Applications International Corp. (SAIC), a strategic think tank. Martin directed studies of targeting, force posture, and arms control in the Office of the Secretary of Defense from 1968 to 1976. For his views on how analytical models influenced nuclear policy, see his "Modeling Nuclear Warfare" in Hughes, *Military Modeling*.

Transferring models to productive role: Thomas, former president of the Military Operations Research Society, wrote "Verification Revisited— 1983" (from which the quotation is taken) and "Models and Wartime Combat Operations" in *Military Modeling*.

"Organized mind-blowing": Bloomfield, "Simulation and Society."

Project Camelot: All quotations are from the Lyndon B. Johnson Library, National Security File: declassified Report on Project Camelot, December 17, 1964; letter from Rex D. Hopper to Joann Galtung, April 5, 1965; confidential cable on June 14, 1965, from Ambassador Ralph Dunganin of Chile to State Department; letter from Joseph A. Califano, Jr., special assistant to the Secretary of Defense, to McGeorge Bundy, special assistant to the President, June 24, 1965.

Rickover quotation: Testimony before the Subcommittee on Department of Defense Appropriations, House Committee on Appropriations, Eighty-ninth Congress, Second Session, Part 6.

McNamara quotation: Summers, *On Strategy*, attributes the quotation to Douglas H. Rosenberg, "Arms and the American Way: The Ideological Dimension of Military Growth," in *Military Force and American Society*, edited by Bruce M. Russell and Alfred Stepan (New York: Harper & Row, 1973).

A *"bitter little story"*: Summers, *On Strategy*.

Chapter 7: Gaming Under Analysis

Fighting "scratch-pad wars": The phrase is from *Counsels of War* by historian Gregg Herken. The book dramatically traces the evolution of nuclear strategy from the Manhattan Project to the Reagan Administration.

John von Neumann: John von Neumann wrote a paper on game theory that was published in Europe in 1928. The theory spread among American economists and mathematicians after von Neumann wrote, with Princeton economist Oskar Morgenstern, *Theory of Games and Economic Behavior*, which was published in 1944. For a look at von Neumann's involvement at Rand, see Kaplan, *The Wizards of Armageddon*.

Shubik on gaming: Shubik, *Game Theory in the Social Sciences*. Some of the book is based on his work at Rand.

"Prisoners' Dilemma": There are many versions of this classic. My interpretation is based on readings in Shubik, ibid; Brams, *Superpower Games*; Rapoport, *Fights, Games, and Debates*.

Ellsberg "Chicken" game: See Ellsberg, "The Theory and Practice of Blackmail" in *Bargaining: Formal Theories of Negotiations*, edited by Oran R. Young, University of Illinois Press, 1975. Ellsberg quotation from "The Crude Analysis of Strategic Choice," *American Economic Review*, May 1961, No. 2, pp. 472–78.

A view of the Cuban missile crisis: Brams, ibid. In this book Brams, a professor of politics at New York University, uses game theory as a way of looking at deterrence games, arms-race games, and verification games, "in which one side tries to hide the truth and the other to discover it."

Soviet matrix: *Operations Research* by E. S. Wentzel, Mir Publishers, Moscow. (Translated from the Russian by Michal G. Edelev. English translation, 1983.) The book was shown to me by Leslie Callahan, who told me that an American analyst spotted the book in a Moscow bookstore and bought every copy on sale. Dog-eared copies circulate through the U.S. gaming community.

Bracken quotation: Bracken, "Deterrence, Gaming, and Game Theory."

Report on STROP: Dalkey, "STROP: A Strategic Planning Model."

Rand's Systems Research Laboratory: Brewer and Shubik, *The War Game.*

Testing players on Mexican-American War: Guetzkow et al., *Simulation in International Relations.* . . . See also Hermann and Hermann, "An Attempt to Simulate the Outbreak of World War I."

Experts rate national power: Guetzkow, "Isolation and Collaboration . . . See also Gutzkow et al., *Simulation in International Relations* . . .

Children playing house: Raser, *Simultation and Society* . . .

First political-military game: Goldhamer and Speier, "Some Observations on Political Gaming." As Bloomfield has noted ("Reflections on Gaming"), other political scientists who conducted classroom games included Lucian Pye, Warner Schilling, and Norman Padelford.

Early Rand history: Kaplan, *The Wizards of Armageddon.* This book, rich in insights about Rand's people, describes in great detail Rand's involvement in the shaping of American strategy.

Discovery of Japanese Pearl Harbor Game: Roberta Wohlstetter, wife of Rand strategist Albert Wohlstetter, in *Pearl Harbor: Warning and Decision*, winner of the Bancroft Prize for 1963, credits Andy Marshall for inspiring her to write the book.

MIT gaming: Bloomfield, ibid.; Mandel, "Political Gaming and Foreign Policy Making During Crises."

U.S. Arms Control and Disarmament Agency games: Bloomfield and Gearin, "Political Games: Experiments in Foreign Policy Research." (When the report was written Gearin, an Army colonel, was working for a doctorate in political science at MIT.)

CONEX games: Interview, Bloomfield; Bloomfield and Gearin, "Games Foreign Policy Experts Play. . . ." The CONEX games were held in September 1968, December 1968, March 1969, and September 1969. The last one, played at MIT in 1969, used the computerized conflict data system Bloomfield and Gearin were developing under contract to the U.S. Arms Control and Disarmament Agency.

President played by executive with presidential pretension; difficulty in getting U.S. intervention: Bloomfield and Gearin, "Political Games. . . ." In a 1985 interview Bloomfield still kept the game President's identity secret. Remaining Bloomfield remarks are from that interview.

Chapter 8: Scenario

SAFE instructions: Brewer and Shubik, *The War Game*. SAFE was developed at Rand by Olaf Helmer and R. E. Bickner.

"Gamesters": Zuckerman, "Judgment and Control in Modern Warfare," *Foreign Affairs*, January 1962. Another British critic, P. M. S. Blackett ("Critique of Some Contemporary Defence Thinking," *Encounter*, April 1961), wondered why, if game theory was so useful for war, the theory would not be even better in more simple endeavors and make its practitioners wealthy poker players. Another critic was C. P. Snow ("Discussions," following "Operational Research and Nuclear Weapons" by Blackett in *Journal of the Royal United Services Institution*, August 1961).

McNamara-Mountbatten meeting: McNamara speaking to Walter Cronkite, CBS Reports, "Hiroshima Plus 40 Years . . . and Still Counting," broadcast July 31, 1985.

Rand on advantages of a first-strike: Herken, *Counsels of War*.

A British view: *The Economist*, June 24, 1961.

STAGE: Wilson, *War Gaming*.

New ideas gamed: Bracken, "Deterrence, Gaming, and Game Theory."

Distinction between crisis management and targeting blurred: Callahan, "Flexible Response Theory . . ."

SPARC: Wilson, ibid.

Players expended entire budget: Brown, "The Role of Modeling . . ."

Greenstein remarks: Interview, Greenstein.

Rand on Soviet cities: Herken, ibid.

Secret advice to Air Force officers: Pringle and Arkin, *S.I.O.P.*

McNamara's SIOP; "scratch-pad war games" replaced: Herken, ibid.

Half-SAFE: Player's Manual of Rules and Procedures for the Half-SAFE Simulation of Strategic Planning and Nuclear War, produced by David P. Burke at the Naval Postgraduate School, Monterey, Calif., February 1980.

SDI: One of the accounts (Greve, San Jose *Mercury-News*) was sent to me by an officer in the Office of the Secretary of Defense with a note saying "this account of the SDI birth is relatively valid." Greve quotes a "key White House analyst" as saying, "None of the things you assume would be considered were considered at all. People just don't believe that the President could make such a momentous decision so impulsively. They think we must have thought through what it would do to treaties and how it might work as a bargaining chip in Geneva, and so on, but I can't find it. Neither can anybody else who's undertaken a history of SDI, and the reason is, it isn't there."

Star War games: "Star Wars," *The New York Times Magazine*, August 4, 1985.

Playing against contractors: "Star Wars in Strategy: The Russian Response," *The New York Times*," December 17, 1985.

Greenstein remarks: Greenstein interview.

Ambiguity: Confidential source.

"Harvard Game": From a scenario given to me by William Martel.

Information on Soviet submarine: Polmar, *Guide to the Soviet Navy*, 1983, plus conversation with Polmar.

Chapter 9: Gaming Guerrilla War

Nhu's formula: Mention of the Bonnett formula is made in Hausrath, *Venture Simulation*, which notes that Bernard B. Fall, the French historian who wrote authoritatively and extensively on Vietnam, quoted Bonnett's formula in *The Two Vietnams: A Political and Military Analysis*. Fall, accompanying U.S. Marines in Vietnam as a war correspondent, was killed by a booby trap in 1967.

Blumstein proposal: Hausrath, ibid.

Fain remarks: Interview, Fain.

Hypothetical 1972 war: Wilson, *War Gaming*.

"American intellectual theoreticans of order": Bell, *The Myth of the Guerrilla* . . .

Adams guerrilla game: Hausrath, ibid.

Army chief of staff questioned counterinsurgency: Summers, *On Strategy*.

Cold War Model: Hausrath, ibid.

A system for upgrading or downgrading effectiveness: Hausrath, ibid.

Tactical changes paid off: Ewell and Hunt, *Sharpening the Combat Edge* . . .

Temper: Brewer and Shubik, *The War Game*; Hausrath, ibid.

Apt and Agile: Wilson, *War Gaming*.

Description of villager survey: See "Vietnam Documents" in Bibliography. In the notes to his authoritative *Vietnam A History*, Stanley Karnow wrote, "Among the most perceptive studies of the Vietcong were those undertaken by the RAND Corporation. Though they had little impact on either American or South Vietnamese government policy, they remain valuable for students of the period."

Questioning of Viet Cong defectors: See "Vietnam Documents" in Bibliography.

Interviewers showing villagers drawing of ladder: Thomas, "Models and Wartime Combat Operations Research."

McNamara quotation: From his deposition, given on March 26, 1984, in the libel suit General William C. Westmoreland filed against CBS, charging that the show *60 Minutes* had libeled him. The show claimed numbers had been doctored to indicate a decrease in enemy forces when, in fact, they were increasing. McNamara's remark about infiltration estimates is from his testimony in the trial, December 6, 1984.

Southeast Asia Analysis Report: Thomas, ibid.

Chapter 10: Vietnam's Prophetic Scenarios

Sigma I-64 and Sigma II-64: All records from the Lyndon B. Johnson Library, National Security Files, declassified (with some sanitizing) under the Freedom of Information Act. Records of Sigma II-64 were "Game Organization and Procedures," "Game Director's Remarks," "Initial Scenario," "Game Messages," and "Critique." The scenario is digested from the "Initial Scenario" handed the team members. "Memorandum for the Senior Participants of Sigma I-64 from the Joint Chiefs of Staff Joint War Games Agency," March 21, 1964, and "Memorandum to General Maxwell D. Taylor from McGeorge Bundy," March 23, 1964, show that participants in that game included Under Secretary of State George W. Ball, U. Alexis Johnson, a State Department careerist who would become the deputy ambassador to South Vietnam; David E. Bell, director of the Agency for International Development, Major General Theodore R. Milton, of the Pacific Command, Carl T. Rowan, director of the U.S. Information Agency, William H. Sullivan from the State Department, Michael Forrestal from the White House, and Air Force Brigadier General John W. Vogt.

Sigma I-64: Discussion of the game is based on "Final Report Sigma I-64," prepared by the Joint War Games Agency, Joint Chiefs of Staff, April 15, 1964. Contents include a letter to participants (eighty-seven names blacked-out) from General Maxwell D. Taylor, chairman of the Joint Chiefs of Staff; Game Director's Remarks, Summary, and Senior Level Critique. The remarks of players are from the critique.

An overt war against the Viet Cong: "Critique of Signa I-64 [*sic*]: Political Considerations," Memorandum for the Record, Seymour Weiss, April 9, 1964, to Blue Senior Team, Blue Action Team, Rear Admiral C. J. Van Arsdall, and McGeorge Bundy.

Navy pilot shot down in June 1964: "Vietnam Rules of Engagement Declassified," *Congressional Record*, March 6, 1985, page S 2633. Senator Barry Goldwater got the Air Force to declassify the secret report and published it in the *Record*.

Tonkin Gulf incident: The best eyewitness account of the nonattack came from Navy pilot James B. Stockdale, who had flown from the *Ti-conderoga* and strafed the North Vietnamese patrol boats in the first attack. He later wrote that when he returned to the aircraft carrier after the second reported attack and was being debriefed, to the question "Did you see any boats?" he answered, "Not a one. No boats, no boat wakes, no ricochets off boats, no boat impacts, no torpedo wakes—nothing but black sea and American firepower." (From *In Love and War* by Stockdale and his wife, Sybil.) Stockdale, shot down on a later mission, was a prisoner of the North Vietnamese for eight years.

Vinh gaming: Edwin W. Besch, a Marine captain who was wounded in Vietnam and returned as a CIA intelligence analyst, wrote about his attempt to sell the landing idea. ("Vinh, The Missed Opportunity," *Shipmate*, November 1984; pp. 18–21.) Retired Marine Colonel James W. Hammond, editor of *Shipmate*, said in an editor's note that when he was head of the command department of the Marine Corps Command and Staff College in 1968 to 1971 "the final or graduation staff problem was an amphibious operation against Vinh."

"Spoken like a true member of the Red Team": Halberstam, *The Best and the Brightest*.

Bundy is sent report on game: Memorandum to McGeorge Bundy from Air Force Major LeRoy A. Wenstrom, a JWGA staff officer, September 14, 1964.

Prophetic scenario: The imagined history was compared to actual events as narrated in Karnow, *Vietnam A History*.

Marines war-gamed Danang landing: Krulak, *First to Fight*. *Sigma I-66*: Information on the game comes from "Final Report Sigma I-66," prepared by the Joint War Games Agency, Joint Chiefs of Staff, November 1, 1966.

Chapter 11: Red and Blue in the White House

Johnson at the sand table: Karnow, *Vietnam A History*.

An "Oriental" playing Ho Chi Minh: Interview, Argo.

Kupperman on White House crises: Quoted in *Information Technology for Emergency Management* (see Congressional Reports in Bibliography), pp. 202–203.

Suggestion for "probabilistic situations" and reference to Freud: Raser, *Simulation and Society . . .*

NCA release authority: See Congressional Research Service, "Authority to Order the Use of Nuclear Weapons," Washington: Government Printing Office, 1977; Department of Defense Directive No 5100.30, December 2, 1971; Department of Defense, Joint Chiefs of Staff, *Publication 1, Department of Defense Dictionary of Military and Associated Terms*, Washington: June 1979, revised. Also, Pringle and Arkin, *S.I.O.P.* and Ford, *The Button . . .*

"calculated ambiguity": This is the suggestion of Richard K. Betts in *Surprise Attack* (Washington: Brookings Institution, 1982). The same phrase could be used for descriptions of the NCA in general.

U.S. nuclear decisionmaking: Bloomfield, "Nuclear Crisis and Human Fraility . . ."

Ivy League game: I have based my report on Ivy League on Pringle and Arkin's *S.I.O.P.*, Fialka's "Nuclear Reaction . . ." and "The Doomsday Exercise," *Newsweek*, April 5, 1982, Department of Defense news release of February 26, 1982, and information from several confidential sources. In order to get official information, including the crucial question of President Reagan's participation, I specifically directed this question to Robert B. Sims, at the time special assistant to the President and deputy press secretary for foreign afairs. (He later became the Assistant Secretary of Defense for public affairs.) Sims replied, "White House involvement in exercises such as those you cited are among the most sensitive subjects here, since knowledge of our procedures could be of great value to a potential adversary." I also checked with Fialka, who broke the story, to see whether he had received any criticism for inaccuracies. "I'm told," he said, "that some sources had second thoughts. But some wanted to tell—to send a message to the Russians. Things have tightened up since."

President's call: Fialka, "Nuclear Reaction . . ."

Leaking of Ivy League: Jack Anderson, "Responding to Europeans' Nuclear Fears," *Washington Post*, January 29, 1982.

Helms on describing decisionmaking process: Quoted in Rostow memorandum to the President, December 6, 1968. Lyndon B. Johnson Library, National Security Files.

Report on WWMCCS: (see Congressional Reports in Bibliography). Representatives of the National Security Council met with the chief executives of several telecommunication firms in March 1982. On September 13, 1982, President Reagan established, by executive order, the National Security Telecommunications Advisory Group, which was to work on ways to improve national security telecommunications. This was undoubtedly a direct result of experiences in Ivy League.

At least one was played at Camp David . . . : The games were played from Friday, September 8 to Monday, September 11, and Friday, September 29, to Sunday, October 1, 1961. A memorandum from Lawrence C. McQuade to Walt Rostow, September 29, 1961, said one game, known as "the ISA [International Security Affairs] Conference," would be played at the Military Assistance Institute in Arlington, Virginia. The other, code-named NATO Planning Conference, was held at Camp David, according to Kaplan (*Wizards of Armageddon*) and Herkin (*Counsels of War*). The latter mentions the nuclear demonstration, crediting Schelling. The McQuade memorandum, scenarios, and other information on the games came from the John F. Kennedy Library, National Security Files (Berlin).

Schelling and "I have a gun" example: Andrews, *War Gaming*.

Kennedy was at the family compound: Presidential appointment book, John F. Kennedy Library.

Meeting in Ball's conference room: Interview, U. Alexis Johnson, Deputy Under Secretary of State for Political Affairs; interview conducted by William Brubeck, senior staff member, National Security Council, November 7, 1964. Oral History files, John F. Kennedy Library.

Navy exercises during Cuban missile crisis: Anderson's speech was made to a Navy League audience in New York City on November 9, 1962. Anderson's argument with McNamara, Allison's *Essence of Decision*.

Study of President's command and control resources: "Defense Command and Control Support to the President," transmitted to McGeorge Bundy, according to a letter, March 6, 1965, from Deputy Secretary of Defense Cyrus Vance. From National Security Files, Lyndon B. Johnson Library.

Helms on presidential decisionmaking: Richard Helms, Director of Central Intelligence, interviewed in his office, April 4, 1969, by Paige Mulhollan. Oral History Collection, Lyndon B. Johnson Library.

Changing from theatrical player to game player: Allison, *The Essence of Decision*.

"Countdown to Looking Glass": Script by Albert Ruben. The drama was produced by L&B Productions and Primedia Productions, on the Home Box Office network in October 1984.

Real decisionmakers: Introduction by Ted Koppel to the first part of "The Crisis Game," November 22, 1983.

Chapter 12: Playing the Crisis Game

"The Crisis Game": The show, produced by ABC News, ran four consecutive nights (Tuesday, November 22, through Friday, November 25, 1983) in the *Nightline* time period, which is 11:30 P.M. to 12:30 A.M., in the East. The quotations of players are from the transcripts of the shows.

Rights of game acquired by Harvard's Kennedy School of Government.: Interview, Moore. As producer, Moore supervised the months of work preceding the show, the creation of a Situation Room that "looked more White Housey" than the real one, and the eighteen hours of taping that were edited down to four. The executive producer was William Lord. The Control Team members were Air Force General William Y. Smith, formerly Deputy Commander in Chief of U.S. forces in Europe; Barry Blechman, vice president of the Roosevelt Center for American Policy Studies, who had been assistant director of the Arms Control and Disarmament Agency in the Carter Administration; Leonard Garment, White House counsel in the Nixon Administration; Arnold Horelick, director of the Rand-UCLA Center for the Study of Soviet International Behavior and, from 1977 to 1981, national intelligence officer for the U.S.S.R. and Eastern Europe; Walter Slocombe, Deputy Under Secretary of Defense for policy planning in the Carter Administration; Richard C. Steadman, Deputy Assistant Secretary of Defense for East Asia and Pacific affairs in the Johnson Administration; Robert C. Tucker, a professor at the Johns Hopkins University School of Advanced International Studies and president of the Lehrman Institute, and Leslie H. Gelb, national security correspondent of *The New York Times* and, from 1977 to 1979, director of the Bureau of Politico-Military Affairs in the Department of State. Gelb was the senior consultant to ABC for the project.

Kirkpatrick on Situation Room: Remarks by Ambassador Jeane J. Kirkpatrick, U.S. permanent representative to the Union Nations, to the Women's Forum, New York, N.Y., December 19, 1984.

1966 list of games: Memorandum for Bundy (blacked-out, but stamped elsewhere "McGeorge Bundy's Office"), from Rear Admiral C. J. Van Arsdall, Chief, Joint War Games Agency, July 12, 1965.

The "most tense situation since World War II": Zumwalt, On Watch. Zumwalt was Chief of Naval Operations at the time of the alert.

Chapter 13: The Perils of Make-believe

Ten bombs on ten cities: McGeorge Bundy, "To Cap the Volcano," Foreign Affairs, October 1969, as quoted in Bloomfield's "Nuclear Crisis and Human Frailty."

Games as dress rehearsal: "Beyond the Hotline: Controlling a Nuclear Crisis," William L. Ury and Richard Smoke, Nuclear Negotiations Project, Harvard Law School, March 1984.

Proposal for high-level games: "Enhancing National Decision Making During Crises," Dr. James W. Browning, Dr. Frank J. Dellermann, and Lloyd H. Hoffmann. The authors were kind enough to give me a copy of this unpublished paper.

"I think this is a terrible problem": From the script for "Countdown to Looking Glass."

Products of other nuclear war games: On Meet the Press, on February 26, 1967, General Earle G. Wheeler, Chairman of the Joint Chiefs of Staff, was asked about Secretary of Defense Robert McNamara's estimates of 90 to 120 million casualties in a nuclear war. In a rare revelation of gaming results by a high official, Wheeler said that the figures "are arrived at by war-gaming a situation. We have run literally thousands of war games of this type over the years. The results are very sensitive to the assumptions and the model, so-called, upon which the war game is based." For a description of the most recent methods for calculating nuclear weapons' effects, see Martel and Savage, Strategic Nuclear War.

Leaders and "real horrors of a nuclear war": Fisher and Deutsch quotations from "Social Scientists Believe Leaders Lack a Sense of War's Reality," The New York Times, September 7, 1982.

"the only games they know how to play"; could not bear implication that "their judgment was faulty": First quotation: Jerome D. Frank, Bulletin of Atomic Scientists, October 1980. Second quotation: Frank, at the first nuclear war conference, under the auspices of the Center for Defense

Information and the Institute for Policy Studies (published in the *Bulletin of Atomic Scientists*, April 1979).

"numb the good sense": Sorensen, *Decision-Making in the White House.*

Leaders during crises: George, "The Impact of Crisis-induced Stress on Decisionmaking."

Pieczenik remarks: Interview, Pieczenik.

A bunch of ambushers: Interview, Atkeson.

Steinhardt on warfare as a mathematical model: Low, "Theater-Level Gaming . . ."

Catalogue of models: Quattromani, *Catalog of Wargaming and Military Simulation Models.*

GAO on model effectiveness: *Models, Data, and War: A Critique of the Foundation for Defense Analyses*, U.S. General Accounting Office, March 12, 1980.

Problems of high-level modeling: Interview, Dupuy.

Officials' dependency on models: Marshall, "Status Report: Theory of Combat and Philosophy of War."

Brehm remarks: Interview, Brehm.

Stand-ins for key leaders: "Command and Control Support to the President," undated report. Lyndon B. Johnson Library, National Security File.

Nifty Nugget "most ambitious test": R. W. Komer, *An Evaluation Report of Mobilization and Deployment Capability Based on Exercises Nifty Nugget-78 and Rex-78*, Office of the Secretary of Defense, June 30, 1980.

"considerable horsepower": Interview, Burcham.

Lindquist remarks: Interview, Lindquist.

"The Army was simply attrited to death": Fialka, "Grim Lessons of Nifty Nugget." This report by Fialka, along with his reports in *The Washington Star* (November 2, 3, 4, 1979), forced some additional information out of the Pentagon, including the Komer report. My information is based on these sources, the interviews cited in this chapter, and civilian and military officers who declined to be identified because, officially, much about the aftermath of Nifty Nugget remains classified.

Pirie remarks: Interview, Pirie.

Emergency laws: They are listed in a tightly circulated document, "A Review of the Adequacy of Principal Statutory Authorities Affecting DoD Surge and Mobilization Capacity," produced for Contractor Systems Research and Applications Corporation in September 1983 by Richard Danzig.

1,100 vehicles unloaded in thirty hours: Thomas, "The Sealift Factor."

Marine storage ships: Meyer, "New Storage Ships . . . ."

Revelations about Proud Spirit: Fialka, "The Pentagon Exercise 'Proud Spirit'. . . ."

Short-war, long-war scenarios: Halstead, "The Atlantic: The Linchpin."

Army's shortages: Perkins, *Global Demands: Limited Forces*.

Chapter 14: In the Theater of Terrorism

Kingfisher game: Summary of Procedure and Analysis, International Seminar on Problems in Political Terrorism and Combatting Terrorism, Center for Strategic Studies, Tel Aviv University and the Office of the Prime Minister's Adviser on Combatting Terrorism, Tel Aviv, July 1979. Obtained from a confidential source. Delegates to the conference included U.S. Ambassador Anthony Quainton, director of the State Department's Office for Combatting Terrorism, and members of the U.S. Delta Force antiterrorist military unit. Other participants were Dr. Frank M. Ochberg, of the National Institute of Mental Health; Brian M. Jenkins, Rand's expert on terrorism, and Robert H. Kupperman, executive director of the Georgetown Center for Strategic and International Studies. Israelis included Tel Aviv University's Ariel Merari, Aharon Yariv, former head of Israeli military intelligence and an adviser on terrorism to the Prime Minister, and representatives of the Israel Saiyeret, which executed the Entebbe raid in 1976. The German delegation included Colonel Ulrich Wegener, commander of the West German antiterrorist team, *Grenzschutzgruppe-9* (Border Protection Group 9). The origins of GSG-9U, as it is also known, trace back to the dawn of modern terrorism: the massacre of Israeli athletes by Black September Arab terrorists at the Munich Olympics in 1972.

The chief member of the media team was Robert Moss, a London *Daily Telegraph* columnist who had developed a reputation as an expert on terrorism, lecturing at the Royal College of Defense Studies in London and the NATO Defense College in Rome. In 1981 he testified on terrorism before the Senate subcommittee on security and terrorism. The game is recounted here by melding various elements in the complex paper flow of the game: "Scenario," "Control Policy and Guidelines," "Events in the Game," and "Teams Comments," each handled separately in the overall, 189-page "Summary of Procedure and Analysis." My narrative viewpoint is that of a journalist drawing from various sources to describe a terrorist crisis after it ends.

Information on Flight 847: Based on news accounts in *Washington Post*, June 17, 1985; *The New York Times*, June 17 and July 7, 1985; *Newsweek*, July 8, 1985.

Symptoms of Legionnaire's disease: Interview, Kupperman.

The 20/20 show: The show "If You Were President" was telecast on August 6, 1981. Information about it is based on a viewing of a tape of the show and the Kupperman interview. The fifty-two hours of videotape were edited down to the one-hour report under the supervision of producer Carol Blakeslee, co-producers Consuelo Gonzalez and Jay La-Monica, and associate producer Linda Maskin.

Chapter 15: Real Problems, Simulated Solutions

Maneuvers: *Washington Post*, August 4 and 21, 1984; October 16, 1984; November 1 and 14, 1984; January 29, 1985; March 22, 1985; April 11, 1985; September 2, 1985. *The New York Times*, October 14, 1984; March 30, 1985; June 4, 5, and 20 1985. *Defence Update International*, No. 46, 1984. "War Games," *The Defense Monitor*, Center for Defense Information, Vol.XIII, No. 7, 1984. *Invasion: A Guide to the U.S. Military Presence in Central America*, National Action Research on the Military Industrial Complex (NARMIC), a project of the American Friends Service Committee, 1985.

Military use of simulation: Allen, "Simulation . . ."; Allen, "Non-Trivial Pursuits." Also, "Cubic Corp. Upgrades Air Combat Training System for Navy Use," *Aviation Week & Space Technology*, December 3, 1984.

Laser "bullets": The system is called MILES, for Multiple Integrated Laser Engagement System. Description based on information from the Army and the developer of MILES, Loral Electro-Optical Systems.

Question of Navy "invalid" Orange play: Lloyd Hoffman, "Operations with the Interim Battle Group Tactical Trainer: Observations and Comments on July Exercise," July 13, 1984, as quoted in Watts, "Diagnostic Observations. . . ." Hoffman, formerly at the National Defense University, is now with Rand.

Uptide: Uptide Report No. 6, Uptide Exercise 3A (Transit Phase). Formerly secret, declassified December 31, 1981.

Comment on Uptide: Correspondence from confidential source, May 31, 1985.

Unsinkable aircraft carriers: Confidential sources and remarks of Representative Newt Gingrich, a Georgia Congressman known for his pro-defense record. He raised the unsinkable carrier issue in a speech at the Workshop on Modeling and Simulation of Land Combat, Callaway Gardens, Georgia, March 28–31, 1982.

"Swiss-cheese" ASW: Allen and Polmar, "The Silent Chase."

Ocean Venture: Christian Eliot, Francisco Figueroa de la Vega, Manuel Ramìrez Gabarrus, "Ocean Venture '81," *Naval Forces*, No. VI/81, Vol. II, pp. 74–76.

Knuth charges: Morton Mintz, "Article Critical of Carriers Stamped 'Secret' by Navy," *Washington Post*, May 4, 1982.

Pentagon whistle-blowers: Interview, Hovin.

Abrams M-1 tank: Coates and Kilian, *Heavy Losses*.

Nuclear snap-on: Seth Bonder, Vector Research, Inc., speaking at a gaming workshop in Leesburg, Virginia. For further reference to conference, see later notes.

Firepower scores: From the proceedings of the Leesburg conference, see fuller reference in subsequent note. Information about the Army models cited is from *Catalog of War Games and Combat Simulations*, prepared by the Office of the Deputy Under Secretary of the Army (Operations Research), second edition, April 1982, and Quattromani, *Catalog. . . .*

Macedonia remarks: Interview, Macedonia.

Validation a satisfied customer and *follow-on contract*: Brewer and Shubik, *The War Game*.

Callahan remarks: Interview, Callahan.

Leesburg Workshop: "Theater-Level Gaming and Analysis Workshop for Force Planning," sponsored by the Office of Naval Research, Leesburg, Virginia, September 22–29, 1977.

Stockfisch criticism: J. A. Stockfisch, *Models, Data, and War: a Critique of the Study of Conventional Forces*, Rand Report, March 1975. His Leesburg remarks are from the proceedings of the conference.

Remarks at conference: The proceedings of the conference.

Euphemism for "how many women and children you kill": *The New York Times*, January 17, 1986.

Low report: Low, "Theater-Level Gaming. . . ."

Information about Carmonette: *Catalog of War Games and Combat Simulations*.

Chapter 16: The Search for the Ultimate Game

Marshall's search: Marshall, "A Program to Improve . . ."

Martel remarks: Interview, Martel. *Their book*: Martel and Savage, *Strategic Nuclear War*.

Greenstein remarks: Interview, Greenstein.

Macedonia remarks: Interviews, Macedonia and Dunnigan.

Information on McClintic Theater Model: Dees, "A Detailed Analysis . . ."; Bettencourt, "Combat Modeling . . ."

Marshall and a nuclear Pearl Harbor: Goldhamer and Marshall, "The Deterrence and Strategy of Total War. . . ." The paper is cited in *The Wizards of Armageddon*, by Fred Kaplan, who got this paper declassified through the Freedom of Information Act.

No bolt from the blue: Interview, Molander.

Wildfire: Information from the Roosevelt Center for American Policy Studies.

Military journal survey of players: Schemmer, "Who Would Start the War? . . ."

Schelling quotation: Ford, *The Button* . . .

Watts remarks: Watts, "Diagnostic Observations . . ."

Goldhamer's report: "Reality and Belief in Military Affairs," a first draft (June 1977) by Herbert Goldhamer, edited by Joan Goldhamer. A report prepared for the Director of Net Assessment, Office of the Secretary of Defense. Rand publication R-2448-NA, February 1979.

Marshall requirements: Marshall, ibid.

Chapter 17: When Ivan Plays Sam

Pentagon invitation: Letter issued by the Defense Nuclear Agency (DNA001-80-R-0002), November 7, 1979. Marshall's Office of Net Assessment often uses the Defense Nuclear Agency for dealing with contractors.

West German gaming: Watts, "Diagnostic Observations . . ."

"The mice had eaten them": Murray, interview with editors of the U.S. Naval Institute *Proceedings*, October 1983.

Makins remarks: Interview, Makins.

"people-in-the-loop" gaming: Martin and Olin, "Improved Methods for Strategic Analysis." The authors worked for SAI when the contest was being planned. When the report was published, Olin was at the Office of Management and Budget.

Game in Moscow: Interview, Bloomfield.

Rand report on automated games: Graubard and Builder, "New Methods . . ."; Graubard and Builder, "Rand's Strategic Assessment Center . . ."

Information on demonstrations: Martin and Olin, ibid; Graubard and Builder, "Rand's Strategic . . ."; interviews, Makins and Resnick.

Questions automata would be asked: Davis et al., "Automated War Gaming . . ."

Davis remarks: Interview, Davis.

RSAC language and "virtues": Davis and Winnefeld, ibid.

Bloomfield remarks: Interview, Bloomfield.

Description of "agents": Davis and Winnefeld, ibid; Schwabe, "Strategic Analysis . . ."; Schwabe and Jamison, "A Rule-Based Policy-Level Model . . ."; Davis and Schwabe, "Search for a Red Agent. . . ." Definition of "Ivan K" is from Davis and Stan, "Concepts and Models of Escalation."

Chapter 18: Across the Threshold

War narrative: Based on information in Schwabe's "Strategic Analysis. . . ." Schwabe derived his analysis from game runs in 1982 using the RSAC Mark II. See also Schwabe and Jamison, "A Rule-Based Policy-Level Model . . ."; Steeb and Gillogly, "Design for an Advanced Red Agent . . ." (Appendix D); Davis and Schwabe, "Search for a Red Agent . . ."

Tritten remarks: Interview, Tritten.

Bibliography

Books

Allison, Graham T. *Essence of Decision*. Boston: Little, Brown, 1971.

Bell, J. Bowyer. *The Myth of the Guerrilla: Revolutionary Theory and Malpractice*. New York: Knopf, 1971.

Bracken, Paul. *The Command and Control of Nuclear Forces*. New Haven: Yale University Pres, 1983.

Brams, Steven J. *Superpower Games*. New Haven: Yale University Press, 1985.

Brewer, Garry D., and Martin Shubik. *The War Game: A Critique of Military Problem Solving*. Cambridge: Harvard University Press, 1979.

Brewer, Garry D., and Peter de Leon. *The Foundations of Policy Analysis*. Homewood, Ill.: Dorsey Press, 1983.

Calder, Nigel. *Nuclear Nightmares*. New York: Penguin, 1981.

Callahan, Leslie G., Jr., ed. *Proceedings of the Workshop on Modeling and Simulation of Land Combat*. Atlanta: Georgia Technical Research Institute, 1983.

Clifford, Lt. Col. Kenneth J. *Progress and Purpose: A Developmental History of the U.S. Marine Corps*. Washington: U.S. Marine Corps, 1973.

Coates, James, and Michael Kilian. *Heavy Losses*. New York: Viking, 1985.

Copeland, Miles. *The Game of Nations*. New York: Simon & Schuster, 1969.

Darryl, William. *Cohesion The Human Element in Combat*. Washington: National Defense University Press, 1985.

Davis, Paul K., and James A. Winnefeld. *The Rand Assessment Center: an Overview and Interim Conclusions about the Utility and Development Options*. Rand publication R-2945-DNA, prepared for Defense Nuclear Agency, March 1983.

Davis, Paul K., and Peter J. E. Stan. *Concepts and Models of Escalation*. Rand publication R-3235, May 1984.

DeLeon, Peter. *Scenario Designs*. Santa Monica: Rand, R-1218.

De Weerd, Harvey A. *Political-Military Scenarios*. Santa Monica: Rand, 1967.

Dunnigan, James F. *Complete Wargames Handbook*. New York: Morrow, 1980.

————. *How to Make War*. New York: Quill, 1983.

————, and Austin Bay. *A Quick & Dirty Guide to War*. New York: Morrow, 1985.

Dupuy, Col. T. N. *Numbers, Predictions and War*. Fairfax, Va.: Hero Books, 1985.

Ewell, Lt. Gen. Julian J., and Maj. Gen. Ira A. Hunt. *Sharpening the Combat Edge: The Use of Analysis to Reinforce Military Judgment*. Washington: Department of the Army, 1974.

Fallows, James. *National Defense*. New York: Vintage Books. 1981.

Featherstone, Donald F. *Battle Notes for Wargamers*. London: Newton Abbot, David & Charles. 1973.

Ford, Daniel. *The Button: The Pentagon's Strategic Command and Control System*. New York: Simon & Schuster, 1985.

Fuchida, Capt. Mitsuo, and Comdr. Masatake Okumiya. *Midway*. Annapolis: Naval Institute Press, 1955.

Garner, William V. *Soviet Threat Perceptions of NATO's Eurostrategic Missiles*. Paris: Atlantic Institute for International Affairs, 1983.

Goldhamer, Herbert, and Andrew W. Marshall, with Nathan Leites. *The Deterrence and Strategy of Total War, 1959–61: A Method of Analysis*. Rand RM-2301, April 30, 1959.

Graubard, Morlie H., and Carl H. Builder. *Rand's Strategic Assessment Center: an Overview of the Concept*. Rand N-1583-DNA for the Defense Nuclear Agency, September 1980.

Guderian, Heinz. *Panzer Leader*. Washington: Zenger Publishing, 1982 (reprint).

Guetzkow, Harold, and Chadwick Alder, Richard Brody, Robert Noel, Richard Snyder. *Simulation in International Relations: Developments for Research and Teaching*. Englewood Cliffs, N.J.: Prentice-Hall, 1963.

Hackett, Sir John et al. *The Third World War August 1985*. New York: Macmillan, 1979.

Halberstam, David. *The Best and the Brightest*. New York: Random House, 1969.

Hausrath, Alfred H. *Venture Simulation in War, Business, and Politics*. New York: McGraw-Hill, 1971.

Herken, Gregg. *Counsels of War*. New York: Knopf, 1985.

Hughes, Wayne P., ed. *Military Modeling*. Alexandria, Va.: Military Operations Research Society, 1984.

Iklé, Fred. *Every War Must End*. New York: Columbia University Press, 1971.

Kaplan, Fred. *The Wizards of Armageddon*. New York: Simon & Schuster, 1983.

Karnow, Stanley. *Vietnam: A History*. New York: Viking, 1983.

Kennedy, Robert F. *Thirteen Days*. New York: Norton. 1971. (With "Afterword" by Richard E. Neustadt and Graham T. Allison.)

Krulak, Victor H. *First to Fight*. Annapolis: Naval Institute Press, Md., 1984.

Kupperman, Robert, and Darrell M. Trent. *Terrorism, Threat, Response, Reality*. Stanford: Hoover Institution Press, 1979.

Lanchester, Frederick W. *Aircraft in War: the Dawn of the Fourth Arm.* London: Constable, 1916.

McHugh, Francis J. *Fundamentals of War Gaming.* Newport, R.I.: U.S. Naval War College, 1966.

Martel, William C., and Paul L. Savage. *Strategic Nuclear War.* Westport, Conn.: Greenwood Press, 1986.

Morschauser, Joseph, III. *How to Play War Games in Miniature.* New York: Walker, 1962.

Perkins, Stuart L. *Global Demands: Limited Forces.* Washington: National Defense University Press, 1984.

Polmar, Norman. *Aircraft Carriers.* New York: Doubleday, 1969.

————. *Guide to the Soviet Navy.* Annapolis: Naval Institute Press, 1983.

Pratt, Fletcher. *Fletcher Pratt's Naval War Game.* Milwaukee: Z&M Enterprises, 1976.

Pringle, Peter, and William Arkin. *SIOP: The Secret U.S. Plan for Nuclear War.* New York: Norton, 1983.

Prins, Gwyn, ed. *The Nuclear Crisis Reader.* New York: Vintage Books, 1984.

Quadripartite Working Group on Army Operational Research. *Catalog of War Games and Combat Simulations.* Washington: Office of the Deputy Under Secretary of the Army, April 1982.

Quattromani, A. F. *Catalog of Wargaming and Military Simulation Models.* 9th ed. Washington, D.C.: Studies Analysis and Gaming Agency, Organization of the Joint Chiefs of Staff, May 1982.

Rapoport, Anatol. *Fights, Games, and Debates.* Ann Arbor: University of Michigan Press, 1960.

Raser, John R. *Simulation and Society: An Exploration of Scientific Gaming.* Boston: Allyn and Bacon, 1969.

Reed, Jean D. *NATO's Theater Nuclear Forces: A Coherent Strategy for the 1980s.* Washington: National Defense University Press, 1983.

Schelling, Thomas. *The Strategy of Conflict.* New York: Oxford, 1960.

Shapiro, Norman Z., and H. Edward Hall, Robert H. Anderson, Mark LaCasse. *The RAND-ABEL Programming Language.* Rand publication R-3274-NA, prepared for the Director of Net Assessment, August 1985.

Shubik, Martin. *Game Theory and Related Approaches to Social Behavior.* New York: Wiley. 1964.

————. *Models, Simulations, and Games—a Survey.* Santa Monica, Calif.: Rand (R3 R-1060), 1972.

————. *Game Theory in the Social Sciences: Concepts and Solutions.* Cambridge, Mass.: MIT Press, 1982.

Sollinger, Jerry M. *Improving US Theater Nuclear Doctrine.* Washington: National Defense University Press, 1983.

Sorensen, T. C. *Decision-Making in the White House.* New York: Columbia University Press, 1964.

Steeb, Randall, and James Gillogly. *Design for an Advanced Red Agent.* Rand publication R-2977-DNA. Prepared for Defense Nuclear Agency, May 1983.

Stockdale, James B., and Sybil Stockdale. *In Love and War*. New York: Harper & Row, 1984.

Summers, Col. Harry G. *On Strategy: A Critical Analysis of the Vietnam War*. Carlisle Barracks, Pa.: Army War College, 1981.

Tidman, Keith R. *The Operations Evaluation Group: A History of Naval Operations Analysis*. Annapolis: Naval Institute Press. 1984.

U.S. General Accounting Office. *Models, Data, and War: A Critique of the Foundation for Defense Analyses*. March 1980.

Vlahos, Michael. *The Blue Sword*. Newport: Naval War College Press, 1980

Walther, Harry J. *Catalog of War Gaming and Military Simulations Models*. Washington: Studies, Analysis and Gaming Agency, June 1975.

Wells, Herbert George. *Little Wars*. New York: Macmillan, 1973.

Wilson, Andrew. *War Gaming*. London: Penguin, 1970. (Published in the United States as *The Bomb and the Computer: Wargaming from Ancient Chinese Mapboard to Atomic Computer*. New York: Delacorte Press, 1969.)

Wohlstetter, Roberta. *Pearl Harbor: Warning and Decision*. Stanford: Stanford University Press, 1962.

Young, J. P. *A Survey of Historical Developments in War Games*. Washington: Department of the Army, Office of the Chief of Military History, MS P-094, 1952. (Also published as Staff Paper ORO-SP-98, Operations Research Office, Johns Hopkins University, March 1959.)

Zumwalt, Elmo R., Jr. *On Watch*. New York: Quadrangle/New York Times, 1976.

Articles and Other Written Sources

Alexander, Joseph H. "Combined Amphibious Operations in Northern Europe." U.S. Naval Institute *Proceedings*, November 1980.

Allen, Thomas B. "It's the Real Thing." *Sea Power*, December 1985.

———. 'Non-Trivial Pursuits." *Sea Power*, April 1984.

———., and Norman Polmar. "The Silent Chase." *The New York Times Magazine*, January 1, 1984.

Arnott, Comdr. Ralph E., and Comdr. William A. Gaffney. "Naval Presence: Sizing the Force." *Naval War College Review*, March–April 1985.

Atkeson, Edward B. (published as Anonymous). "In Pursuit of the Essence of War." *Army*, January 1984.

———. "The 'Fault Line' in the Warsaw Pact: Implications for North Atlantic Strategy." *Orbis*, Spring 1986.

Bettencourt, Lt. Col. Vernon M., Jr., "Combat Modeling Evaluation at the United States Military Academy." Proceedings of the SIMOWAGA Conference, December 1984.

Blackett, P. M. S. "Critique of Some Contemporary Defence Thinking." *Encounter*, April 1961.

Bloomfield, Lincoln P. "Reflections on Gaming." *Orbis*, Winter 1984.

———. "Nuclear Crisis and Human Frailty: Making it through the 80's and 90's." Unpublished paper.

————, and Cornelius J. Gearin. "Games Foreign Policy Experts Play: The Policy Exercise Comes of Age." *Orbis*, Winter 1973.

————. "Political Games: Experiments in Foreign Policy Research." *Technology Review*. October–November 1974.

Bracken, Paul. "Unintended Consequences of Strategic Gaming." *Simulations & Games*, September 1977.

————. "Deterrence, Gaming, and Game Theory." *Orbis*, Winter 1984.

Brewer, Garry D and Bruce G. Blair. "War games and national security with a grain of SALT." *Bulletin of the Atomic Scientists*, June 1979.

Brown, Thomas A. "The Role of Modeling in Force Sizing," in Hughes, ed., *Military Modeling*.

Browning, James W. "Rehearsing the Critical Decisions." U.S. Naval Institute *Proceedings*, February 1984.

————, and Frank J. Dellerman, Lloyd H. Hoffmann. "Enhancing National Decision Making During Crises." Unpublished paper written in April 1985.

Bunn, Matthew, and Kosta Tsipis. "The Uncertainties of Preemptive Nuclear Attack." *Scientific American*, November 1983.

Callahan, Maj. Leslie G., Jr. "Robot Generals." *Combat Forces*, April 1953.

Callahan, Lt. Col. Leslie G., Jr. "Communication Between the Military and the Scientist." Industrial College of the Armed Forces thesis, March 1964.

————, ed. "The Need for Multidisciplinary Modeling Language in Military Science and Engineering." *Proceedings of the Workshop on Modeling and Simulation of Land Combat*, March 1982.

————. "Flexible Response Theory and Doctrine in a Nuclear Age." General Accounting Office publication, August 1976.

Cambareri, Carmen S. "Scuttle the *Graf Spee!*" U.S. Naval Institute *Proceedings*, June 1983.

Dahlberg, Susan, and Tom McNevin. "The name of the game: Half-SAFE." *Bulletin of the Atomic Scientists*, April 1982.

Dalkey, Norman C. "STROP: A Strategic Planning Model." Rand RM-4817-PR (Project Rand), Air Force Grant 49(638)-1700, July 1966.

Davis, Paul K. "Game-structured Analysis as a Framework for Defense Planning." The Rand Paper Series (P-7051), January 1985.

————, and William L. Schwabe. "Search for a Red Agent to be Used in War Games and Simulations." Paper for Thinking Red Wargaming Workshop, National Defense University, April 23–25, 1985.

————., and Peter J. E. Stan, Bruce W. Bennett. "Automated War Gaming as a Technique for Exploring Strategic Command and Control Issues." Rand Note for Director of Net Assessment, N-2044-NA, November 1983.

Dees, Capt. Robert F. "A Detailed Analysis of the McClintic Theatre Model," in Callahan, ed., *Proceedings of the Workshop on Modeling and Simulation of Land Combat*, March 1982.

De Weerd, Harvey A. "Political-Military Scenarios." Rand paper P-3535, February 1967.

Dunnigan, James F. "Experience with Commercial Wargames." *Proceedings of the Workshop on Modeling and Simulation of Land Combat*, March 1982.

Dupuy, Col. T. N. "Criticism of Combat Models Cites Unreliability of Results." *Army*, March 1985.

Ellsberg, Daniel. "The Crude Analysis of Strategic Choices." *American Economic Review*, May 1961.

———. "The Theory and Practice of Blackmail." Rand publication, 1968.

Fialka, John J. "Nuclear Reaction: U.S. Tests Response to an Atomic Attack." *The Wall Street Journal*, March 26.

———. "The Pentagon's Exercise 'Proud Spirit': Little Cause for Pride." U.S. Army War College *Parameters*, March 1981.

———. "Grim Lessons of Nifty Nugget." *Army*, April 1980.

Fulton, Richard P. "Wargaming in the United States Army." *Wargamer's Digest*, February 1978.

George, Alexander L. "The Impact of Crisis-induced Stress on Decision-making." Symposium on Medical Aspects of Nuclear War, National Academy of Science, Washington, September 20–22, 1985. (Paper revised in October 1985.)

Goldhamer, Herbert, and Hans Speier. "Some Observations on Political Gaming." *Journal of World Politics*, October 1959.

Graubard, Morlie H., and Carl H. Builder. "New Methods for Strategic Analysis: Automating the Wargame." *Policy Sciences*, 15 (1982).

Greve, Frank. "'STAR WARS,' How Reagan's Plan caught many insiders by surprise." *San Jose Mercury-News*, November 17, 1985.

Guetzkow, Harold. "Isolation and Collaboration: A Partial Theory of Internation Relations." *Journal of Conflict Resolution*, (No. 1) 1957.

Guy, Thomas G. "War in the Fourth Dimension—Is the Navy Ready for It? How Can the Navy Prepare for It?" *Naval War College Review*, January–February 1983.

Halstead, Ambassador John G. H. "The Atlantic: The Linchpin." U.S. Naval Institute *Proceedings, Sea Link Supplement*, December 1984.

Hamburg, David A. "Crisis Management." *Bulletin of the Atomic Scientists*, June–July 1984.

Hermann, Charles F., and Margaret G. Hermann. "An Attempt to Simulate the Outbreak of World War I." *American Political Science Review*, LXI, No. 2 (1967).

Hoffman, Lloyd H. Jr. "Defense War Gaming," *Orbis*, Winter 1984.

Kapper, Frank. "The Simulation of Crisis" and "Sun Tzu, the Spring Offensive & the Home Hobbyist." *Defense 81*, May 1981.

Lawrence, Lt. Gen. Richard D. "Playing the Game." *Defense 86*, January–February 1986.

Low, Lawrence J. "Theater-Level Gaming and Analysis Workshop for Force Planning: Summary, Discussion of Issues and Requirements for Research," May, 1981.

Mandel, Robert. "Political Gaming and Foreign Policy Making During Crises." *World Politics*, July 1977.

Marshall, A. W. "A Program to Improve Analytic Methods Related to Strategic Forces." *Policy Sciences*, 15 (1982).

Marshall, Donald S. "Status Report: Theory of Combat and Philosophy of War." *Proceedings of the Workshop on Modeling and Simulation of Land Combat*, March 1982.

Martin, J. J., and Douglas C. Olin. "Improved Methods for Strategic Analysis." *Policy Sciences*, 15 (1982).

Meyer, Deborah O. "New Storage Ships Are 'Incalculable' USMC Readiness Asset, Kelley Says." *Armed Forces Journal International*, April 1985.

Nicolosi, Anthony S. "The Spirit of McCarty Little." Naval Institute *Proceedings*, September 1984.

Paine, Christopher. "Nuclear combat: the five-year defense plan." *Bulletin of the Atomic Scientists*, November 1982.

Pilisuk, Marc. "Games Strategists Play." *Bulletin of the Atomic Scientists*, November 1982.

Poole, J. B. "Simulations Business." *Army Quarterly and Defence Journal*, January 1978.

Prina, L. Edgar. "The 'New' NWC: Professionalism . . ." *Sea Power*, July 1976.

Reinman, Robert A. "National Emergency Telecommunications Policy: Who's In Charge?" National Security Affairs Monograph Series, National Defense University Press, 1984.

Schemmer, Benjamin F. "Who Would Start a War? A Civilian? The Military Advisor? Will It Be a Woman?" *Armed Forces Journal International*, December 1981.

Schwabe, William L. "Strategic Analysis as Though Nonsuperpowers Matter." Rand publication prepared for Defense Nuclear Agency, June 1983.

————, and Lewis M. Jamison. "A Rule-Based Policy-Level Model of Nonsuperpower Behavior in Strategic Conflicts." Rand R-2962-DNA for Defense Nuclear Agency, December 1982.

Schwartz, Benjamin L. "Continuous Analytical Models of BMD Effectiveness." Paper for Joint National Meeting of The Institute of Management Sciences and Operations Research Society of America, Boston, April 29–May 1, 1985.

Snyder, Glenn H. "'Prisoner's Dilemma' and 'Chicken' Models in International Politics." *International Studies Quarterly*, March 1971.

Steeb, Randall, and James Gillogly. "Design for an Advanced Red Agent for the Rand Strategy Assessment Center." Rand Document R02977-DNA, prepared for the Defense Nuclear Agency, May 1983.

Steinbruner, John D. "Launch under Attack." *Scientific American*, January 1984.

Taylor, Theodore C. "Tactical Concentration and Surprise—in Theory." *Naval War College Review*, July–August 1985.

TEMPER as a Model of International Relations: An Evaluation for the Joint War Games Agency, DoD (New York: Simulmatics Corp. AD-653-606, December 1966. (See also Mathematica. "Final Report: Review of TEMPER Model." (Washington D.C.: Defense Communication Agency, Contract DCA 100-6-C-083, September 1966.)

Thomas, Clayton J. "Models and Wartime Combat Operations Research," in Hughes, ed., *Military Modeling*.

Thomas, Vincent C. "The Sealift Factor." *Sea Power*, January 1985.

Vlahos, Michael. "Wargaming, an Enforcer of Strategic Realism: 1919–1942." *Naval War College Review*, March–April 1986.

Watts, Lt. Col. Barry D. "Diagnostic Observations on Theater-Level War Gaming." Presentation at Thinking Red Wargaming Workshop, National Defense University, April 23–25, 1985.

Vietnam Documents*

Davison, W. P., and J. J. Zasloff. "A Profile of Viet Cong Cadres." RM-4983 ISA/ARPA, June 1966.

Denton, Frank H. "Some Effects of Military Operations on Viet Cong Attitudes." Rand RM-49 66-ISA/ARPA, November 1966.

Hoeffing, Oleg. "Bombing North Vietnam: An Apprasial of Economics and Political Effects (Draft), August 1966."

Pearce, R. Michael. "Evolution of a Vietnamese Village—Part I; the Present, After Eight Months of Pacification." Memorandum RM-4552-ARPA, April 1965.

Simulmatics Corporation. "Improving the Effectiveness of the ChiHo Program: A Summary." September 1967.

———. "A Socio-Pyschological Study of Regional/Popular Forces in Vietnam (Final Report)." September 1967.

Zasloff, J. J. "Political Motivation of the Viet Cong: The Vietminh Regroups." June 1966.

Congressional Documents

"Problems in the Acquisition of Standard Computers for the World Wide Military Command and Control System." Report to the House Committee on Appropriations, December 29, 1970.

"The World-Wide Military Command and Control System—Major Changes Needed in its Automated Data Processing Management and Direction." Report to the Congress (LCD-78-117), September 21, 1978.

"Emergency Management Information and Technology." Hearings, 97th Congress, first session, September 29–30, 1981, House Committee on Science and Technology, Subcommittee on Investigation and Oversight.

"NORAD Computer Systems are Dangerously Obsolete." Twenty-third Report by the House Committe on Government Operations, March 8, 1982.

*All from National Security Files, Lyndon B. Johnson Library.

"War Powers Resolution." Hearing, 98th Congress, first session, Senate Foreign Relations Committee, September 21, 1983.
"The Climatic, Biological, and Strategic Effects of Nuclear War." Hearing, 98th Congress, second session, House Committee on Science and Technology, September 12, 1984.

Interviews

Col. R. W. Argo, U.S. Army (Ret.), December 13, 1985.
Maj. Gen. Edward B. Atkeson, U.S. Army (Ret.), Alexandria, Va., May 1, 1985.
Lincoln Bloomfield, Cambridge, Mass., June 20, 1985.
William K. Brehm, Arlington, Va., January 25, 1985.
Col. Jerry J. Burcham, U.S. Army (Ret.), Arlington, Va., January 25, 1985.
Col. Leslie G. Callahan, U.S. Army (Ret.), Georgia Institute of Technology, Atlanta, Ga., October 11, 1985.
Paul K. Davis, Santa Monica, Calif., October 18, 1985.
James Dunnigan, New York, N.Y., June 7, 1985.
Col. T. N. Dupuy, U.S. Army (Ret.), Fairfax, Va., March 29, 1985.
Janice B. Fain, McLean, Va., May 29 1985.
Sherman Greenstein, Rosslyn, Va., January 25, 1985.
Robert Hardy, Arlington, Va., February 4, 1985.
Maj. Gen. Fred Haynes, U.S.M.C. (Ret.), Arlington, Va., January 24, 1985.
Mark Herman, New York City, August 23, 1985.
Paul Hovin, Washington, D.C., February 1, 1985.
Frank Kapper, Pentagon Annex, Va., May 16, 1985.
Richard L. Kugler, Washington, D.C., August 26, 1985.
Robert Kupperman, Washington, D.C., April 24, 1985.
William G. Lese, Jr., the Pentagon, May 23, 1985.
Col. Gary E. Lindquist, U.S. Army (Ret.), Arlington, Va., January 25, 1985.
Raymond M. Macedonia, Wilmington, Mass., June 21, 1985.
Christopher Makins, Washington, D.C., February 13, 1985.
William Mandel, Washington, D.C., May 14, 1985.
Roger Molander, Washington, D.C., February 13, 1985.
Bill Moore, Washington, D.C., January 15, 1985.
Steve Pieczenik, Bethesda, Md., June 4, 1985.
Robert B. Pirie, Alexandria, Va., May 30, 1985.
Joel Resnick, McLean, Va., January 30, 1986.
Brig. Gen. Edwin H. Simmons, U.S.M.C. (Ret.), Marine Museum, Washington (D.C.) Navy Yard, March 21, 1985.
Comdr. James J. Tritten, U.S. Navy, Pentagon, October 25, 1985.
Alvin D. Ungerleider, Alexandria, Va., January 5, 1986.

Index